CAMBRIDGE LIBRARY COLLECTION

Books of enduring scholarly value

Earth Sciences

In the nineteenth century, geology emerged as a distinct academic discipline. It pointed the way towards the theory of evolution, as scientists including Gideon Mantell, Adam Sedgwick, Charles Lyell and Roderick Murchison began to use the evidence of minerals, rock formations and fossils to demonstrate that the earth was older by millions of years than the conventional, Bible-based wisdom had supposed. They argued convincingly that the climate, flora and fauna of the distant past could be deduced from geological evidence. Volcanic activity, the formation of mountains, and the action of glaciers and rivers, tides and ocean currents also became better understood. This series includes landmark publications by pioneers of the modern earth sciences, who advanced the scientific understanding of our planet and the processes by which it is constantly re-shaped.

The Minerals of New South Wales, Etc.

Throughout the nineteenth century, Britain remained hungry for minerals to fuel her industrial and economic growth. Archibald Liversidge (1846–1927) found his knowledge and research to be in high demand. He had studied at the Royal College of Chemistry, and then obtained an exhibition to Cambridge, where he founded the Cambridge University Natural Sciences Club. At just twenty-seven years old Liversidge was appointed Reader in Geology at the University of Sydney, where he revolutionized the study of minerals and their potential applications. First published in 1876, and reprinted here from the enlarged, third edition of 1888, his chemical audit of the minerals of New South Wales became a key text for students of this field. Divided into two sections that address metallic and non-metallic minerals in turn, and incorporating a detailed map and substantial appendix, this work is of enduring interest and importance to geologists, chemists and historians of science.

Cambridge University Press has long been a pioneer in the reissuing of out-of-print titles from its own backlist, producing digital reprints of books that are still sought after by scholars and students but could not be reprinted economically using traditional technology. The Cambridge Library Collection extends this activity to a wider range of books which are still of importance to researchers and professionals, either for the source material they contain, or as landmarks in the history of their academic discipline.

Drawing from the world-renowned collections in the Cambridge University Library, and guided by the advice of experts in each subject area, Cambridge University Press is using state-of-the-art scanning machines in its own Printing House to capture the content of each book selected for inclusion. The files are processed to give a consistently clear, crisp image, and the books finished to the high quality standard for which the Press is recognised around the world. The latest print-on-demand technology ensures that the books will remain available indefinitely, and that orders for single or multiple copies can quickly be supplied.

The Cambridge Library Collection will bring back to life books of enduring scholarly value (including out-of-copyright works originally issued by other publishers) across a wide range of disciplines in the humanities and social sciences and in science and technology.

The Minerals of
New South Wales, Etc.

Archibald Liversidge

CAMBRIDGE
UNIVERSITY PRESS

CAMBRIDGE UNIVERSITY PRESS

Cambridge, New York, Melbourne, Madrid, Cape Town,
Singapore, São Paolo, Delhi, Tokyo, Mexico City

Published in the United States of America by Cambridge University Press, New York

www.cambridge.org
Information on this title: www.cambridge.org/9781108039055

© in this compilation Cambridge University Press 2011

This edition first published 1888
This digitally printed version 2011

ISBN 978-1-108-03905-5 Paperback

THE MINERALS

OF

NEW SOUTH WALES.

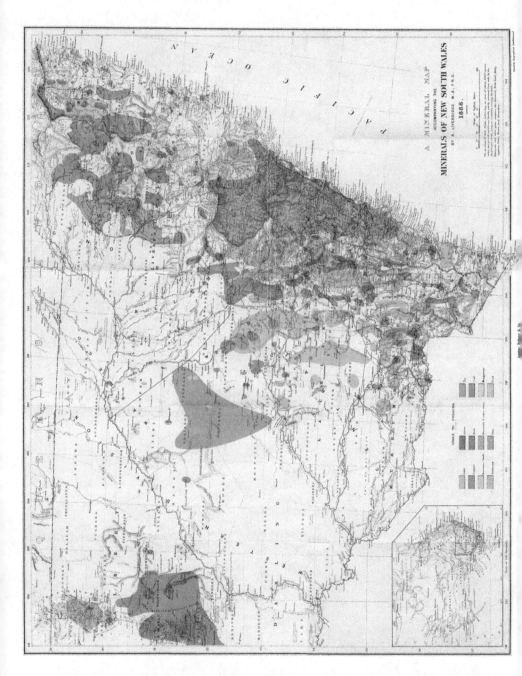

The material originally positioned here is too large for reproduction in this reissue.
A PDF can be downloaded from the web address given on page iv of this book,
by clicking on 'Resources Available'.

THE MINERALS

OF

NEW SOUTH WALES,

ETC.

BY

A. LIVERSIDGE, M.A., F.R.S.

PROFESSOR OF CHEMISTRY AND MINERALOGY IN THE UNIVERSITY OF SYDNEY.

With Map.

LONDON:

TRÜBNER & CO., LUDGATE HILL.

1888.

Ballantyne Press
BALLANTYNE, HANSON AND CO.
EDINBURGH AND LONDON

CONTENTS.

———

PART II.—NON-METALLIC MINERALS.

CLASS I.—CARBON AND CARBONACEOUS MINERALS.

CLASS II.—SULPHUR.

CLASS III.—SALTS.

CONTENTS.

APPENDIX.

THE

MINERALS OF NEW SOUTH WALES.

INTRODUCTION.

THE following account of the minerals of New South Wales originated in a paper which was read before the Royal Society of New South Wales in December 1874, and appeared in the Society's *Transactions* for that year. A second and enlarged edition of it appeared in the "Mineral Products of New South Wales," published by the Mining Department in 1882. In this, the third edition, an effort has been made to bring the matter up to date, in order that it may serve, as far as possible, as a record of the progress made in our knowledge of the mineralogy of New South Wales during the first hundred years of the colony's history. Since 1874 every opportunity open to me has been taken advantage of to correct and add to it; special attention has been paid to the chemical composition of the minerals; but on account of the great length of time required to make complete analyses, and the difficulty of obtaining specimens sufficiently pure for the purpose, the number of minerals analysed is by no means equal to my wishes.

In addition to my own, I have incorporated the analyses of minerals made by others, and notably those made by Mr. W. A. Dixon, F.I.C., and by the Government Analyst for the Mining Department, and published in the Annual Reports of the Department of Mines, Sydney.

I may perhaps state that the descriptions of the minerals are given almost entirely from specimens which I have either collected myself or which have come under my own personal observation. It is much to be regretted that no systematic examination of the minerals and rocks of New South Wales has been undertaken similar to that performed in other colonies. The amount of exact information upon the chemical composition of the various minerals occurring in New South Wales which has yet been published is extremely small, and by no means equal to what might naturally be expected from a colony so rich and prosperous, and so well endowed with mineral wealth.

Great difficulty has been experienced in identifying certain of the

A

localities from the changes which the names of places have, in many
cases, undergone—numbers of localities I have had to reject altogether
on this account, and some uncertain ones probably still remain ; but
every effort has been made to eliminate errors of the kind as far as
possible.   Some mistakes have doubtless crept in, for in such a work
as this it is almost impossible that some should not occur, although I
have done my best to keep the number down to as few as possible.
Too often it is the practice to intentionally mislead, especially if the
collector fancies that the mineral is likely to be of commercial value ;
this is done, of course, with the object of preventing the information
leaking out in any way, and the finder being forestalled in making
application for a mineral lease or the right to work the deposit.

Some of the localities have been taken from papers published by
the late Rev. W. B. Clarke, M.A., F.R.S., the late Mr. Stutchbury,
who was for some time Government Geologist, from some of the reports of the earlier explorers, and from the publications of the Mining
Department.

The Mineral Map, although based upon the one issued by the
Department of Mines in 1885, is practically a new one, prepared
afresh from the information gathered together in this volume.

At the end of the book I have reproduced some papers, which
were out of print, since they may be of interest in connection with
the mineralogy and geology of New South Wales.

I know that the work has many imperfections, but I feel that those
who are aware of the difficulties met with in preparing a book of this
kind will be ready to make allowance for its shortcomings.   In order
that it might appear in 1888, the centenary of the settlement of
Australia, I have had it printed and published during my absence from
the colony, and under conditions which rendered undivided attention to
it impossible and access to books of reference difficult.

I have, however, received valuable help in the preparation for the
press and in the correction of proofs from Mr. Felix Oswald, and I
have much pleasure in acknowledging the assistance which he has
rendered me.

A. LIVERSIDGE.

LONDON, *December* 1887.

DIAGRAM SHEWING THE VALUE OF THE ANNUAL PRODUCTION OF GOLD, COAL, COPPER AND TIN IN NEW SOUTH WALES, FROM 1881. TO 1886 INCLUSIVE.

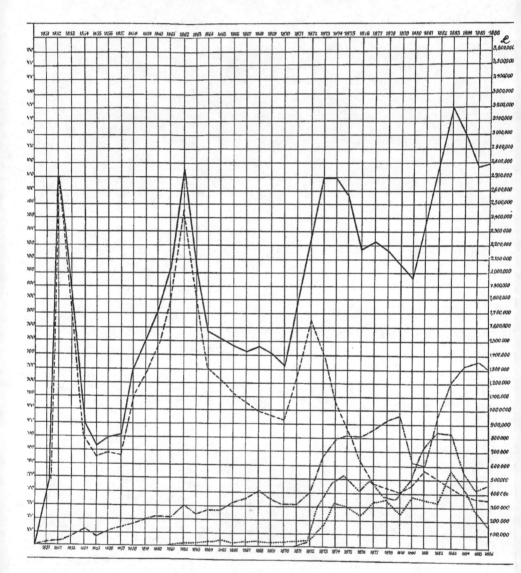

INDEX TO LINES.

Total Minerals thus	———————————
Gold .... , ...	— — — — — — — —
Coal	— · — · — · — · — · —
Copper ——— · — · — · ———	··
Tin	—— · · · —— · · · ——

The vertical divisions indicate the years, and the horizontal divisions denote the value of the Minerals.

PART I.

METALLIC MINERALS.

GOLD.

ONLY one true mineral species of gold has up to the present been found in New South Wales, and that is—

NATIVE GOLD.

Crystallises in the cubical system. Well-developed crystals are very rare, and are never of large size, seldom exceeding $\frac{1}{4}$ inch in diameter, and the faces are usually more or less cavernous; the most common form are the octahedron and rhombic dodecahedron; single and detached crystals are seldom found—they are usually attached end to end, forming strings, wires, and branching or arborescent forms. A beautiful branching tree-like group of large rhombic dodecahedral crystals, weighing some 20 oz., was formerly to be seen in the Australian Museum collection; but the specimen has been stolen, so that it is unfortunately lost to science, for no goniometrical measurements were made, and not even a cast or drawing seems to have been retained. Occasionally elongated crystals of rhombic dodecahedra are met with, arranged in columnar masses very similar to groups of basaltic columns. Some very perfect crystals were obtained in the early days of gold-mining from the Louisa Creek. As with other minerals, the smaller crystals are usually the most perfect.

A beautiful group of gold crystals is to be seen in the Museum of Science and Art at Edinburgh—perhaps one of the finest in existence. A model of this rare and very valuable nugget has been kindly made for me by the late Professor Archer, the Director of the Museum.

As will be seen from the woodcut, the crystals are for the most part imperfect octahedra and elongated cubes; some have imperfectly developed faces of the rhombic dodecahedron, joined end to end in an arborescent form.

Professor Archer was under the impression that the specimen came from New South Wales, but the exact locality is no longer known. This notice may, perhaps, draw attention to the specimen, and be the means of eliciting some information as to its history.

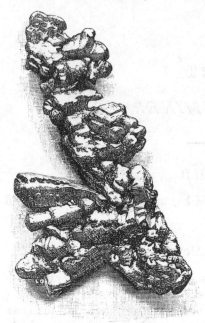

FIG. 1.
Group of Gold Crystals in Edinburgh Museum.

It is much to be regretted that more of such specimens have not been preserved. At the present day they are extremely scarce, and even in the early days of the gold discoveries they were never abundant. Unfortunately most of them very quickly find their way into the melting-pot, and of the few which have been preserved, probably even fewer are to be found in Australia than elsewhere.

Filiform, reticulated, and spongy shapes are common; but more so are irregular plates, scales, and strings, which interpenetrate the matrix in every direction. In one or two specimens from the "Uncle Tom Mine," Lucknow, I have observed capillary crystals or filaments of gold resembling the artificial "moss gold," or the better known "moss copper;" in this mine the gold occurs with mispickel and calcite, the matted or moss-like filaments being met with in small cavities in the former mineral.* Sometimes, as observed by Mr. C. S. Wilkinson at the Cowarbee Mine, about forty miles north-west of Wagga Wagga, the plates are so exceedingly thin that they form mere films like gold-leaf, and in this particular instance the films run both between and across the laminæ of the red-coloured schistose rock in which they occur. Then, again, gold occurs in New South Wales, as elsewhere, so finely divided and equally diffused throughout the matrix as to be invisible even by the aid of a lens.

In alluvial deposits gold occurs in more or less rounded and water-worn flattened grains, scales, and pebbles or nuggets. The largest nuggets discovered in Australia have been found in Victoria; none at all to compare with them in size have been found in New South Wales.

* "On the Formation of Moss Gold and Silver" (A. Liversidge, *Trans. Roy. Soc. of N. S. W.*, 1876).

EXAMPLES OF NEW SOUTH WALES GOLD MASSES AND NUGGETS.

No. 1. Found in July 1851 by a native boy, amongst a heap of quartz, at Meroo Creek, or Louisa Creek, River Turon, fifty-three miles from Bathurst, and twenty-nine miles from Mudgee, New South Wales, where there is now a township known as Hargraves. It was in three pieces when discovered, though generally considered as one mass. The aboriginal who discovered these blocks "observed a speck of some glittering yellow substance upon the surface of a block of quartz, upon which he applied his tomahawk, and broke off a portion." One of the pieces weighed 70 lbs. avoir., and gave 60 lbs. troy of gold; the gross weight of the other two about 60 lbs. each. These three pieces, weighing 1¾ cwt., contained 106 lbs. troy, of gold, and about 1 cwt. of quartz. In the same year another nugget, weight 30 lbs. 6 oz., was discovered in clay, twenty-four yards from the large pieces; and in the following year, near to No. 6, there were found two nuggets, weighing 157 oz. and 71 oz.

Gross weight (troy), 106 lbs., or 1272 oz.

The following account of the discovery of the above "hundred-weight of gold," as it was termed, is quoted in Stirling's "Gold Discoveries of 1862," from the *Sydney Morning Herald* of 18th July 1851:—

"Bathurst is mad again. The delirium of golden fever has returned with increased intensity. Men meet together, stare stupidly at each other, talk incoherent nonsense, and wonder what will happen next. Everybody has a hundred times seen a hundredweight of flour; a hundredweight of sugar or potatoes is an everyday fact; but a hundredweight of gold is a phrase scarcely known in the English language. It is beyond the range of our ordinary ideas—a sort of physical incomprehensibility; but that it is a material existence our own eyes bore witness on Monday last.

"Mr. Suttor, a few days previously, threw out a few misty hints about the possibility of a single individual digging four thousand pounds' worth of gold in one day, but no one believed him serious. It was thought that he was doing a little harmless puffing for his own district and the Turon Diggings. On Sunday it began to be whispered about town that Dr Kerr (Mr Suttor's brother-in-law) had found a hundredweight of gold. Some few believed it; but the townspeople generally, and amongst the rest the writer of this article, treated the story as a piece of ridiculous exaggeration and the bearer of it as a jester, who gave the Bathurstonians unlimited credit for gullibility. The following day, however, set the matter at rest. About two o'clock in the afternoon two greys, in tandem, driven by W. H. Suttor, Esq.,

M.C., made their appearance at the bottom of William Street. In a few seconds they were pulled up opposite the *Free Press* Office, and the first indication of the astounding fact which met the view was two massive pieces of the precious metal, glittering in virgin purity as they leaped from the solid rock. An intimation that the valuable prize was to reach the town on that day having been pretty generally circulated in the early part of the morning, the townspeople were on the *qui vive*, and in almost as little time as it has taken to write it 150 people had collected around the gig conveying the time's wonder, eager to catch a glimpse of the monster lump said to form a portion of it. The two pieces spoken of were freely handed about amongst the assembled throng for some twenty minutes. Astonishment, wonder, incredulity, admiration, and the other kindred sentiments of the human heart were depicted upon the features of all present in a most remarkable manner, and they were by no means diminished in intensity when a square tin box in the body of the vehicle was pointed to as the repository of the remainder of the hundredweight of gold. Having, good-naturedly, gratified the curiosity of the people, Mr. Suttor invited us to accompany his party to the Union Bank of Australia to witness the interesting process of weighing. We complied with alacrity, and the next moment the greys dashed off at a gallant pace, followed by a hearty cheer from the multitude.

"In a few moments the tin box and its contents were placed on the table of the board-room of the bank. In the presence of the manager, David Kennedy, W. H. Suttor, I. J. Hawkins, Esqs., and the fortunate proprietor (Dr. Kerr) the weighing commenced, Dr. Machattie officiating, and Mr. Ferrand acting as clerk. The first two pieces already alluded to weighed severally 6 lbs. 4 oz. 1 dwt. and 6 lbs. 13 dwts., besides which were sixteen drafts of 5 lbs. 4 oz. each, making in all 102 lbs. 9 oz. 5 dwts. From Dr. Kerr we learned that he had retained upwards of 3 lbs. as specimens, so that the total weight found would be 106 lbs. (one hundred and six pounds), all disembowelled from the earth at one time. And now for the particulars of this extraordinary gathering, which has set the town and district in a whirl of excitement.

"A few days ago an educated aboriginal, formerly attached to the Wellington Mission, and who had been in the service of W. J. Kerr, Esq., of Wallawa, about seven years, returned home to his employer with the intelligence that he had discovered a large mass of gold amongst a heap of quartz upon the run whilst tending his sheep. Gold being the universal topic of conversation, the curiosity of this sable son of the forest was excited, and, provided with a tomahawk, he had amused himself by exploring the country adjacent to his employer's land, and had thus made the discovery. His attention was first called to the lucky spot by observing a speck of some glittering yellow substance upon the surface of a block of quartz, upon which he applied his tomahawk and broke off a portion—at that moment the splendid prize stood revealed to his sight. His first care was to start off home and disclose his discovery to his master, to whom he presented whatever gold might be procured from it. As might be supposed, little time was lost by the worthy doctor. Quick as horseflesh would carry him, he was on the ground, and in a very short period the three blocks of quartz, containing the hundredweight of gold, were released from the bed, where, charged with unknown wealth, they had rested perhaps for thousands of years, awaiting the hand of civilised man to disturb them. The largest of the blocks was about a foot in diameter, and weighed 75 lbs. gross. Out of this piece 60 lbs. of pure gold was taken. Before separation it was beautifully encased in quartz. The other two were something smaller. The auriferous mass weighed as nearly as could be guessed from 2 to 3 cwts. Not being able to move it conveniently, Dr. Kerr broke the pieces into small fragments, and herein committed a very grand error—as specimens the glittering blocks would have been invaluable. Nothing yet known of would have borne comparison, or, if any, the comparison would have been in our favour. From the description given by him, as seen in their original state, the world has seen nothing like them yet.

"The heaviest of the two large pieces presented an appearance not unlike a honeycomb or sponge, and consisted of particles of a crystalline form, as did nearly the whole of the gold. The second larger piece was smoother, and the particles more condensed, and seemed as if it had been acted upon by water. The remainder was broken into lumps of from 2 to 3 lbs. and downwards, and were remarkably free from quartz or earthy matter. When heaped together on the table they presented a splendid appearance, and shone with an effulgence calculated to dazzle the brain of any man not armed with the coldness of stoicism.

"The spot where this mass of treasure was found will be celebrated in the golden annals of these districts, and we shall therefore describe it as minutely as our means of information will allow. In the first place, the quartz blocks formed an isolated heap, and were distant about 100 yards from a quartz-vein, which stretches up the ridge from the Murroo Creek. The locality is the commencement of an undulating tableland, very fertile, and is contiguous to a never-failing supply of water in the above-named creek. It is distant about fifty-three miles from Bathurst, eighteen from Mudgee, thirty from Wellington, and eighteen from the nearest point of the Macquarie River, and is within about eight miles of Dr. Kerr's head station. The neighbouring country has been pretty well explored since the discovery, but, with the exception of dust, no further indications have been found.

"These particulars were kindly furnished by Mr. Suttor and Dr. Kerr, and may therefore be relied on as correct."

No. 2. A model of what is said to be the first large nugget found in New South Wales is to be seen in the Australian Museum, Sydney. Found in Ophir Creek.

Several other large nuggets appear to have been found in this creek, but none of them approaching to the above in size and value.

No. 3. A nugget weighing 26 oz. was found at Bingera in 1852.

No. 4. Found by a party of four, on 1st November 1858, at Burrandong, near Orange, New South Wales, at a depth of 35 feet; when pounded with a hammer it yielded 120 lbs. of gold, for which £5000 were offered. Melted at the Sydney Mint, when it weighed 1286 oz. 8 dwts.; after refining, 1182 oz. 7 dwts.; loss, 8 per cent.; fineness, 87·4 per cent.; the standard weight of gold being 1127 oz. 6 dwts. Value, £4389, 8s. 10d. The gold was mixed with quartz and sulphide of iron (mundic). Assay, 87·40 per cent. gold = 20 car. $3\frac{7}{8}$ car. grs.

Gross weight (troy), 107 lbs. 2 oz. 8 dwts.; or, 1286 oz. 8 dwts.

No. 5. Found at Kiandra, Snowy River, New South Wales, October 1860.

Gross weight (troy), 33 lbs. 4 oz.; or, 400 oz.

No. 6. "The Brenan Nugget." Found in Meroo Creek, Turon River, New South Wales, embedded in clay; measures 21 inches in circumference. It was found twenty-four yards from No. 1. Sold in Sydney, 1851, for £1156.

Gross weight (troy), 30 lb. 6 oz.; or, 364 oz. 11 dwts.

No. 7. Found at New Chum Hill, Kiandra, Snowy River, New South Wales, July 1861.

Gross weight (troy), 16 lbs. 8 oz.; or, 200 oz.

No. 8. Found at Kiandra, Snowy River, New South Wales, March 1860.

Gross weight (troy), 13 lbs. 4 oz.; or, 160 oz.

No. 9. Found, in 1852, at Meroo Creek, Turon River, New South Wales, close to No. 1. This was called "The King of the Waterworn Nuggets."

Gross weight (troy), 13 lbs. 1 oz.; or, 157 oz.

No. 10. Found in 1860, at the Tooloom Diggings, New South Wales; nearly solid gold.

Gross weight (troy), 11 lbs. 8 oz.; or 140 oz.

No. 11. Found at Kiandra, Snowy River, New South Wales, March 1860.

Gross weight (troy), 7 lbs. 9 oz. 18 dwts.; or 93 oz. 18 dwts.

No. 12. Found in 1852, at Louisa Creek, New South Wales; a solid lump of gold.

Gross weight (troy), 6 lbs. 10 oz.; or 82 oz.

No. 13. Found by two boys, in July 1861, at Gundagai (new diggings), New South Wales.

Gross weight (troy), 5 lbs. 4 oz. 7 dwts.; or 64 oz. 7 dwts.

No. 14. Found in 1857, at Louisa Creek, New South Wales; gold and crystallised quartz.

Gross weight (troy), 4 lbs. 2 oz.; or 50 oz.

No. 15. Found at New Chum Hill, Kiandra, New South Wales, in July 1861.

Gross weight (troy), 3 lbs. 6 oz.; or 42 oz.

No. 16. Found at Summer Hill Creek, New South Wales. The earliest nugget found in New South Wales after the gold discovery there by Hargraves. 13th May 1851.

Gross weight (troy), 1 lb. 1 oz.; or 13 oz.

In the *Annual Reports of the Mines Department* the discovery of the following nuggets are reported:—

Nos. 17 to 23. A nugget weighing 22 oz. 18 dwts. 12 grs. was found in 1874 on "M'Guiggan's Lead," about nine miles from Parkes; the metal was of dark colour and free from gangue; also one of 134 oz., and other smaller ones of 7 oz., 25 oz., 35 oz., 37 oz.; and in 1876 one of 36 oz.

No. 24. In 1874 a nugget of 65 oz. was found on Woods' Flat, about twelve miles from Cowra.

No. 25. At the same place, and in the same year, another of 50 oz. is reported.

No. 26. One weighing 64 oz. 3 dwts. was unearthed in the Canadian Lead, near Gulgong, November 1876, at a depth of 140 feet; it was stated to have been so completely invested with a coating of iron oxide as to be superficially unrecognisable as gold.

Nos. 27 to 29. A nugget weighing 19 oz. 12 dwts. was found early in 1876 at the "Wapping Butcher Mine," the Terrace, near Parkes; also others of 16 oz. 10 dwts. and 18 oz., together with a large number of smaller nuggets.

Nos. 30 and 31. A nugget of 43 oz., together with one of 23 oz., was discovered on the Nundle Gold-field in 1879.

No. 32. One of 32 oz. 15 dwts. was found in October 1879, in Broad Gully, in the Braidwood District, together with several smaller ones in the same year.

No. 33. A nugget weighing 28 lbs. was found on the Whipstick Flat, Kiandra; recorded by Mr. Lamont Young, F.G.S., in the *Annual Report of the Mines Department* for 1880, but no date is given.

Nos. 34 to 45. At Temora the following were found during 1880:—99 oz., 84 oz., 76 oz., 72 oz., 68 oz., 64 oz., 63 oz., and one of 59 oz. 1 dwt.—this measured 7 inches by $2\frac{1}{4}$ inches wide, with a thickness of about 1 inch, and described as not waterworn, but jagged, and with a half-turn or twist in it. During the same year others of 46 oz. 18 dwts. 20 grs., 40 oz., 28 oz., 24 oz., 16 oz., and 14 oz. were met with.

No. 46. At Nerrigundah, at the foot of Mount Dromedary, a small one of 13 oz. 15 dwts. was found in 1880.

No. 47. On March 16th, 1882, a nugget was found at Temora, weighing 153 oz. 17 dwts., at a depth of about 14 feet.

No. 48. Nugget found at Cadia, near Orange, New South Wales, near the surface, in September 1882. Weight, 70 oz.

No. 49. In March 1882 a nugget weighing 29 oz., with other smaller ones, was obtained by J. Ward, in a blind gully leading into the Upper Meroo; also, in April a nugget was found at the Pyramul.

No. 50. A nugget weighing 12 oz. was found on the surface at Mount Browne, Milparinka Division.

Nos. 51 to 53. During the year 1883 three nuggets were found in the direction of Pine Bowl and Two-mile Flat, Gulgong Division. The first nugget weighed 19 oz. 6 dwts.; the second 19 oz. 16 dwts., and the third 12 oz.

Nos. 54 to 65. During the year 1884 twelve nuggets have been found in the Gulgong Division, varying from 3 to 39 oz.

Nos. 66 to 86. The nuggets which have been obtained during the year 1884 are as follows:—One 5 oz.; one 6 oz. 7 dwts.; two of 8 oz. each; one 8 oz. 12 dwts.; one 10 oz.; one 12 oz.; one 15 oz.; one 16 oz.; one 17 oz.; one 18 oz.; one 20 oz.; one 25 oz.; one 28 oz.; one 32 oz.; one 38 oz.; one 45 oz.; one 90 oz.; two pieces weighing together 98 oz.; and one 175 oz. 15 dwts. All of these nuggets were obtained from alluvium between Upper and Lower Temora.

Nos. 87 to 95. The following nuggets were found in the Mudgee Mining District during the year ended the 31st December 1884, besides smaller ones:—

37 oz.	12 dwts.	at Rat's Castle, near Mudgee.
7 „	10 „	at Piambong, seventeen miles from Mudgee.
12 „	0 „	„ „ „ „
7 „	10 „	„ „ „ „
39 „	0 „	at Rat's Castle, near Piambong.
7 „	0 „	at Mudgee Gold-mining Co.
5 „	0 „	„ „ „ „
6 „	0 „	„ „ „ „

No. 96. At Spring Creek, on the Macquarie River, near Burrendong, a nugget was found weighing 62 oz.

Nos. 97 and 98. At Apple Tree Flat, Mudgee Division, Rochester and party obtained a nugget weighing 102 oz., and another party a nugget weighing 20 oz.

No. 99. A vermiform nugget was found near the surface at Cadia, near Orange, in 1885. Weight about 21 oz.

Nos. 100 and 101. In 1887 a nugget weighing 350 oz., worth about £1400, was found at.Maitland Bar, in the Mudgee District; and one of 225 oz. was found by some Chinamen at Hargraves, near Mudgee, in August 1887. The celebrated mass of gold known as "Dr. Kerr's Nugget" was also found near Hargraves in 1851.

For the accounts of No. 1 and Nos. 4 to 16 I am indebted to Mr. Brough Smyth's "Gold-fields and Mineral Districts of Victoria." They are arranged in the order of their weights, the others according to the date of their discovery.

Colour.—Most of the New South Wales gold is usually of fairly deep yellow, being rather lighter than Victorian, and not so light as much of the Southern Queensland gold; but occasionally specimens of very pale and of very dark gold are met with. The quantity of silver present greatly affects the colour of the metal.

Specific gravity.—In specific gravity it varies considerably, the mean being about 17·5. A specimen of Braidwood gold had a specific gravity of 18·28.

Composition.—No specimens of actually pure gold have been met with. There is always more or less silver present, and usually traces of copper, bismuth, iron, and other metals.

The tables on pp. 15, 16, 17, have been compiled from the table of forty-eight assays of New South Wales gold exhibited prior to transmission to the Paris Exhibition, December 1854, by F. B. Miller, F.C.S.; the table of assays made at Sydney Mint in 1856, and also on August 9, 1860; the assays of samples characteristic of the New South Wales gold-fields exhibited at the Philadelphia Exhibition, 1876; the assays of samples of New South Wales gold exhibited at the Paris Exhibition, 1878; the assays of specimens (2 oz.) of alluvial gold exhibited at the Sydney Exhibition, 1879; and a table of assays of New South Wales gold exhibited at the Colonial and Indian Exhibition, London, 1886.

The value assigned to the gold in the earlier tables is less than that in the later ones; hence the average value given in the last column is in some cases lower than the market value of gold at the present day.

COMPOSITION OF NEW SOUTH WALES GOLD.

Assays, made at the Sydney Branch of the Royal Mint, of Forty-eight Specimens of New South Wales Gold, from the Collection exhibited in the Australian Museum, prior to Transmission to the Paris Exhibition.—*December* 1854.

Locality.	Pure Gold in 1000 parts.	Silver.	Copper and Iron.	External Character of Specimens.
Western District.				
TAMBAROORA.—Dirt Hole Creek	945·90			Dull gold, in rounded grains like coarse sand.
,, Dirt Hole Road Creek	952·45			Light and brilliant, small grain gold, with small nuggets.
,, Hayes' Flat	950·10			{ Bright nuggetty gold, presenting very irregular shapes; little waterworn.
,, Golden Gully	942·65			
,, Bald Hill's Creek	947·00			
,, Oaky Creek	946·00	{ 47·35 to 76·4	} Trace to 12·5	Larger waterworn nuggets, dull in colour.
,, Lower Turon	944·55			Bright scaly gold, of uniform character.
,, Macquarie River	946·10			Bright gold, consisting of small elongated and flattened pieces, with irregular nuggets.
,,	947·30			Small nuggets or grains, moderately waterworn and dark-coloured.
,, Upper Pyramul	922·85			
,, Lower Pyramul	948·75			} Brilliant, light scaly gold.
,, Junction of Pyramul and Macquarie	945·10			
,,	944·55			Dull scaly gold, with small rounded nuggets.
,,	946·45			Rough-grained gold.
TURON RIVER (SOFALA).—Erskine Flat	923·80			Dull scaly gold, of uniform character.
,, Green Wattle Flat	916·05			Nuggetty gold, showing marks of crystallisation; moderately waterworn.
,, Little Oaky Creek	926·10	{ 42·9 to 83·7	,, 1·3	Rough-grained gold.
,, Big Oaky Creek	931·60			Small rounded nuggets of dull colour.
,, Nuggetty Gully	956·40			Small nuggets, moderately waterworn.
,, Golden Point	929·50			Scaly dull-coloured gold.
,, Paterson's Point, E 1	925·60			Fine scaly gold, of uniform character, not bright.
,, ,, E 2	923·05			Scales and rounded nuggets.
MEROO RIVER (AVISFORD).—Devil's Hole Creek	957·95			{ Bright scaly gold, with waterworm nuggets.
,, Nuggetty Gully	961·40	{ 38·6 to 50·05	,, 1·05	Small waterworn nuggets, light and bright.
,, Richardson's Point	958·45			{ Light and brilliant small scales.
,, Gifford's Point	949·65			
,, Deep Crossing Place	952·15			Do. with large scales.

Locality				Description
BURRENDONG.—Long Point (Macquarie, below junction of Ophir Creek) .	934·85	...	...	Dark-coloured scaly gold.
" Devil's Hole Creek. "Dry Diggings" .	917·90	56·9	" 0·3	Nuggetty gold, with marks of crystallisation.
" Mookerawa Creek .	942·80			Dull, dirty scales and waterworn nuggets.
OPHIR CREEK .	940·60	59·3	Other metals 0·1	Nuggetty gold, much waterworn.
BROWN'S CREEK.—22 miles south of Bathurst .	932·35	...		Dark, rough grains, mixed with blackish impurities.
South-Western District.				
ADELONG CREEK.—5 miles below source .	936·85			Rough nuggetty gold.
" 8 miles below source .	916·45	51·15 to 66·65	" 0·15 to 1·3	Nuggetty; smaller and more waterworn than last.
" 11 miles from source .	945·20			Fine granular gold, light in colour.
" 25 miles from source .	948·60			Fine, bright scaly gold.
	932·00			
Southern District.				
ARALUEN.—Major's Creek, Southern Arm .	935·10			Bright granular gold.
" Bell's Paddock .	895·90	·50-5· to 103·45	...	Dull granular, with rough nuggets.
" Major's Creek, Western Arm .	949·20			Dark-coloured, rounded grains, larger than last.
" In broken granite, 10 feet below surface .	915·05			Bright granular gold.
Northern District.				
HANGING ROCK (NUNDLE).—Oakanville Creek .	936·80			Rough nuggetty gold.
" Same Creek, 3 miles farther down .	937·60			Rough scaly gold.
" Cordillera Gold Co.'s property on the River Peel .	906·65	62·95 to 93·25	...	
" Gully leading to the Peel, north of Oakanville Creek .	930·15			Small dark-coloured nuggets, moderately waterworn.
BINGARA.—Nugget weighing 4 oz. 3 dwt. .	874·25	125·25	...	A porous, spongy kind of nugget, containing dusty impurities in the pores.
.	894·45			Small rounded grains and nuggets of brightish colour.
ROCKY RIVER .	943·70	56·3	...	Very small granular gold, light and brilliant.

The average value of the above was found to be 80s. 6d. per oz., the value of standard gold being 77s. 10½d. per oz.

SAMPLES of GOLD characteristic of the Gold-fields of New South Wales, exhibited by the Mining Department, and assayed at the Royal Mint, Sydney. From the New South Wales Official Catalogue, Philadelphia Exhibition, 1876.

Locality.	Description of Gold.	Weight of Sample.	Loss in Melting, per cent.	Gold and Silver in 1000 parts after Melting.		Value per oz., after Melting, at £3, 17s. 10½d., Standard.		
		ozs.		Gold.	Silver.	£	s.	d.
West District.								
Sofala	In fine scales, and coarse plates and grains	2·50	1·54	923·0	72	3	18	9½
Bathurst	Fine scales and coarse grains, with some spongy and stringy	2·00	2·00	923·5	71	3	18	10
„	Fine scales, plates, and coarse grains	2·00	1·47	918·0	76	3	18	4½
Hargraves	Fine dust and coarse grains	2·00	1·23	920·5	70	3	18	6½
„	Scaly, with some grains	2·00	1·15	961·0	33	4	1	9½
Tambaroora	Fine and coarse, scaly and grains	2·00	1·31	940·0	54	4	0	1
„	Fine scales and grains	2·00	1·55	943·5	50	4	0	5
„	Reef gold—reticulated	2·00	2·77	944·5	51	4	0	6
„	Coarse waterworn grains or nuggets	2·00	2·00	935·5	54	3	19	8½
Hill End	Fine dust and coarse grains	2·00	2·47	945·5	47	4	0	7
„	Scaly, with coarse spongy grains	2·53	1·41	945·5	50	4	0	7
„	Fine scales and coarse crystalline gold	2·00	2·18	947·0	47	4	0	8½
„	Scaly and coarse filiform gold	2·00	1·97	942·5	49	4	0	4
Mudgee	Fine scales and coarse grains	2·50	1·93	941·0	56	4	0	2½
„	Coarse grains with some scales	2·00	2·04	926·0	68	3	19	0
„	Fine and coarse scales	2·00	1·77	937·0	58	3	19	10½
Gulgong	Coarse spongy grains and some scales	2·00	1·78	938·0	58	3	19	11½
„	Dust and coarse scales	2·00	1·78	916·5	79	3	18	3
„	Coarse pieces—filiform and spongy	2·00	1·78	925·0	70	3	18	11
„	Scaly, with some grains	2·00	1·59	946·0	48	4	0	7½
Carcoar	Fine scales, very porous, with some magnetic iron	2·00	10·92	878·0	119	3	15	2
„	Fine and coarse filiform gold of a dark colour	2·00	2·94	960·0	36	4	1	8½
Orange	Scaly	2·00	2·67	943·0	51	4	0	4½
„	Fine dust—"gunpowder gold"	2·00	2·53	930·5	62	3	19	4
Stony Creek	Scaly	2·00	1·56	942·0	54	4	0	3½
South District.								
Braidwood	Plates and fine scaly	2·00	1·79	959·0	34	4	1	7½
Araluen	Fine dust—"gunpowder gold"	2·00	2·19	951·5	42	4	1	0½
Adelong	Fine scaly and coarse filiform	2·00	2·63	944·0	52	4	0	5½
„	Scaly	2·00	1·27	941·0	53	4	0	2
„	Coarse filiform, with some scaly	2·50	1·69	946·0	50	4	0	7½
Tumut	Fine and coarse, with some very spongy	2·00	6·28	927·5	70	3	19	1½
Young	Scaly dust gold	2·00	2·39	957·0	36	4	1	5½
„	Fine dust—"gunpowder gold"	2·00	1·52	943·0	49	4	0	4½
Nerrigundah	Strings, scales, and plates	2·50	1·64	980·5	15	4	3	4½
Kiandra	Scales and plates, with some grains and threads	2·00	3·15	927·0	63	3	19	1
Goulburn	Coarse grains and reticulated	2·00	6·87	975·0	22	4	2	11½
Bombala	Very fine scaly dust—"gunpowder gold"	2·00	2·63	963·0	34	4	1	11½
Cooma	Filiform crystalline, and some scaly	2·00	3·17	938·0	56	3	19	11½
„	„ „ „ „	2·00	4·22	924·0	70	3	18	10
North District.								
Nundle	Fine scaly and coarse filiform	2·00	3·33	919·5	73	3	18	6
„	Scales, plates, and coarse filiform; of a brownish colour	2·00	3·28	902·5	90	3	17	1½
Tamworth	Spongy, filiform, and crystalline, some with a little quartz attached	2·00	3·28	912·0	83	3	17	10½
„	„ „ „ „	2·00	3·31	914·0	80	3	18	0½
„	Fine dust and shotty grains	2·00	3·31	899·5	93	3	18	10½
Armidale	Scales, with some threads	2·00	3·30	948·0	44	4	0	9
„	Fine scales	2·00	1·91	888·5	105	3	16	0

GOLD—NORTHERN DISTRICT.

Locality.	No. of Samples.	Loss per cent. in Melting.	Average Loss.	Gold in 1000 parts after Melting.	Average of Gold.	Silver in 1000 parts after Melting.	Average of Silver.	Value per Ounce after Melting.	Average Value.
								£ s. d. £ s. d.	£ s. d.
Armidale	4	1·91 to 3·30	2·65	749·5 to 948·0	868·9	44·0 to 242·0	123·5	3 4 8 to 4 0 9	3 14 4½
Bingera	3	5·137	...	874·25 „ 908·5	894·17	80·0 „ 125·25	98·4	3 12 9 „ 3 15 3	3 14 0
Boonoo Boonoo	1	...	...	854·0 „ 659·0	756·5	298·0 „ 337	317·5	...	...
Copeland	1	3·748	...	900·0	...	90·0	...	3 13 10	...
Fairfield	1	...	...	872·0	...	121·0	...	...	...
Glen Innes	1	1·535	...	943·5	...	50·0	...	3 19 0	...
Grafton	2	2·82	...	900·0 to 918·5	909·25	75 to 95	85·0	3 16 4 to 3 16 10	3 16 7
Nundle	10	2·642 to 3·33	3·084	898·5 „ 937·6	921·37	62·95 „ 93·25	74·15	3 15 1 „ 3 18 6	3 16 9½
Peel River	1	...	...	929·0	...	67·0	...	...	...
Richmond River	2	...	...	947·0 to 952·5	949·75	40·0 to 45·0	42·5	3 19 1 „ 3 19 4	3 19 2½
Rocky River	4	1·423	...	876·0 „ 962·0	928·9	56·3 „ 115·0	73·8	3 13 9	...
Stroud	1	1·83	...	850·0	...	14·3	...	3 12 9	...
Tamworth	6	1·716 to 4·24	3·162	899·5 to 935·5	915·5	60·0 to 93·0	79·0	3 17 10½ to 3 18 10½	3 18 2½
Tenterfield	2	2·00 „ 2·73	2·36	886·5 „ 890·5	888·5	10·0 „ 10·6	10·3	3 13 11 „ 3 15 9	3 14 10
Tibooburra	1	·423	...	973·5	...	20	...	4 2 4	...
Timbarra	2	...	...	708·0 to 899·0	803·0	97·0 to 280·0	188·5	...	...
Uralla	2	1·215	...	945·0 „ 955·0	950·0	40·0 „ 50·0	45·0	3 19 0 to 4 0 3	3 19 7½

GOLD—WESTERN DISTRICT.

Locality.	No. of Samples.	Loss per cent. in Melting.	Average Loss.	Gold in 1000 parts after Melting.	Average of Gold.	Silver in 1000 parts after Melting.	Average of Silver.	Value per Ounce after Melting.	Average Value.
								£ s. d. £ s. d.	£ s. d.
Bathurst . . .	8	1·47 to 2·033	1·83	827·0 to 930·0	905·5	59·0 to 164	84·25	3 14 0 to 3 19 11	3 17 10¾
Brown's Creek .	1	...	...	932·35	...	...	...	...	...
Burrendong . .	3	...	...	917·90 to 942·80	931·85	56·9	...	...	...
Carcoar . .	3	2·94 to 10·92	6·93	878·0 ,, 960·0	922·16	36·0 to 119·0	71·16	3 15 2 to 4 1 8½	3 17 11
Gulgong . .	7	1·26 ,, 1·78	1·64	916·5 ,, 949·0	933·35	45·0 ,, 79·0	60·7	3 17 6 ,, 4 0 7½	3 19 1½
Hargraves . .	8	·97 ,, 1·33	1·17	915·0 ,, 961·0	934·87	33·0 ,, 83·0	54·25	3 15 11 ,, 4 1 9½	3 19· 6
Hill End . .	11	1·41 ,, 3·00	2·245	940·5 ,, 947·0	943·63	45·0 ,, 50·0	49·0	3 18 2 ,, 4 0 8½	3 19 9½
Ironbarks . .	1	2·420	...	942·0	...	55·0	...	3 18 3	...
Meroo River (Avisford)	5	...	...	949·65 to 961·40	955·92	38·6 to 50·05	44·32	...	...
Mudgee . .	7	1·55 to 8·86	2·97	887·0 ,, 941·0	924·8	55·0 ,, 105·00	68·8	3 13 5 to 4 0 2½	3 18 4
Ophir . .	1	...	...	915·0	...	82·0	...	...	3 .
Ophir Creek .	1	...	...	940·6	...	59·3	...	...	...
Orange . .	10	2·13 to 3·93	2·74	830·5 to 952·0	922·35	10·0 to 151·0	63·8	3 11 2 to 4 1 1	3 17 9½
Parkes . .	7	1·54 ,, 2·48	2·01	898·0 ,, 926·0	919·1	65·0 ,, 96·0	73·8	3 16 0 ,, 3 18 9	3 16 11¾
Pyramul . .	2	1·12	...	947·0 ,, 954·5	950·75	43·0 ,, 45·0	44·0	3 19 6 ,, 4 1 3	4 0 4½
Sofala . .	9	1·37 to 2·4	1·79	920·0 ,, 943·5	927·0	52·0 ,, 72·0	66·4	3 16 2 ,, 4 0 5	3 17 11¾
Stony Creek .	3	1·56 ,, 2·19	1·87	939·0 ,, 942·5	941·16	50·0 ,, 54·0	51·3	3 18 0 ,, 4 0 3½	3 19 9
Tambaroora .	23	1·31 ,, 2·77	1·83	922·35 ,, 954·0	944·34	42·0 ,, 76·4	52·61	3 18 3 ,, 4 0 6	3 19 7½
Tuena . .	2	...	...	987·3 ,, 943	940·15	54·0 ,, 55·0	54·5	3 18 2	...
Turon . .	2	...	...	918·0 ,, 928·0	923·0	68·0 ,, 78·0	73·0	...	...
Turon River .	8	...	...	916·05 ,, 956·4	929·01	42·9 ,, 83·9	63·4	...	...
Wellington .	2	...	...	855·0 ,, 946·5	900·75	45·0 ,, 135·0	90·0	3 11 5 to 3 18 6	3 14 11½
Windeyer . .	3	·995	...	946·0 ,, 959·0	953·3	37·0 ,, 53·0	43·3	4 0 5	...

GOLD—Southern District.

Locality.	No. of Samples.	Loss per cent. in Melting.	Average Loss.	Gold in 1000 parts after Melting.	Average of Gold.	Silver in 1000 parts after Melting.	Average of Silver.	Value per Ounce after Melting.	Average Value.
Adelong	20	1·27 to 2·63	1·76	931·7 to 954·0	942·64	40·0 to 65·6	52·54	£ s. d. £ s. d. 3 18 0 to 4 0 7½	£ s. d. 3 19 5½
Adelong Creek	5	...	...	932·0 ,, 948·6	941·82	51·15 ,, 66·65	58·9	...	...
Araluen	17	1·34 to 2·19	1·65	895·0 ,, 958·5	925·15	35·0 ,, 105·10	72·4	3 16 5 to 4 1 0½	3 18 5½
Bombala	1	2·63	...	963	...	34	...	4 1 11½	...
Braidwood	8	1·79 to 2·33	2·03	928·0 to 959·0	944·9	34·0 to 67·0	48·8	3 18 7 to 4 1 7½	3 19 9¾
Burrangong	1	...	...	948	...	43	...	...	...
Cooma	3	3·17 to 4·22	3·69	924·0 to 933·0	932·16	55·0 to 70·0	60·3	3 17 3 to 3 19 11½	3 18 8
Delegate	1	...	...	971	...	27	...	...	...
Emu Creek	1	...	...	971	...	27	...	...	...
Forbes	1	2·647	...	921·5	...	75	...	3 16 5	...
Goulburn	2	2·51 to 6·87	4·69	948·5 to 975·0	961·7	22·0 to 43·0	32·5	4 0 9 to 4 2 11½	4 1 10¾
Kiandra	10	3·07 ,, 11·307	5·29	924·0 ,, 937·7	927·95	63·0 ,, 73·4	66·31	3 11 3 ,, 3 19 1	3 13 2
Mitta Mitta	1	...	...	895·7	...	104·3	...	...	...
Monaro	1	...	...	972·0	...	20·0	...	4 0 7	...
Murrumburrah	1	1·906	...	947·0	...	45·0	...	3 19 0	...
Nerrigundah	4	1·395 to 1·64	1·517	972·5 to 983	979·6	10·0 to 15·0	13·75	4 1 5 to 4 3 4½	4 2 4½
Temora	1	2·166	...	957·5	...	30·0	...	3 19 7	...
Tumberumbah	2	2·955	...	945·5 to 946	945·75	45·0 to 55·0	50·0	3 18 1 to 3 18 3	3 18 2
Tumut	1	6·28	...	927·5	...	70·0	...	3 19 1½	...
Urana	1	...	...	974·5	...	20·0	...	4 0 7	...
Young	3	1·11 to 2·39	1·67	943·0 to 957·0	949·3	36·0 to 49·0	44·3	4 0 4½ to 4 1 5¼	4 0 10½

The following three tables are extracted from the Report on the
Southern Gold-fields by the late Rev. W. B. Clarke, M.A.:—

ASSAYS OF GOLD MADE AT SYDNEY MINT, 1856.

Locality.	In 1000 parts.		Copper (with trace of Iron).	Remarks.
	Gold.	Silver.		
SOUTHERN DISTRICT.				
Araluen	934·90	65·1	0·0	...
,,	895·50	104·3	0·2	...
,,	915·20	84·8	0·0	...
,,	935·10	...	...	Bright granular gold.
,,	949·20	50·80	...	Dark-coloured grains.
,,	895·90	105·10	...	Dull granular, and rough nuggets.
In broken granite 10 ft. below surface	915·05	...	...	Bright granular gold.
Adelong	936·70	62·3	1·0	...
,,	946·40	53·1	0·5	...
,,	931·70	65·6	2·7	...
,,	936·85	...	...	Rough, nuggetty.
,,	946·45	...	... }	Smaller, more waterworn, nuggetty.
,,	945·20	...	... }	
,,	948·60	...	...	Light-coloured, fine, granular.
,,	932·00	...	...	Fine, bright, scaly gold.
,,	941·00	58·18	...	...
Mitta Mitta	895·70	104·30	...	...
Omeo	852·25	147·75	...	...

ASSAYS OF GOLD MADE AT SYDNEY MINT, 9th August 1860.

KIANDRA—New South Wales.

No.	Weight of Gold Dust in oz.	Loss per cent. in Melting.	Gold in 10,000 parts.	Silver.	Copper.	Net Value per oz.	Remarks.
						£ s. d.	
1	200·00	5·345	9,277	723	...	3 11 5–465	Rough, nuggetty.
2	215·08	5·375	9,258	734	8	3 11 3–347	,, ,,
3	63·94	11·307	9,335	656	9	3 7 4–647	,, ,,
4	92·48	4·520	9,264	717	19	3 11 11–367	,, ,,
5	67·59	4·348	9,247	731	22	3 11 11–692	,, ,,
6	42·17	5·620	9,377	623	...	3 12 0–192	Coarse, dull, granular.
7	31·88	4·925	9,262	727	11	3 11 8–320	Mixed, granular.
Mean	101·877	5·920	9,288	701·5	9·85	3 11 1–290	

The following examples of gold from other sources are given for the
purpose of comparison:—

TASMANIA.

Locality.	Gold in 100 parts.	Silver.	Iron.	Copper.	Tin, Lead, Cobalt, Nickel.	Remarks.
Black Boy Flat . .	94·76	5·04	...	...	...	Bright, granular.
,, ,, . .	94·95	4·66	0·08	trace.	{ traces T.L.N.	} Granular.
Nook, Fingal . .	92·55	7·10	0·17	trace	trace T.	Rough and fine.
Fingal	90·89	8·02	...	trace.	1·000	Waterworn nuggets.

COMPOSITION OF GOLD FROM VARIOUS PLACES.

Locality.	Specific gravity.	Gold.	Silver.	Iron.	Copper.	Bismuth.	Lead.	Silica.	Total.	Analyst.
Queensland—										
Gilbert River .	...	89·920	9·688	0·070	0·128	...	0·026	...	99·832	R. Smith.
Paddy's County	...	92·800	6·774	0·114	0·048	trace.	0·048	...	99·684	R. Daintree.
Cornwall, Ladock	...	92·34	6·06	trace.	...	...	...	1·60	100·000	A. Church.
Ashantee . .	17·55	90·055	9·940	trace.	trace.	...	...	...	99·995	,,
Scotland—										
Wanlockhead .	16·50	86·60	12·39	0·35	...	...	...	...	99·340	,,
Sutherlandshire	16·62	79·22	20·78	...	...	...	...	...	100·000	,,
Australia . .	...	99·28	0·44	0·20	0·07	...	...	0·01	100·000	Northcote.
Bathurst, N.S.W.	...	95·68	3·92	0·16	...	...	...	...	99·760	Henry.

Danas' "Descriptive Mineralogy," p. 5.

Locality.	Specific gravity.	Gold.	Silver.	Iron.	Copper.	Silica.	Total.	Analyst.
Wales—								
Clogau, quartz vein, No. 2 .	17·26	90·16	9·26	trace.	trace.	0·32	99·74	D. Forbes.
	15·62	89·93	9·24	trace.	...	0·74	99·81	,,
Mawddach River, Gwyn Fynydd, wash gold	15·79	84·89	13·99	0·34	...	0·43	99·65	,,
Cornwall, St. Austell Moor .	16·52	90·12	9·05	...	...	0·83	100·00	,,
Ireland, Wicklow, wash gold {	15·07 } 14·34 {	91·01	8·85	...	...	0·14	100·00	,,
Sutherlandshire—								
Kildonan Valley . . .	15·799	81·11	18·45	...	...	0·44	100·00	,,
,, ,, . . .	15·799	81·27	18·47	...	...	0·36	100·00	,,
Venezuela	...	93·58	3·69	1·60	0·65	...	99·53	Williams.
West Africa, gold grains . .	14·63	89·40	10·07	...	0·53	...	100·00	K. Wibel.
,, ,, . .	16·20	87·91	11·40	...	0·69	...	100·00	,,
,, gold dust . .	...	97·23	2·77	...	...	...	100·00	,,
,, ,, . .	...	96·40	3·60	...	...	...	100·00	,,
,, ,, . .	...	92·03	5·82	...	2·15	...	100·00	,,
,, gold dust washed from clay .	...	97·81	2·19	...	...	...	100·00	,,

Watts' "Dictionary of Chemistry," vol. vii., p. 572.

		Gold.	Silver.
Transylvania—Vöröspatak	. . . . .	60·49	38·74
South America— { Antioquia	. . . .	64·93	35·07
{ Marmato	. . . .	73·45	26·48
British Columbia—Stephen's Creek	. . .	79·50	19·70
Wales—Welsh Gold-Mining Co.	. . .	76·40	22·78
Scotland— { Sutherland	. . . . .	79·22	20·78
{ Wanlockhead	. . . .	86·60	12·39
California—Mariposa	. . . . .	81·00	18·70
Russia—Borushkoi	. . . . . .	83·85	16·15
Australia	. . . .	87·78	6·07
Africa—Ashantee	. . . . . .	90·055	9·940

Dr. Ure's "Dictionary of Arts," &c., vol. ii. pp. 686, 687.

The average fineness of Californian gold is stated at ·880. Canadian gold usually contains from 100 to 150 parts of silver to the 1000, but the Nova Scotian gold much less. In some of the above, notably that from Vöröspatak and Antioquia, the amount of gold present is less than two-thirds.

The average fineness of Victorian gold is about 23 carats, that is to say, it contains about 96 per cent. gold and 3½ per cent. of silver, with about ½ per cent. of other metals. Farther north, in New South Wales, the average fineness is 22 carats 1⅞ grains, or 93½ per cent. gold and 6 per cent. silver. Still farther north, in Queensland, the average fineness is but little more than 21 carats, or 87·25 per cent. gold and 12 per cent. silver. Maryborough gold only contains 85 per cent. gold and as much as 14 per cent. silver (F. B. Miller, F.C.S., *Trans. Roy. Soc. N. S. W.*, 1870). But beyond this the northern gold again becomes richer; the gold from the Palmer River alluvial workings has a greater fineness of gold, with only small quantities of silver and other metals. The gold from Mount Morgan has a fineness of 997, so that it is practically pure.

Vein Gold.—The greater portion of the gold found *in situ* in New South Wales occurs in quartz-veins running through the older and metamorphic rocks belonging to the Silurian, Devonian, and Carboniferous periods. Calcite is occasionally the vein-stuff. Gold is said to have been found in crystallised felspars, a most unusual matrix.

The rocks in which auriferous veins are most commonly met with are the various argillaceous slates and chloritic and talcose schists; also in granite, as at Braidwood and Bowenfels, porphyries, and other similar metamorphic rocks; in eisenkiesel at Carcoar. The most productive auriferous quartz-veins have been found in connection with diorites, hornblendic granites, porphyry, Silurian slates, schists, and with serpentine. The walls and "country" of such veins are also usually auriferous to greater or less distances.

As examples of the richness of portions of gold-veins the following may be cited:—A telegram from Hill End, on February 1st, 1873, stated that at Beyers & Holtermann's mine 102 cwt. of gold had been raised in 10 tons of stuff. From the same mine a slab of vein-stuff and gold weighing 630 lbs. was exhibited which was estimated to contain about £2000 worth of gold. Many other similarly rich blocks were also shown.

The Mint returns for the gold from 415 tons of vein-stuff from this mine were 16,279·63 oz., value £63,234, 12s., in 1873.

Krohmann's Company, also at Hill End, raised in 1873 436 tons 9 cwt. of stuff, for which the mine returns were 24,079 oz. 8 dwts. of gold, value £93,616, 11s. 9d.

Gold-reefs in New South Wales have not yet been worked to any great depth. At the United Gold-mine, Adelong, they were in 1881 getting good stone from a depth of 874 feet. In 1884 the Great Victoria Company was working at 1060 feet, and the Williams Mine was at about the same depth. The Consols Mine, Grenfell, had a depth of 940 feet in 1882 ; and Krohmann's Mine, Hill End, was 830 feet deep in 1881.

Associations.—The most common minerals which are found with vein-gold are iron pyrites, which is never quite free from, and is sometimes exceedingly rich in, gold ; and iron oxide, which is for the most part derived from the decomposition of various pyrites. It is found in association with antimonite at Sandgate, co. Sandon, New England; at the Eleanora Mine, Hillgrove, in the same county. In some cases the antimonite serves as the matrix of the gold, but in most of the specimens which have come under my notice the gold is held by quartz intimately mixed with the antimonite. This association of gold and antimonite is extremely rare not only in New South Wales but elsewhere. At the New Reform Gold-mining Company, Lucknow, native gold occurs with native arsenic in calcite, and in mispickel, where the mispickel contains in some places over 2000 oz. of gold per ton; with mispickel at Carcoar and near Scone, and at Moruya with silver sulphides too; also in calcite at the Crow Mountains, Barraba; at Solferino, in the Garibaldi Reef; and, it is stated, near Gunnedah ; at Tuena, at Lake Cowal, at Humbug Creek, at Grenfell, and at Merimbula. In 1886 large masses of gold were obtained from calcite at Armstrong's Mine, Ti Tree Creek, Oaky, ten miles south of Barraba, in the Peel and Uralla mining district. The reef was about 16 inches wide and had been followed to a depth of 80 feet. With calcite and serpentine with mispickel at Dungog; with pyrrhotine and calcite at Hawkin's Hill ; with galena and zinc-blende at Grenfell; with galena, zinc-blende, magnetite, molybdenite, chlorite, and scheelite at the Williams Mine, Adelong; calcite, talc, asbestos, hornblende, and serpentine at the Floreston Mine, near Gundagai. On the calcite and hornblende the gold is sometimes in the form of paint-like films.   Gold has been found in a thin vein of talc in quartz near Bathurst; with serpentine at Lucknow; with steatite, cuprite, malachite, tenorite, and other copper ores, notably in the Canobolas and in the Winterton Mine, Mitchell's Creek, near Bathurst, where it is also associated with barytes in well-developed, although small, crystals, and with mimetite, a chloro-arseniate of lead. Beautiful specimens of native gold in malachite and red oxide of copper have been yielded by the Kaiser Mine, Mitchell's Creek, near Bathurst. Gold is also found with mimetite in the Adelong district, with carbonate of lead from about twelve miles from Emmaville; with bismuth ore from Kingsgate and Comstock Lode, near Yarrow; with barytes from near Glanmire; it is reported with tinstone in the cliffs at Eden, and with native arsenic at Solferino.   At Hill End it was found with muscovite mica.

Gold and native copper have been found together in quartz-veins, and in rocks through which the veins pass.

In alluvial deposits gold is associated in New South Wales with a very large number of minerals; and it is remarkable that certain of them, such as platinum, osmo-iridium, sapphire, ruby, oriental emerald, and diamond, have not yet been found *in situ*.   Amongst other minerals we have tinstone, titaniferous iron, magnetic iron, chrome iron, brookite, rutile, anatase, emerald, beryl, topaz, zircon, hyacinth, spinelle, garnet, red and brown hæmatite, pyrites, binoxide of manganese, galena, blende, tourmaline, magnesite, and many more of less value.   Quite recently alluvial gold and metallic copper have been discovered together in some new ground opened at the head of Whet Creek, near Mount Misery, Nundle, a specimen of which was forwarded to me by Mr. D. A. Porter, of Tamworth, on April 13, 1882.   The particles of metallic copper are much smaller than those of the gold; the latter, however, do not exceed a square millimetre in area.   The gold is not much waterworn, and under the microscope is seen to be distinctly crystallised in parts.

The grains of copper, although of more or less spherical form with mammillated surfaces, are in some instances partially crystallised.

Mr. Porter's assay of the sample gave him the following results:—

Gold	23·0
Copper	61·0
Iron oxide	10·0
Loss	6·0
	100·0

The iron oxide in the above is in the form of titaniferous iron and magnetite; smaller quantities of other minerals, usually found with alluvial gold, are also present.

The alluvial deposits are of various ages; those which yield payable gold are of Permian, Cretaceous, Tertiary, and Quaternary age, and are often deeply buried by overflows of igneous rocks. Some are being worked to a depth of 200 feet.

Gold is found in small quantities in the tin-drifts of New England, especially in the older drifts—conglomerates or " cements," as they are termed by the miners.

Gold in the Coal-Measures.—With reference to this, the Rev. W. B. Clarke made the following remarks in the fourth edition of his " Sedimentary Formations of New South Wales," p. 9 :—

" This (*i.e.*, the occurrence of gold in the Carboniferous rocks) is thus referred to in a communication to me from Mr. Daintree, F.G.S., in a letter dated Maryvale, North Kennedy, January 22, 1870 :—

" ' I believe if the Peak Downs district were carefully mapped, it would be incontestably proved that *payable* drift gold is there found in the Carboniferous conglomerates.'

" He then gives a section of the shaft and drive then being worked at the Springs, about twelve miles from Clermont, and adds :—' The miners use the Carboniferous sandstone, the Glossopteris bed at bottom, and take the cement several inches from its junction with the Glossopteris bed for their wash-dirt. The surface of the Glossopteris bed is unbroken, dips southerly at an angle of about 5°, and the cement lies conformably on it ; and little patches of mud deposit in the cement, similar in appearance to the Glossopteris sediment, lie in the same plane as that bed, and I have no doubt the cement is conformable to the Glossopteris bed of the same period of deposit. Small fragments of coal were taken from the adjoining shaft, and I have no doubt, with the necessary time given to the work, Carboniferous fossils may ultimately be found in the conglomerates themselves—so putting the matter beyond reach of dispute.'

" A similar instance of such an occurrence was examined by myself in the coal-measure drift of Tallawang, in the county of Phillip, in the year 1875, and recognised as payable by C. S. Wilkinson, Esq., F.G.S., the present Geological Surveyor, in his report to the Minister of Mines, December 1876, in which place there is mention of other notices by myself of like association. The localities are similar in geological structure ; for, almost in the words of Mr. Daintree, which Mr. Wilkinson never read, the latter says :—' These conglomerates are associated with beds of sandstone and shale, containing Glossopteris, the fossil plant characteristic of our coal-measures.'—*Annual Report of the Department of Mines* for year 1876, p. 173.

" I made a section of the deposits which I found resting on hard shales (probably Devonian) in which numerous shallow shafts have produced alluvial gold. The bottom of the beds above the base exhibited a brecciated fragmentary deposit, well seen a mile or two away, on the road to Cobbora—above which sandstones, flinty shale, coarse grits, the red shales of Mount Victoria and Blackheath occur; and, nearer the top, Vertebraria and Glossopteris and charcoal are met with. One of the beds was of quartz-pebbles, cemented by ferruginous matter, precisely like many detrital fragments in other gold-fields, and specially resembling that above Govett's Leap, in which I obtained gold in 1863."

Mr. Clarke had previously ascertained that the Hawkesbury sandstone on the north shore of Sydney Harbour and at Govett's Leap contained traces of gold, and had also detected gold in the coal-measures of the southern part of the colony, near Shelley's Flat, Shoalhaven; * and the late Sir Thomas Mitchell also found gold in a quartz-pebble from the Carboniferous conglomerates in the year 1855, at Wingello, on the road from Braidwood.

Gold is also found in the coal-measures in Tasmania, and in New Zealand.

In connection with the above, it is interesting to note that the Carboniferous limestone near Bristol, England, contains gold and silver. Messrs. W. W. Stoddart and Pass found appreciable quantities of both metals in the limestone at Walton, near Clevedon.

The analysis of the dried limestone gave :—

Alumina	·8777
Oxide of iron	4·8000
Carbonate of lime	94·3000
Silica	·0200
Silver	·0023
Gold	a trace
	100·000

An assay was made by Mr. J. P. Merry, of Swansea; he found in one sample 94 grains of silver to the ton, and another sample contained very nearly an ounce. The quantity of gold varied from 3 to 5 grains per ton.—See Dr. Ure's "Dictionary of Arts," &c., vol. iv. p. 419.

The Rev. W. B. Clarke mentions that gold is found at the mouth of the Richmond River distributed in the sand, and covering pebbles on the sea-beach; a similar distribution is found in the sand of Shell Harbour. The black sand found in places along the coast between the Richmond and Tweed Rivers is all more or less auriferous, and after it has been concentrated by the action of storms it is sufficiently rich to pay to work. The gold is in exceedingly fine particles. Other spots give similar indications, and some specimens of gold were brought up from the sea-bottom by the sounding apparatus of H.M.S. Herald off Port Macquarie.

Distribution.—From the fact that gold is so widely scattered over

* "Southern Gold-fields," W. B. Clarke, pp. 43, 44, and 245. Sydney, 1860.

nearly the whole of New South Wales, it would be almost an endless task to attempt to enumerate the names of all the localities at which it has been found; it must, therefore, suffice to refer to the names of the principal gold-fields already cited in the tables which show the proportion of silver contained by gold from various parts of the colony, and to the mineral map accompanying this volume, which roughly shows the approximate area of the various gold-fields. The proclaimed goldfields cover an area of some 70,000 square miles; the workable area is probably far greater.

The Discovery of Gold.—It is not my present intention to express any opinion upon the long-disputed question as to who was the original discoverer of gold in Australia; but it may not be out of place to quote certain statements which have been made from time to time, so that each may judge for himself.

The first mention of the occurrence of gold in New South Wales was made as early as the month of August 1788. The alleged discovery by a convict of the name of Dailey, however, proved to be without foundation, as he afterwards confessed that he had filed down a yellow metal buckle, and had mixed with it some gold filed from a guinea, and some earth to give it a natural appearance.—*Vide* Captain Hunter's *Journal*, p. 84, published 1793. Mr. John White, Surgeon-General to the settlement, also gives a similar account of the matter in his *Journal*, published in 1790.

"Some convicts who were employed cutting a road to Bathurst are said to have found gold in a considerable quantity, and were only compelled to keep silence on the point by menaces and floggings, 1814." —Heaton's "Australian Dictionary of Dates," p. 109. These statements were probably true, since the last portions of the road pass through what has since proved to be gold-bearing country.

"A convict flogged in Sydney on suspicion of having stolen gold, which he stated he had found in the bush, 1825."—*Ibid.*

The *Evening News* of Sydney for 7th August 1875 contains the following statement with respect to the original discovery of gold:—

"We are in a position to show that gold was discovered, and we believe officially reported to the Government, upwards of fifty-two years ago, viz., on the 16th February 1823. On that date Mr. Assistant-Surveyor James M'Brian discovered the precious metal at a spot on the Fish River, about midway between O'Connell Plains and Diamond Swamp, a little to the north of the old Bathurst road, and about fifteen miles east of Bathurst. We have now before us an extract from Mr. M'Brian's field book, which book is preserved in the Surveyor-General's Office. It reads as follows:—'February 15, 1823. At 8 chains 50 links to river, and marked gum-tree. At this place

I found numerous particles of gold in the sand and in the hills convenient to the river.'"

It is stated in a Sydney paper that Mr. Cohen, a silversmith of Sydney, purchased a piece of auriferous quartz from a labouring man in December 1829.

Mr. Davison mentions in his book on "The Discovery and Geognosy of Gold Deposits in Australia," London, 1860, that a servant of Mr. Low's had, in 1830, found a specimen of gold several ounces in weight on the Fish River, nearly in the same locality as Mr. Assistant-Surveyor M'Brian.

In reference to the early discovery of reef gold Mr. Wilkinson makes the following remarks (*Annual Report of the Department of Mines,* 1877, p. 202):—

"In one of the reefs in diorite, near the summit of Diamond Hill, it is said that gold in quartz was discovered in 1823. Mr. J. Willard Low, of Sidmouth Valley, informed me that in that year, in his presence, his father (Mr. Robert Low) and Lieutenant W. Lawson, while collecting some specimens of quartz crystals from the reef, found one specimen of quartz containing a piece of gold of the size of a pea. To make sure that it was gold, these gentlemen are said to have had the specimen tested. It is also interesting to observe that on the Fish River, about $2\frac{1}{2}$ miles north from this spot, Mr. Assistant-Surveyor M'Brian, when engaged on the survey of the river, on the 15th February 1823, stated that he discovered gold."

Count Strzelecki found gold, associated with pyrites, in 1839, in the Vale of Clwydd.

The two following letters were published in the *Sydney Morning Herald* of 17th May 1851, and are of very great interest in connection with this question as to the first discovery of gold:—

"*To the Editors, Sydney Morning Herald.*

"GENTLEMEN,—Whilst reading this afternoon the leading article headed ' Gold,' in your number of to-day, I felt convinced that Count Strzelecki must be entitled to more credit as a discoverer of gold ore in this colony than had therein been accorded to him ; for the belief was strong in my mind that previously to 1840 he had himself informed me of its existence in the country west of the Blue Mountains.

"Searching this evening amongst my old letters, I have luckily met with one addressed to me by the Count in 1839, which I think proves, at all events, that its existence was then fully believed in by him, and had been at least *scientifically discovered by and known to him ;* and this, as far as his fame as a geologist is concerned, is; I conceive, the gist of the matter, and of more consideration than if, by accident or otherwise, he had actually picked up a specimen of the precious metal.

" In justice to a highly accomplished and much-esteemed gentleman and man of science, to whom the colonists are much indebted for his arduous and gratuitous researches and labours in the field of Australian geology, I shall be glad if you will publish the extract from his letter to me.—I am, Gentlemen, your obedient servant,

" THOMAS WALKER.

"FORT STREET, *May* 15, 1851."

———

" WELLINGTON, 16*th October* 1839.

" MY DEAR SIR,—I write you this from Wellington, and on my knee, as it happens that in the place the epistolary fit has taken hold of me there is no table, but in compensation plenty of petrified bones, which I excavate here with my hands—bones, may be, of hippopotamus, or some other species which once was in this part of the world, and is no more. I find the Wellington caves far superior to the Boree ones, and most interesting, but frightfully absorbing my time. I say frightfully, because, thinking of what little I have seen of the colony, and what still remains to be explored, I shudder.

" The distances, too, extend themselves most provokingly under my pursuits; for instance, the distance between Wellington and Sydney, 180 miles, but it was in 420 miles I accomplished it, in true zig-zag rambling, scrambling, and occasionally starving; but seeing much, and surveying barometrically a great track, and securing for mineralogy and geognosy a pretty considerable number of notes. This I accomplished every inch on foot, carrying a weight of 40 lbs.

" You may take it for granted that between Sydney and the ' Dividing Range,' in the direction of Bathurst, and in the width of sixty miles, there are no metals except iron, no minerals of any consequence but alum in its native state; carburet of iron (black lead), and plenty of coal. Not far from Mount Hay there is a thermal spring of chalybeate water, strongly impregnated with carbonic acid—most beneficial to health impaired by dyspepsia or nervous affection, but, as fate would have it, threatening to kill by the exhausting fatigue of the journey whomsoever should attempt to get at it.

" On this side the Dividing Range the variety of rocks and embedded minerals augment—indications most positive of the existing silver and gold veins are met with. The want of means, however—that is, time and men—did not allow me to trace them to their proper sources. Why has the Government not sent heretofore a man of science and mineralogical and mining acquirements to lay open these sources of health still hidden beneath, and which may prove as beneficial to the State and individuals as the rest of the branches of colonial industry?—Believe me, yours most truly,

"P. E. DE STRZELECKI.

"Thomas Walker, Esq."

The following extract from a letter written by Count Strzelecki to Captain P. King, R.N., also dated from Wellington, but ten days later, viz., 26th October 1839, and quoted by Judge Therry in his book entitled "Thirty Years' Residence in New South Wales," is another account in Count Strzelecki's own words of his share in the discovery of gold:—

"I have specimens of excellent coal, some of fine serpentine with asbestos, curious native alum and brown hæmatite, fossil bones, and plants, which I digged out from Boree and Wellington caves, but particularly a specimen of native silver in hornblende rock, and *gold in speck in silicate*, both serving as strong indications of the existence of these precious metals in New South Wales. It was beyond my power to trace these veins or positively ascertain their gauge. I would have done so with pleasure, *pro bono publico*, but my time was short, and so were the hands. I regret that the Government, having reserved all the mines for its benefit, did not send here a scientific man, truly miner and mineralogist, to lay open these hidden resources, which may prove as beneficial to the State and individuals as the rest of the branches of the colonial industry."

The reasons why Count Strzelecki did not follow up his discovery are also given by himself as follows:—

"I was warned of the responsibility I should incur if I gave publicity to the discovery, since, as the Governor argued, by proclaiming the colonies to be gold regions the maintenance of discipline would be impossible. These reasons of State policy had great weight with me, and I willingly deferred to the representations of the Governor-General, notwithstanding that they were opposed to my private interests."

Gold is said to have been found in 1843–44 by a shepherd named Macgregor, in the Wellington Valley; and in 1849 another shepherd boy was reported to have found a gold nugget in the Pyrenees, Victoria.

With reference to the important part which the Rev. W. B. Clarke played in the discovery of gold in Australia, I cannot do better than

quote the words of Professor Archibald Geikie, F.R.S., who, in his
" Life of Murchison," says :—" Count Strzelecki appears to have been
the first to ascertain the actual existence of gold in Australia ; but, at
the request of the colonial authorities, the discovery was closely kept
secret. The first explorer who proclaimed the probable auriferous veins
of Australia on true scientific grounds, that is, by obtaining gold *in situ*
and tracing the parent rocks through the country, was the Rev. W. B.
Clarke, M.A., F.G.S., who, originally a clergyman in England, has
spent a long and laborious life in working out the geological structure
of his adopted country, New South Wales. He found gold in the
Macquarie Valley and Vale of Clwydd in 1841, and exhibited it to
numerous members of the Legislature, declaring at the same time his
belief in its abundance. While, therefore, geologists in Europe were
guessing, he, having actually found the precious metal, was tracing its
occurrence far and near on the ground."

The Rev. W. B. Clarke gave the following evidence before a Select
Committee of the Legislative Council, 24th September 1852 (*vide
Parliamentary Papers*) :—

" *Q.* Have you any objection to state to the Committee when your
attention was first directed to the existence of gold in this country ?
A. It was in 1841, when I crossed the Dividing Range to the westward
of Parramatta, in endeavouring to satisfy myself as to the extent of the
Carboniferous formation in that direction, that I first became aware of
the existence of gold in Australia, by detecting it at the head of the
Winburndale rivulet, and in the granite westward of the Vale of Clwydd.
" *By Mr. Holroyd.*—*Q.* Did you go farther to the westward ? *A.*
No ; I had satisfied myself as to the object of my journey, and returned
home. At that time I knew nothing of the history of gold ; but since
then I have obtained every information I could upon the subject.
There are many persons living who know that I, very shortly afterwards, began to speak of the abundance of gold likely to be found in
the colony, and that as early as 1843 I mentioned it generally. On the
9th April 1844 I also spoke to the then Governor, Sir G. Gipps, and
exhibited to him a sample, but without any result as to further inquiry.
The matter was regarded as one of curiosity only, and considerations
of the penal condition of the colony kept the subject quiet, as much as
the general ignorance of the value of such an indication. In that year
I exhibited the gold, and spoke of its probable abundance, to some of
the then members of the Council ; and one of them, the late Mr
Robinson, replied to me, ' You ought to have been a miner,' but took
no further notice of it. The only one who seemed to take much interest in the subject was His Honour Mr. Justice Therry. I am able
to fix the date of the time when I spoke to Sir G. Gipps by the recol-

lection that I spent that day with him at Parramatta, and that it was the day on which a certain great meeting of squatters was held in Sydney.

"*Q.* What was the character of the gold you found? *A.* It was imbedded in a matrix of quartz, and also, as it is generally found in granite, in small flakes. I did not find alluvial gold.

"*Q.* Did you make it known to any of your scientific friends in England? *A.* Not at the time at which it was found, but I have written to my friends often since; and Sir R. Murchison has quoted from one of my letters to him in an article published by him in the *Quarterly Review* of September 1850. The editors of the *Illustrated Australian Magazine*, published at Melbourne 1851 (October), state also that they had seen letters written by me to my friends in England ten years ago, which proved that I knew the country to be auriferous (p. 211). I do not mention these facts for the sake of speaking of myself, but to substantiate my claim to have declared the auriferous character of this country many years ago, before the present gold workings began, and in consequence of the jealousies which have arisen respecting my knowledge and investigations of it.

"*By the Chairman.—Q.* How much gold was there in the specimens you found in 1843? *A.* The weight of one specimen was about a pennyweight; it was what might be termed a fair sample.

"*Q.* Did you find any other specimens afterwards? *A.* I had no opportunity of revisiting the localities; my official duties prevented me, and when I had opportunities of again going away on detached duties, it was altogether in other directions. It was always my intention, had occasion allowed, to make a close investigation of that district.

"*By Captain King.—Q.* Did you ever hear that Count Strzelecki had found gold at Bathurst? *A.* No; I never heard of his having found gold at all until last year, 1851 (June, I believe), when I read a letter published by Mr. Walker in the *Herald* newspaper, in which Strzelecki stated that he had found indications of veins of gold and silver near Wellington. There is no mention whatever of gold in his ' Physical Description,' which was published in 1845; and in the Geological Report of his Journey to Mount Kosciusko and Gipps Land, printed in the *Parliamentary Papers*, the only allusion he makes to gold is in his notice of auriferous pyrites, which he says was too insignificant to be regarded commercially.

"*By Mr. Holroyd.—Q.* Did you obtain your specimens from the creek or were they brought to you? *A.* The gold of which I have spoken as having first led me to the knowledge of the existence of the metal in New South Wales I obtained myself.

"*Q.* Did you break off any more quartz? *A.* No; I was not looking for gold; my object at that time was different. I was not then aware that other persons had found gold in various places of the western country.

"*By the Chairman.—Q.* Were you aware of its containing gold until you returned home? *A.* I knew it was gold, but I did not at first see what it indicated.

"*By Mr. Holroyd.—Q.* You did not prosecute the investigation any further? *A.* Not at that time; I merely regarded it as a mineralogical discovery."

In 1844 Sir R. Murchison pointed out the similarity of the Blue Mountain Chain of Australia, the Cordillera, to that of the Ural, and predicted the occurrence of gold. His prognostications, 1844–46–47, appear* to have been the first published. Colonel Helmerson, a member of the Imperial Academy of Sciences, St. Petersburg, who was well acquainted with the Ural Gold-fields, also expressed at this time a similar belief in the existence of gold in Australia.

In the Report of the Commissioners of the International Congress of Australian Statistics, held in London in 1861, it is stated that:—
"The first known discovery of the precious metal was made by Count Strzelecki in 1839, and was mentioned by him to some personal friends and to Sir George Gipps, the then Governor of the colony of New South Wales. It was again discovered and specially noticed by the Rev. W. B. Clarke, of Sydney, in 1841. The attention of the colonial public, however, was not attracted to the subject until the existence of an extensive gold-field throughout Australia was announced by Mr. E. H. Hargraves in 1851. A long time previous to this announcement, namely, in 1844, and without being aware of the finding of specimens of the precious metal by Count Strzelecki and the Rev. W. B. Clarke, Sir R. Murchison publicly asserted the high probability of the existence of gold in Australia. This bold induction was based on his knowledge of the geological formation of that country. And the wonderful results of gold-mining in Victoria and New South Wales afford a proof of scientific sagacity almost unparalleled in the history of science.

"JAMES MACARTHUR, ⎫
"EDWARD HAMILTON, ⎬ New South Wales.
"STUART A. DONALDSON, ⎭
"M. A. MARSH, Queensland.
"WILLIAM WESTGARTH, Victoria.
"EDWARD STEPHENS, South Australia.
"JAS. A. YOUL, Tasmania.
"J. E. FITZGERALD, New Zealand.

"OFFICES OF THE CONGRESS:
"SOMERSET HOUSE, LONDON, 18*th July* 1860."

---

* Royal Geographical Society's volume for 1845; *Trans. Royal Geographical Society of Cornwall,* 1846; *Report of the British Association,* 1849.

Simpson Davidson, in his " Gold Deposits in Australia," p. 27, says :—
" During all the time (apparently from 1847 to 1849) of my being at
Goodgood, the very crystallinic character of the mica schist continued
to attract the attention not only of myself, but also of the shepherds,
who were continually bringing specimens to me to ask if it were not
gold, or an indication of it; and amongst others whom I had lately
engaged as a shepherd was one by name Thomas Appleby. This man
had seen better days, and had had a great deal of experience in the
colony.   He was, besides, gifted with strong, natural good sense, and
intemperate habits alone had reduced him to the necessity of servitude
in this humble capacity. Appleby was always disposed to look for
gold at Goodgood, and I think it likely he may have lived in the
Western Districts, about the Wellington Valley, since he was not only
acquainted with the fact of a shepherd in that neighbourhood having
found gold during a number of years past, and of having effectually
concealed the fact from the authorities, but he described very correctly
the manner in which the fortunate shepherd got his gold, by breaking
up ' white flint, just such as this, sir,' as Appleby one day said, while
picking up at the same time the quartz pebbles which were scattered
about in tolerable abundance on the Goodgood Run, in addition to the
compact quartz-veins to which I have already alluded.

" Appleby was not the first man who mentioned to me the secret of
the gold-finding shepherd, for the fact of a shepherd habitually finding
gold was known, I venture to say, to every other shepherd in the colony
of two years' standing. The tradition had passed from shepherd to
shepherd ; and whilst the Government and the men of science, as it
afterwards appeared, either were or affected to be ignorant of the
circumstances, the facts were universally spoken of at this time in the
pastoral districts, though they might be but little heard of amongst the
Sydney citizens. But Appleby described the manner in which the
lucky shepherd obtained his gold more circumstantially and more correctly than any other person I met with, and I think that he must
either have collected his information from the immediate neighbourhood
of Wellington, or it may have been from an actual personal acquaintance
with the gold-collecting shepherd himself."

Page 275.—" It should also be stated that the Mr. Smith who is
mentioned purchased the gold, which it appears he sent to Sir R.
Murchison in England, for he never discovered any gold in Australia
himself.   Mr. Smith is chiefly known in the colony as having
exhibited to the Colonial Secretary a lump of gold found by a
shepherd about the year 1846 in the very neighbourhood where Mr.
Hargraves washed out the first gold on Summer Hill Creek. This
shepherd only found one piece of gold, and could never find any
more (on p. 356 Mr. Davison states that at the time it was supposed

by most people to have been melted down from stolen jewellery); but another shepherd—the more notorious Macgregor—had collected at various times numerous pieces near Wellington, about fifty miles distant from the former place, and I presume that some of these may have been the specimens which came into possession of Mr. Smith and Mr. Phillips, and were by them forwarded to Sir Roderick Murchison in 1848, since neither of these persons claim to be actual gold-finders."

And at page 340.—" Although the existence of gold in New South Wales was known for many years past to scientific men, yet it is generally admitted that Macgregor was the first person who found it in remunerative quantities. In the scramble for notoriety which occurred several years subsequent to Macgregor's success his claims were overlooked or set aside by those who laboured through the press and elsewhere to enforce their own demands, and he, being a man of humble position and of unobtrusive habits, made no endeavour at the time to establish a priority so justly his due. Macgregor, now a wealthy man, was formerly a shepherd in Mr. Montefiore's establishment at Wellington. His flock fed over land situated on Mitchell's Creek, and possessing a geological turn of mind, and from the nature of his occupation, abundant leisure to prosecute research, he was led to break up and examine portions of a quartz-ridge which traversed his sheep-run. During this investigation he met with a metal (amongst several others) which he supposed to be gold, and forwarded a sample of it to Sydney. The result proved the correctness of his opinion, and thenceforth he devoted the whole of his available time to the accumulation of the precious metal. The shepherd was ordinarily a prudent man, but becoming enamoured of a young woman, he revealed to her the secret of his wealth, and produced ample proofs of its reality. From this moment ceased the monopoly which he had enjoyed undisturbed for some years; the circumstances connected with his discovery gradually became known to the public, and the local excitement was intense. The quartz-ridge and its neighbourhood were visited by hundreds eager in the pursuit, all of whom were enabled to bear away an auriferous fragment. Dr. Curtis communicated the facts to Sir George Gipps, but failed to direct official notice to the locality; and ultimately Macgregor left the district (to which he is yet an occasional visitor) in search of other gold-fields. The excitement of the good people of Wellington is at present little less than it was in Macgregor's time, from the fact of these identical lands being now in the market. They consider, and with probability, that an opportunity will now be afforded for testing the auriferous capabilities of the immediate vicinity of the township. Three sections of 640 acres each are to be submitted for sale on the 29th of April instant at Wellington, and the result is looked forward to with impatience. Copper and other ores have been also found here, in addition to which the lands are of the highest character, probably the best in the country for agricultural purposes, being watered by Mitchell's Creek."

Again on page 348.—"By inquiring on the spot I have learnt that Macgregor had collected altogether gold of the value of about two hundred pounds sterling previously to the discovery of gold in placer-deposits. This sum may appear small, but considering that it was entirely obtained by breaking the surface quartz with a hammer while following the occupation of sheep-tending, I should think that it not improbably represented a thousand separate instances of gold-finding between the year 1840 and 1850."

Mr. Davison also mentions that in June 1849 there appeared an article in a Sydney journal headed "Port Phillip a Gold-field," with a circumstantial account of some youth having found a lump of gold between Melbourne and the Pyrenees. The statement was a good deal doubted at the time, but the account was perfectly true.

The above statement was made, Mr. Davison says, while he and Mr. E. Hammond Hargraves were detained by the weather in Sydney Harbour, on board the barque *Elizabeth Archer*, then bound for the gold-fields of California.

In a pamphlet on "Gold and Gold-fields," by James Wyld, London, occurs the following statement (p. 32) :—"Mr. Francis Forbes, of Sydney, about two years ago published and circulated in New South Wales a paper, in which he affirmed in the strongest manner, on scientific data, the existence of gold formations in New Holland. Mr. Forbes, not being listened to nor encouraged in his researches, went to California, where he died in 1850."

On 23d June 1875 some articles and letters referring to the discovery of gold appeared in the *Parkes Gazette*, in which it is stated that Mr. John Phillips announced the discovery of gold in 1847. A letter, dated from Jermyn Street, 16th July 1855, from Sir Roderick Murchison to Sir Charles Hotham, is cited, which states that "Mr. Phillips is the person who first announced to me that he had detected it (gold) in your government (1847). I so stated the fact in my letter of 1848 to the Colonial Secretary (Lord Grey), when I urged upon H. M. Government to take the initiative in developing the auriferous resources of the region."

Mr. Austin brought to Sydney a nugget of gold, worth £35, which he had found in the Bathurst district, January 1851.—Heaton's "Dictionary of Dates."

In the *Quarterly Journal of the Geological Society*, 1852, vol. viii. p. 134, there is an abstract of a paper by Sir R. I. Murchison on "The Anticipation of the Discovery of Gold in Australia."

"From 1841 to 1843 Sir Roderick published descriptions of the auriferous phenomena of the Ural Mountains on different occasions, as read before the Geological Society and the British Association. In 1844 he compared the eastern chain of Australia, about to be described by Strzelecki, with the Ural Mountains. In 1846 (a year before the Californian discovery) he addressed Sir C. Lemon, the President of the Royal Geological Society of Cornwall, on the subject, and recommended any Cornish tin-miners who were unemployed to emigrate to New South Wales, and dig for gold in the *débris* and drift in the flanks of what he had previously termed the "Australian Cordilleras," in which he had recently heard that gold had been discovered in small quantities, and in which, from similarity with the Ural Mountains, he anticipated it would be certainly found in abundance. Having received letters from residents in Sydney and Adelaide, saying that, in consequence of his writings, they had sought and obtained gold, specimens of which they sent, the author wrote to Earl Grey, Minister for the Colonies, in November 1848, referring to his anticipation as being about to be realised in a manner which might operate a great change in the colony. The author spoke of a geological discovery recently communicated to him by the Rev. W. Clarke, F.G.S., viz., the existence of many fossils of known Silurian species, including *Pentamerus Knightii*, and many shells and corals on the flanks of the dividing range of New South Wales. This discovery is important, for it completes the resemblance of the Australian Cordilleras (along which Devonian and Carboniferous fossils had been found) with the Ural Mountains; the two chains being thus shown to be zoologically, as well as lithologically, similar, and both to possess the same auriferous constants."

The following extract is from a lecture upon the geology of Australia by the late Prof. Beete Jukes, F.R.S. : *—

"Some of Sir. R. Murchison's observations, having found their way to the Australian papers, a Mr. Smith, engaged at that time in some ironworks at Berrima, was induced by them in the year 1849 to search for gold, and he found it. He sent the gold to the Colonial Government, and offered to disclose its locality on payment of £500. The Governor, however, not putting full faith in the statement, and being, moreover, unwilling to encourage a gold fever without sufficient reasons, declined to grant the sum, but offered, if Mr. Smith would mention the locality, and the discovery was found to be valuable, to reward him accordingly. Very unwisely, as it turns out, Mr. Smith did not accept this offer, and it remained for Mr. Hargraves, who came with the prestige of his Californian experience, to remake the discovery, and to get the reward from Government on their own conditions."

* "Lectures on Gold, delivered at the Museum of Practical Geology," p. 32. London, 1852.

The early discoveries, or alleged discoveries, prior to 1840, may be roughly summarised as follows:—Convict Dailey, 1788; Convicts, 1814–1825; Surveyor M'Brian, Mr. Robert Low, Lieutenant Lawson, 1823; Mr. Cohen's purchase, 1829; Mr. Low's servants, 1830; Count Strzelecki, 1839. The Rev. W. B. Clarke reports finding gold in granite, *in situ*, on 23d and 24th February 1841, near Hartley, at the head of the Wimburndale rivulet.

The Rev. W. B. Clarke says in his " Plain Statements," 1851, p. 7, that Messrs. Macgregor, Stewart, Trappitt, and others found gold prior to Mr. Hargraves. There is considerable evidence to prove that gold was several times obtained in Victoria and publicly exhibited in Melbourne in 1848 and 1849.

To Mr. Hargraves, in 1851, was reserved the satisfaction of showing that gold existed in great quantities in various parts of the colony, and that it could be readily obtained from alluvial deposits by means of the cradle.

There is some interesting matter relating to the early discovery of gold in Australia in an Imperial Blue-book, " Correspondence relative to the recent Discovery of Gold in Australia," presented to both Houses of Parliament. London, February 3, 1852.

RETURN showing the Quantity and Value of Gold produced in the Colony of New South Wales (from the *Annual Reports of the Department of Mines*, Sydney).

Year.	Quantity.	Value.	Year.	Quantity.	Value.
	ozs.	£		ozs.	£
1851	144,121	468,336	1870	240,858	931,016
1852	818,751	2,660,946	1871	323,610	1,250,485
1853	548,052	1,781,172	1872	425,130	1,643,582
1854	237,910	773,209	1873	361,785	1,395,175
1855	171,367	654,594	1874	270,823	1,040,329
1856	184,600	689,174	1875	230,883	877,694
1857	175,949	674,447	1876	167,412	613,190
1858	286,798	1,104,175	1877	124,111	471,418
1859	329,363	1,259,127	1878	119,665	430,033
1860	384,054	1,465,373	1879	109,650	407,219
1861	465,685	1,806,172	1880	118,600	441,543
1862	640,622	2,467,780	1881	149,627·06	566,513
1863	466,111	1,796,170	1882	140,469	526,521
1864	340,267	1,304,926	1883	123,806	458,509
1865	320,316	1,231,243	1884	107,199	395,292
1866	290,014	1,116,404	1885	103,736	378,665
1867	271,886	1,053,578	1886	101,417	366,294
1868	255,662	994,655			
1869	251,492	974,149	Total . .		36,469,138

DIAGRAM SHOWING THE ANNUAL PRODUCTION OF GOLD RAISED IN NEW SOUTH WALES TO 31 DECEMBER 1886.

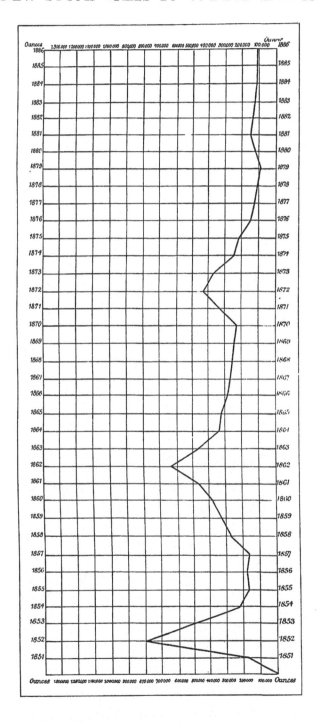

The Horiontal divisions indicate the Year and the Vertical divisions denote the Quantity of Gold.

SILVER.

NATIVE SILVER.

Native silver crystallises in the cubical system; specific gravity, 10·1 to 11·1. It does not appear to have been found in any quantity in New South Wales.

The Rev. W. B. Clarke mentions in his "Southern Gold-fields," published in 1860, that silver had been met with in the southern districts in two or three places in the form of small fragments and arborescent crystals. The same author mentions finding a thin plate of flexible silver having a specific gravity of 10.

Count Strzelecki makes the following mention of the occurrence of native silver on Honeysuckle Range, from Piper's Flat,* in New South Wales, in 1839 :—" Silver (native) in very minute and rare spangles, disseminated in primitive greenstone, . . . deserving further researches." Native silver is said to occur also in the Boorook Mines, with the chloride, sulphide, and other ores of silver; and at Calton Hill, Dungog, Hunter, and Macleay District; and Warril Creek, to the north of Kempsey.

Some years ago a small plate of silver was found by Mr. C. Suttor, junior, in the quartz of a vein containing galena, on the Mount Grosvenor Estate, near Bathurst.

At the Umberumberka Mine, about 1½ mile from Silverton native silver was obtained at a depth of about 240 feet, where the vein is about 4 feet wide. It is in the form of thin scales, and is associated with galena and siderite, the layers or scales of silver being attached for the most part to the galena.

Silverton is situated in the Barrier Ranges, and is about 822 miles west of Sydney, and fifteen miles from the South Australian border; but the silver-bearing country is said to extend some distance round, thirty or forty miles, and from Mundi Mundi on the west to Mount Gipps on the east.

ARGENTITE.—Silver Glance.

Chem. comp.: Sulphide of silver, AgS; silver, 87·1; S, 12·9 = 100. Crystallises in the cubical system; specific gravity, 7·19 to 7·36.

This ore has been found with iron pyrites in quartz, also in limestone on the Clarence and Manning Rivers. It occurs also at two or three places near Bathurst, co. Bathurst; at Copper Hill; at Brownlea; on the Page and Isis Rivers, Upper Hunter, co. Brisbane; and at Burnaby Creek, co. Argyle. With cobalt, zinc, and iron at Broulee,

* "Physical Description of New South Wales and Van Diemen's Land," by P. E. de Strzelecki. London, 1845.

Moruya, co. Dampier; at Teesdale, co. Bathurst; on the Queanbeyan River, co. Murray; at Burra Creek, co. Selwyn; on the Yass River and Burrowa Creek, co. King; at Buckinbah, co. Gordon; at Tacking Point, co. Macquarie; and on the Crookwell River, co. Georgiana. With gold, lead, and zinc at Gulgong; with carbonate of lead at Peel-wood; with galena and iron pyrites at Shellmalleer; on the Molonglo River, near its junction with the Murrumbidgee; and at the junction of Murrumbidgee Creek with Mountain Creek, co. Murray. In nearly all cases the silver sulphide occurs, mixed more or less intimately with galena, so that properly it should probably be termed argentiferous galena. It occurs with galena on Brookes' Creek, Upper Gundaroon, co. Murray, and Adelong, co. Wynyard; with fluorspar and galena at Woolgarloo; with galena at Wellingrove, in the Glen Innes District; at Grenfell, co. Monteagle; and Araluen, co. St. Vincent. The silver-bearing lodes at Yarrahappini, Warrell, Macleay District, run through granite and Devonian rocks; the vein-stuff consists of quartz, containing mispickel, zinc blende, iron pyrites, and galena, but up to the present these veins have not proved payable.

Silver-bearing veins occur at Boorook, about twenty miles to the north-east of Tenterfield, distant some thirty miles by the road; the veins are said to run through slate and "felspar porphyry," and the silver minerals are associated with quartz, oxide of iron, and iron pyrites, often rich in both silver and gold. In the upper portions of the lodes the silver seems to be mainly present as chloride; below, at a depth of 110 feet or 120 feet, it changes for the most part into silver sulphide.

At the Golden Age Mine the vein is composed of a porous quartz, with chlorite, clay, and much oxide of iron, and the ore is principally the chloride down to 80 feet, below which it changes to sulphide, mixed with argentiferous pyrites and zinc blende. The rock or country is described as a blue shale, or soft slate, in parts fossiliferous.

Some of the ore from the Boorook Mines contained as much as 800 oz. of silver and 8 oz. of gold to the ton; most of the gold is in the free state.

ANTIMONIAL SILVER ORE.

The compositions of the antimonial silver compounds hitherto met with have not yet been worked out. Some of the silver ore from Boorook is in part an antimonial one, mixed with the chloride, sulphide, and perhaps arseniate of silver; associated with this mixed ore are found native gold, iron oxide, iron pyrites, copper pyrites, chlorite, quartz, and other minerals. An arsenical compound of silver and antimony occurs at Moruya, probably fahlerz. Stephanite and polybasite are said to occur at Umberumberka.

The following analyses, made by Mr. W. A. Dixon for the Mining Department in 1879, will serve to show the composition of some of the mixed silver ores.

No. 1. Stone from Mr. J. Moffat's property, Boorook. Gold, 1 dwt. 14 grs. per ton; silver, 1 oz. 11 dwts. 5 grs. per ton.

No. 2. Stone from the Grand Junction Reef, Boorook. Gold, 3 oz. 16 dwts. per ton; silver, 42 oz. 4 dwts. 10 grs. per ton. Calculated into their proximate constituents, these analyses give the results in Nos. 3 and 4.

	No. 1.	No. 2.
Silica	97·710	91·765
Iron	·791	2·631
Zinc	traces	·477
Lead	·023	1·492
Copper	traces	·047
Antimony	·125	·132
Arsenic	·000	traces
Silver	} ·004 {	·129
Gold		·011
Sulphur	·324	1·552
Water	·762	1·008
Oxygen and loss	·261	·756
	100·000	100·000

	No. 3.	No. 4.	
Silica	97·710	91·765	
Sulphide of antimony	·174	·184	
Gold	·000	·011	
Sulphide of silver	·004	·148	
,, lead	·026	1·722	
,, zinc	·000	·715	
,, copper (Cu_2S)	·000	·060 } Copper pyrites,
,, iron (Fe_2S_3)	} ·485 {	·082 } 0·142
,, ,, (FeS_2)		1·774	
Oxide of iron	·823	2·507	
Water	·762	1·008	
	99·984	99·996	

KERARGYRITE.—Horn Silver, Silver Chloride.

Chem. comp.: Silver chloride, $AgCl$; silver, 75·3; chlorine, 24·7 = 100·0.

Specific gravity, 5·31 to 5·55.

Crystallises in the cubical system, often has the appearance of and cuts like wax. When crystallised, usually as small cubes and octahedra, grey, greenish, brown, or black in colour; colourless when pure and unexposed. Fuses in the candle flame. Said to occur in a vein near Braidwood, and within two or three miles of Queanbeyan.

Occurs at Boorook, especially in the upper portion of the veins, from which some well-developed crystals have been obtained.

During the past year or two very large quantities of silver chloride and other similar argentiferous minerals have been obtained from the Silverton District and Barrier Ranges.

"The loose masses found on the surface have usually a dirty green or brownish colour, and are known by the miners as 'slugs;' on cutting them or driving into them with a pick they present the usual characteristics of masses of silver chloride, being very tough and horn-like, and yielding a smooth shining surface. They sometimes weigh many pounds, but are usually much less.

"One specimen, from a vein 17 feet wide at the 312-feet level, Broken Hill Mine, Barrier Range, which was examined yielded the following result :—

Chloride of silver	81·67
Bromide of silver	10·19
Oxide of iron, alumina, silica, &c.	8·14
	100·00

"The vein-stuff is sometimes earthy, at others it consists largely of chlorite, and in other instances it contains bluish chalcedonic translucent quartz; at times it is mainly ferruginous.

"At the Broken Hill Mines also, at the 150-feet level, the silver chloride was found associated with chrysocolla, i.e., hydrous copper silicate. At the 100-feet and 212-feet levels the vein had a porphyritic structure, and the silver chloride was mixed with earthy grey copper sulphide (redruthite), and was associated with small crystallised red garnets; at the 212-feet level cuprite and quartz were also present in addition to the copper sulphide and other minerals.

"Large quantities of cerussite, i.e., lead carbonate, and galena seem to be present throughout most of the veins, so that smelting should be easy.

"In some cases the vein has proved to be very rich in silver: 48 tons of the ferruginous part of the vein at Broken Hill Mine yielded 37,000 oz. of silver, although by the method of smelting followed the loss must have been, as was alleged, very great. In another case 1300 oz. of silver per ton are said to have been obtained. In some cases the silver chloride is very well crystallised; the octahedra are quite one-eighth of an inch in length and fairly distinct, being, however, generally attached to one another in incipient branching forms, of a greenish shade, with a horn-like translucency.

"In some cases where the silver chloride is crystallised on a velvet black or brown hæmatite the effect is very fine, and the appearance is really very beautiful. This form is also found at the Broken Hill Mine, and associated with the silver chloride are small yellow imperfect hexagonal prisms of silver iodide.

"In one vein at the Broken Hill the vein-stuff is a white earthy

mineral resembling kaolin, and the vein is known as the Kaolin Vein
in consequence. The greenish crystals and plates of silver chloride
crystallised on some of the slicken-side surfaces of the kaolin have a
very pretty effect.

"Amongst the richest mines, in addition to the Broken Hill Mines,
are:—Christmas Mine, Lubra Mine, War Dance and Gipsy Girl Mine,
Thackeringa, North May Bell Mine, Silverton, Day Dream, Hen and
Chickens Mine, where the silver chloride occurs with azurite or blue
copper carbonate.

"Selected specimens, of course, assay very high; one piece of the
vein from the Lubra yielded 8493 oz. of silver per ton, and a yield of
16,000 oz. has been obtained from surface slugs." *

Chloride of silver—"slug" from Silverton, Barrier Range:—

Analysis.

Moisture	·48
Chloride of silver	72·23
Iron carbonate	4·26
Alumina	2·33
Carbonate of lime	6·40
Carbonate of magnesia	3·75
Insoluble in acids	9·70
Undetermined	·85
	100·00

(*Mines Report*, 1885.)

BROMARGYRITE.—Silver Bromide.

Chem. comp.: AgBr.   Silver, 57·4; bromine, 42·6 = 100·00.

Crystallises in minute cubes and octahedra, usually of a yellow,
green, or grey colour; cuts readily.   Very similar to silver chloride.
Said to occur in the Silverton Mines.

EMBOLITE.—Silver Chloro-bromide.

Chem. comp.: AgClBr.   The proportion of silver varies from 61
to 72 per cent.   Crystallises in the cubical system, usually in small
cubes of a yellow or green colour.
Specific gravity, 5·31 to 5·81.
Found at Winter & Morgan's Mine, Sunny Corner, Mitchell's
Creek, Bathurst.   Also in the Silverton Mines.

IODARGYRITE.—Silver Iodide.

Chem. comp.: AgI = silver 46, iodine 54 per cent.   Crystallises
in the hexagonal system, usually in scales and foliated masses, of a
lemon or greenish yellow; also grey.
Found at Silverton and Barrier Ranges.

* A. Liversidge, *Journal of the Royal Society of New South Wales*, 1886.

Chlor-iodide of silver occurs in ironstone at the 150-feet level, Broken Hill Mine, Barrier Range :—

Analysis.

Silver iodide	3·58
Silver chloride	3·78
Silver bromide	trace
Antimony	trace
Copper	trace
Iron oxide	45·56
Alumina	32·96
Silica	14·38
Potash	trace
	100·26

Ferruginous carbonate of lead, containing silver iodide from Broken Hill, Barrier Range :—

Analysis.

Moisture at 100° C.	·300
Silver iodide	·223
Lead carbonate	56·960
Copper carbonate	·780
Iron oxide and alumina	15·950
Silica	22·350
Carbonate of lime	2·200
Sulphuric anhydride, phosphoric acid, magnesia, &c.	1·237
	100 000

(*Mines Report*, 1885.)

ARGENTIFEROUS ORES.

On account of the mixed nature of many of the silver-bearing veins, the following general list of argentiferous localities is appended :—

In co. Argyle, cerussite and galena with a trace of antimony and mispickel,—antimonite and cervantite with lead in ferruginous quartz at Marulan; galena, cerussite, and copper sulphide in porous quartz at Slate Creek, Locksley; and earthy cerussite at Tarrago. Co. Ashburnham, earthy carbonate of copper ten miles west of Molong; and silver ore at Gumble Flat, Molong. Co. Auckland, silver ore near Merimbula and Perico. Co. Bathurst, ferruginous barytes near Bathurst; ore at Blayney; galena at Bull-Dog Range; barytes near Glanmire; iron pyrites and mispickel at Mount Grosvenor; porous ferruginous quartz with silver chloride,—cubical iron pyrites,— at Ophir; bismuth ore near Orange. Co. Beresford, galena at Cooma; and five miles from Cooma. Co. Bligh, quartz with silver chloride near Denisontown. Co. Buller, silver chloride,—iron pyrites,—blende, galena, and iron pyrites from Boorook. Co. Clive, mispickel,—earthy cerussite,—galena, pyrites, and copper carbonate,—iron pyrites and zinc blende in dark claystone at Bolivia; fahlerz and pyrites at Tenmile, parish of Park; iron pyrites and garnets in altered claystone,—

lead oxide,—fahlerz in quartz,—copper pyrites and mispickel at Pye's
Creek; metallic bismuth in ferruginous quartz at Hamilton's, Pye's
Creek; galena, zinc blende, and copper pyrites three miles north of
Pye's Creek; fahlerz and galena,—fahlerz in claystone on Severn
River, Tenterfield; galena, cerussite, and anglesite at Campbell's Reef,
Severn River; galena in felspathic rock at the Severn River Silver-
mine; cerussite and galena eleven miles from Tenterfield. Co.
Dampier, iron pyrites near Moruya. Co. Dudley, mispickel and
galena thirty-seven miles north-east of Kempsey. Co. Durham, mispickel with calcite and serpentine from Dungog. Co. Georgiana,
brown iron ore at Abercrombie; silver ore at Back Creek, near
Rockley; ferruginous cerussite at Costigan's Mount; fahlerz,—ferruginous quartz and galena near Rockley; galena, cerussite, and carbonate of copper on New Sewell's Creek, Rockley; copper pyrites in
ferruginous quartz on Thompson's Creek; galena and copper pyrites,
—black sand (chiefly magnetic iron), Trunkey; brown iron ore and
copper carbonate at Tuena; cerussite,—copper carbonate and cerussite
near Tuena. Co. Gipps, galena, cerussite and copper carbonate on
the Melrose Road, Condobolin. Co. Gough, galena at Cadell's, Deepwater; mispickel from head of Deepwater River, at Emmaville, Kingsgate, and Parish Parkes; mispickel, galena, and iron pyrites at Emmaville; galena with mispickel and arseniate of iron six miles south of
Emmaville; cerussite twelve miles from Emmaville; zinc blende and
pyrites from the Grampians; bismuth ore at Glen Innes, the Comstock
lode (near Yarrow), and Kingsgate; blende and pyrites in felspathic
clay at Parish Parkes; ore from Gordon's Reef, near Strathbogie;
mispickel and pyrites on the Mann River; fahlerz in claystone eight
miles from Mann River. Co. Goulburn, ore at Jingellic Creek. Co.
Harden, iron pyrites and galena at Muttama Reef; lead ore at
Mylora. Co. Hardinge, mispickel at Bundarra; galena and zinc
blende ten miles south-west of Tingha. Co. Inglis, antimony ore
at Bendemeer. Co. King, fine-grained galena at Yass; lead ore at
Good Hope Mine. Co. Mouramba, ferruginous chlorite rocks at
Nymagee. Co. Murray, galena and zinc blende near Boro; ferru-
ginous barytes and protoxide of lead at Captain's Flat, Molonglo.
Co. Napier, silver chloride at Narrangarie Silver-mine, which is between
Bong Bong and Coolah Creek. Co. Raleigh, blende, galena and
pyrites on Nambucca River; galena from Coast Range north of Macleay
River; ore on Warrell Creek. Co. Roxburgh, iron pyrites at Butler's
Creek and Clear Creek; lead ore and ironstone at Peelwood; ferru-
ginous cerussite at Sunny Corner. Co. St. Vincent, cerussite from
Nowra Ranges. Co. Wallace, galena and ferruginous cerussite,—
ferruginous cerussite with copper carbonate and lead protoxide from
Mount Trooper, Snowy River. Co. Wellington, porous brown iron ore

(friable and stalactitic) from Lewis Ponds Creek. Co. Westmoreland, copper pyrites on Blackman's Creek, Oberon; brown iron ore from O'Connell District; ironstone on Widow's Creek, O'Connell District; iron pyrites and galena on Native Dog Creek; green carbonate and red oxide of copper at Wiseman's Creek, where the copper ores are richer in silver than the lead ores. Co. Woore, galena and copper ore from Monaro District. Co. Wynyard, galena and magnetic pyrites at the head of Adelong Creek; ironstone from Wagga Wagga. Co. Yancowinna, silver chloride with galena, opaline quartz, zinc blende, pyrites, &c., at South Broken Hill; cerussite at Broken Hill; ore from Mount Gipps, Poolamacca, and Thackaringa; iron pyrites, galena, and other minerals at Umberumberka. Argentiferous mispickel thirty miles from Uralla. Galena and arsenical silver ore from the Melrose Ranges. Argentiferous galena from Everton, near Burrowa; Scrubby Rush; Sounding Rock, near Trunkey; and Judd's Creek, near Burraga.

The following is an abstract of Mr. Wilkinson's report upon the Sunny Corner silver lodes (*Mines*, 1886):—

"The primary formation of the Sunny Corner District, Mitchell's Creek, consists of Siluro-Devonian sandstones and shales, upheaved and penetrated by elvanite and quartz-porphyry, chiefly in a north and south direction. After the eruption of the igneous rocks, displacements took place on at least two different occasions, resulting in the opening of irregular fissures from a few inches to forty feet wide, in which were deposited gold and silver bearing sulphides of iron, copper, lead, zinc, arsenic, and quartz. A dyke of the elvanite, in the Sunny Corner Mine, has been split in two, and the fissure filled with clay, showing that the fracture took place after the intrusive rock had solidified. Then, again, the sedimentary formation has in some cases been dis- placed from its contact with the igneous rock, forming, where the original line of junction was uneven, the irregular cavities now filled with lode-stuff; but where the line of junction was even the lodes pinch out, though a well-defined fissure joint continues. In the Great Western Mine the surface of the rock has been grooved by the friction of one rock upon the other, and in the fissure argentiferous sulphides of lead, copper, zinc, and iron have been found in patches. The fissure then opened again and became filled with clay, in one place twenty feet thick, and somewhat resembling a decomposed felspathic basalt rock. Then shrinkage cracks formed in the clay lode were filled with cerus- site, probably derived from the decomposition of the galena in the breccia-lode. The sliding movement of the upper formation has taken place towards the north-west, for the ore deposits occur chiefly upon the north-western slopes of the intrusive formation. The larger ore deposits occur more in the higher levels than in the lower, which is perhaps due to the fissure walls being more regular in the latter than in the former. In the slopes above the south drive of the Great Western Mine native silver and native copper have been found in the joints, where the lode has been much oxidised; these metals have, no doubt, been precipitated from solution from the decomposed sulphides in the brecciated portion. Sulphides of iron, copper, lead, and zinc, in small crystalline masses, occur disseminated through the elvanite, which has been bored to a depth of 200 feet. I am of opinion that it is from thermal water permeating this intrusive rock that the metalliferous deposits in the fissure lodes have been derived. Towards the walls of the Paddy Lackey Reef the quartz becomes seamed with thin greenish layers containing iron pyrites. Galena also occurs, and gold can be seen in fine specks generally surrounding a nucleus of other mineral. The old reefs on the Big and Little Hills traverse a series of shales, sandstones, and coarse grits, which have been much tilted. From these reefs the alluvial gold in the bed of Mitchell's Creek has been derived."

The following analyses (*Mines Report*, 1886) are given to show the general nature of the ore on Mitchell's Creek :—

No. 1 is an analysis of an average sample of the lode-stuff from the New Nevada lode, Mitchell, taken from the whole width of the lode at the end of the north drive. It yielded 13 oz. 12 dwts. of silver per ton and a trace of gold, under 2 dwts. a ton.

No. 2 is an analysis of a similar average sample taken from a slope in the south drive near the 90-feet shaft, where the lode-stuff is more oxidised, and contains less sulphides than on the lower level. It yielded 2 oz. 14½ dwts. of silver per ton and a trace of gold.

	No. 1.	No. 2.
Oxides of iron and alumina	24·18	26·31
Insoluble siliceous matter (gangue)	46·63	51·02
Lead	9·34	6·25
Copper	5·06	4·02
Bismuth	·88	trace
Antimony	·43	trace
Arsenic	·90	trace
Sulphur	12·42	9·32
Manganese	trace	trace
Chromium	...	trace
Undetermined, loss, &c.	·16	4·08
Total	100·00	100·00

Respecting the silver deposits in the west Mr. Wilkinson, F.G.S., says (*Mines Report*, 1884) :—"I have examined eighty-one lodes, and there are a few others, the Day Dawn, Ophir, Black Prince, &c., that I did not see; but those above described include all the principal ones, and from them it will be seen—

" 1. That the geological formations which contain the argentiferous lodes of the Barrier Range silver-field are mica schists, clay slates, and sandstones, traversed by numerous quartz-reefs and intrusive masses and dykes of coarsely crystalline granite (pegmatite) and diorite. Nearly all the lodes occur in the mica schists; and they have been found over a tract of country seventy miles long and thirty miles wide, which has been only partly prospected, so that many more lodes will probably be discovered. But the metalliferous formations are known to occupy a much larger area, and extend to Kooringbury on the north, and on the east as far as the Eight-mile Tank, on the road to Silverton, about thirty-eight miles from Wilcannia.

"2. That the lodes, with the exception of those of the Broken Hill and Pinnacles, which are chiefly composed of ferruginous quartzite, all consist either of brown iron ore (gossan), containing argentiferous carbonate of lead and galena in bunches, and sometimes chloride and chloro-bromide of silver and carbonate of copper, or rarely of argentiferous carbonate of lead and galena alone; quartz is sometimes, though not always, present; and in one instance baryta occurs. It is evident that the oxides, carbonates, and chlorides have resulted from the decomposition of the sulphides, and perhaps arsenides, of iron, lead, silver, and copper, &c., which will be met with in their original condition below the water-level. Sulphide of lead (galena) and, in two instances, iron pyrites are even found above the water-level. I did not notice any distinct sulphide of silver, iodide of silver, or antimonial ores in the lodes; however, I have collected certain samples of ore for analysis, but they have not yet reached Sydney. Mr. J. Cosmo Newbery, C.M.G., Superintendent of the Technological Museum, Melbourne, reports having found 'chloride, bromide, and iodide of silver, with brown iron ore, carbonate and sulphide of lead, oxide and sulphide of antimony, and traces of bismuth' in the ore from the Christmas Mine. It is stated that 12 cwts. of this ore treated at the Victorian Pyrites Smelting Company's Works yielded 2575 oz. of silver. In one mine the water-level has been reached at a depth of 133 feet, in another at 72 feet, but no lode has been mined below the water-level.

"3. That the lodes, without exception, are very inconstant in thickness, both in longitudinal and vertical extent, and many of them thin out entirely within a few yards. A surface plan of the numerous lodes would resemble the shrinkage cracks upon the surface of a dried piece of cross-grained wood; in fact, the lode fissures were shrinkage cracks formed by the contraction of the rock mass after the intrusion of the igneous rocks.

"Some of the lodes appear to have been formed along an original joint in the strata, which is indicated by a well-defined wall in the lodes; and these will, I believe, continue to great depths, though varying in thickness in places. There is, therefore, a probability that silver and lead mining in this district will be a permanent industry."

EXAMPLES OF SILVER AND GOLD BEARING MINERALS.

These examples are not chosen solely for their richness in gold and silver, but rather to show how large a variety of minerals contain these metals. In the *Annual Reports of the Department of Mines*, from which they are taken, numerous other assays are given, to which the reader is referred for further details. In some cases the gold and silver mineral could probably be separated by mechanical means previous to smelting, but in other cases they are too intimately associated, and are lost in the ordinary dressing and smelting processes, *i.e.*, they are carried off with the gangue and baser metal. The presence of the precious metals should always be sought for, since, if the information yielded by these assays be trustworthy, the annual loss to the colony by the careless and imperfect treatment of the ores of the common metals must be very great. The presence of free gold and of distinct silver minerals was probably quite apparent to the eye in most of the very rich ones, although that is not stated.

Description.	Locality.	Per Ton.				Other Metals.
		Silver.		Gold.		
		Oz.	Dwts.	Oz.	Dwts.	
Brown iron ore	Abercrombie	4	1½	0	16	...
" porous	Lewis Ponds Creek	98	3½	0	13	...
" stalactitic	" "	151	9½	1	4½	...
Ironstone	Wagga Wagga	700 to 4,300 oz.		...		...
Ironstone, with quartz fragments	Peelwood	138	0			...
Iron pyrites	Boorook	50 and 60 oz.		0	15	...
" " cubical	Conoblas, near Orange	20	3½	0	10½	...
Iron pyrites in quartz	Ophir	17	12			...
" "	Clear Creek	8	16½	0	1½	...
Iron pyrites and galena	Butler's Creek	6	16	4	7	...
" "	Muttama Reef	4	0½	7	19	...
" "	Native Dog Creek	4	11	7	12½	...
Iron pyrites and mispickel	Mount Grosvenor	4	10	0	16	...
Iron pyrites and garnets in altered claystone	Pye's Creek, Bolivia	155	3	0		...
Black sand (principally magnetic iron)	Trunkey	10	4	8	3	...
Cerussite	Broken Hill	305	8½	...		...
"	Melrose Ranges	58	16	...		...

EXAMPLES OF SILVER AND GOLD BEARING MINERALS—*continued.*

Description.	Locality.	Per Ton.				Other Metals.
		Silver.		Gold.		
		Oz.	Dwts.	Oz.	Dwts.	
Cerussite	Nowra Ranges	26	18½	...		...
,,	Near Tuena :	41	13	...		Lead, 63·84 per cent.
Earthy cerussite	Bolivia District	5,281	10⅝	...		...
,, ,,	Tarrago	196	16	...		...
Ferruginous cerussite	Costigan's Mount, Tuena	1	3	60	18½	Lead, 43·5 per cent.
,, ,,	Milparinka	1,019	4	...		...
,, ,,	Sunny Corner	62	7½	0	9½	...
Cerussite and quartz	Twelve miles from Emmaville	1	19½	1	1½	...
Cerussite and galena	Eleven miles from Tenterfield	...		65	6½	Lead, 71·2 per cent.
Cerussite and galena, with a trace of antimony and mispickel	Carrington Mine, Marulan	2	17	197	4	Lead, 24·8 per cent.
Ferruginous cerussite, with copper carbonate and protoxide of lead	Near Mount Trooper, Snowy River	208	12½	...		Lead, 51·98 per cent.
Lead oxide	Pye's Creek	35	8½	...		...
Galena	Cooma	37	11	...		Lead, 74 per cent.
,,	Five miles from Cooma	30	4	...		...
,,	Melrose Ranges	136	0	...		...
,,	Three miles north of Pye's Creek	180	17⅝	...		...
Galena in felspathic rock	Severn River Silver-mine	...		4	2⅙	Lead, 40·4 per cent.
Galena and quartz	Cadell's, Deepwater	300	10½	...		...
Galena and ferruginous cerussite	Mount Trooper, Snowy River	20	8			
Galena, cerussite, and anglesite	Campbell's Reef, Severn River	...		5	1	Lead, 35·0 per cent.
Galena, cerussite, and carbonate of copper	Melrose Road, Condobolin	70	4½	...		Lead, 60 per cent.
Galena, zinc blende, and copper pyrites	Three miles north of Pye's Creek	39	4			...
Galena and magnetic pyrites	Head of Adelong Creek	28	6½	0	8	...
Galena, with a little mispickel and arseniate of iron	Six miles south of Emmaville	31	17	...		...
Mispickel	Emmaville	43	14			...
,,	Kingsgate	3	5	0	8	Bismuth, 2·6 per cent.
,,	,,	92	14	...		Bismuth, 9·2 per cent.
,,	Near Vegetable Creek	11	17	0	4	...
Mispickel in quartz	Head of Deepwater River	66	3	...		...
Mispickel, with calcite and serpentine	Dungog	1	12½	21	4½	...
Mispickel and pyrites	Mann River	13	1	...		...

Mineral	Locality					Remarks
Copper pyrites	Blackman's Creek, Oberon	52	…	…	…	…
Copper pyrites and mispickel in quartz	Pye's Creek, Bolivia	68	5	…	…	Copper, 19·65 per cent.
Earthy carbonate of copper	Ten miles west of Molong	26	19	…	…	Copper, 15·80 per cent.
Blue and green carbonates of copper	Abercrombie Ranges	16	6½	…	…	…
Green carbonate and red oxide of copper	Wiseman's Creek	15	2	…	…	Copper, 20·3 per cent.
Carbonate of copper and cerussite	Near Tuena	22	17	…	…	…
Fahlerz in quartz	Pye's Creek, Bolivia	1,238	1	…	…	…
Fahlerz in quartz	Near Rockley	16	10½	0	8	Copper, 27·4 per cent.
Fahlerz in claystone	Eight miles from Mann River	64	10	…	…	…
Fahlerz and galena	Severn River, Tenterfield	44	13	0	10½	(Copper, 5 per cent.; zinc, 20 per cent; lead, a small quantity.)
Fahlerz and Pyrites	" Ten-mile, Parish Park, co. Clive	196	16½	…	…	…
Copper ore	" Monaro, co. Woore	474	9½	2	19½	…
Zinc blende and pyrites	Lower Lode, the Grampians	18	15½	…	…	…
,, ,, with pyrites in felspathic clay	Parish Parkes, co. Gough	61	13	…	…	…
,, ,, galena, and pyrites	Boorook	27	7	…	…	…
Bismuth ore	Nambucca River	120	0	2	9	Bismuth, 35·6 per cent.
,, ,, ,,	Comstock Lode, near Yarrow	57	3	4	1½	Bismuth, 69·3 per cent.
Metallic bismuth in ferruginous quartz	Glen Innes	9	16	…	…	…
Bismuth, carbonate of bismuth, molybdic oxide and sulphide	Kingsgate	123	0	…	…	…
Bismuth ore	Hamilton's, Pye's Creek	57	3	…	…	…
,, ,, ,,	Tenterfield, from 4-foot reef	35	10½	1	4	Bismuth, 60·09 per cent.
Bismuth ore	Glen Innes	0	8½	1	12½	Bismuth, 72·7 per cent.
,, ,, ,,	New England	8	0	2	9	Bismuth, 79·5 per cent.
Antimonite and cervantite with lead	Carrington Mine, Marulan	2	17	164	3	Antimony, 16·75 per cent.; lead, 22·27 per cent.
Ferruginous chlorite rocks	Nymagee	5	3	…	…	…
Ferruginous barytes	Bathurst	2	0	trace		…
Ferruginous barytes and lead protoxide	Captain's Flat, Molonglo	10	12	14	14	…
Mixed minerals	Addison Lode, Boorook	682	15	…	…	…
,, ,,	Golden Age	398	10	…	…	…
,, ,,	Pye's Creek, Bolivia	2,075	0	…	…	…
,, ,,	Webb's, near Emmaville	200 to 800 oz. 4	9½	trace		…
,, ,,	Mount Gipps	to 3,240	10½	…	…	Lead up to 40·27 per cent.

D

RETURN showing the Quantity and Value of Silver produced in the Colony of New South Wales (from the *Annual Report of the Department of Mines,* Sydney).

Year.	Quantity.	Value.	Year.	Quantity.	Value.
		£			£
1862	266 tons ore	say 5,320	1876	69,179 oz. 0 dwts.	15,456
1863	28 ,,	1,080	1877	31,409 ,, 0 ,,	6,673
1864	13 ,,	130	1878	60,563 ,, 0 ,,	13,291
1865	736 oz.	184	1879	83,164 ,, 0 ,,	18,071
1866	...	...	1880	91,419 ,, 0 ,,	21,878
1867	...	...	1881	57,254 ,, 0 ,,	13,026
1868	...	...	1882	38,618 ,, 0 ,,	9,024
1869	753 oz. 0 dwts.	199	1883	77,065 ,, 18 ,,	16,488
1870	13,868 ,, 0 ,,	3,801	1884	93,660 ,, 5 ,,	19,780
1871	71,312 ,, 0 ,,	18,681	1885	794,174 ,, 0 ,,	159,187
1872	49,545 ,, 0 ,,	12,663	1886	1,015,433 ,, 10 ,,	197,544
1873	66,998 ,, 0 ,,	16,278			
1874	78,027 ,, 0 ,,	18,880		Total . .	580,428
1875	52,553 ,, 0 ,,	12,794			

Most of the silver produced in New South Wales was formerly obtained in the refining of gold at the Mint.

PLATINUM.

NATIVE PLATINUM.

Crystallises in the cubical system.

Specific gravity, 16–19.

Reported to occur with gold in the Shoalhaven River, co. Dampier; in the Ophir gold district, co. Wellington; in the form of small grains at Bendemeer, co. Inglis; and at Calton Hill, Dungog, in the Hunter and Macleay District, co. Durham. A small nugget, weighing 268 grs. (also stated as weighing $1\frac{1}{3}$ oz.), and having a specific gravity of between 15 and 16, was obtained from Wiseman's Creek, co. Westmoreland, with alluvial gold.

Platinum has been obtained in small quantity in the washings from the Aberfoil River, about fifteen miles from Oban, associated with metallic tin, gold, iridosmine, cassiterite, sapphires, topaz, &c.; also in the Sara River (*F. A. Genth,* Oct. 2, 1885).

A small quantity of platinum occurs in the sand along the sea-coast, near the Richmond River. An assay of some by Mr. W. A. Dixon* gave—

Gold . . . . 1 dwt. 5 grs. per ton.
Platinum . . . . Less than 5 grs. ,,

* Vide *Annual Report of the Department of Mines,* 1878, p. 43.

Analyses of Australian Platinum.

	No. 1.	No. 2.
Platinum	59·80	61·40
Gold	2·40	1·20
Iron	4·30	4·55
Iridium	2·20	1·10
Rhodium	1·50	1·85
Palladium	1·50	1·80
Copper	1·10	1·10
Iridosmine	25·00	26·00
Osmium	·80	...
Sand	1·20	1·20
	99·80	100·20

(St. Claire Deville and Debray, "Ann. Chem. et Phys.," III. lvi. p. 449.)

The above was probably found in New South Wales.

RHODIUM.

It is stated, in the "Catalogue of Natural and Industrial Products of New South Wales," exhibited by the Paris Exhibition Commissioners, 1854, that the iron ore from Nattai contains both rhodium and nickel, together with traces of antimony and gold; but no evidence is given in support of this statement.

OSMIUM AND IRIDIUM.

OSMO-IRIDIUM OR IRIDOSMINE.

This compound of osmium and iridium is very commonly met with in the auriferous and other drifts of New South Wales, in the form of minute grains and scales. The area over which it is found may be regarded as roughly corresponding with those of the alluvial gold deposits.

I have observed it in fair quantities in the gem sand at Bingera, co. Murchison; Mudgee, co. Phillip; Bathurst, co. Bathurst; and other places.

The iridosmine grains from the Aberfoil River seem to be present both as *newjanskite*, in the white flat scales, and as *sisserskite*, in lead or greyish-white-coloured scales; some have an imperfect hexagonal form, but most of them are irregular (*F. A. Genth*, October 1885).

Its presence in alluvial gold is occasionally a source of trouble at the Mint, for minute grains are often mechanically enclosed by the gold after melting, which, by their hardness, speedily destroy the dies during the operation of coining.

The following analysis is given of a specimen of this alloy, but no particulars are afforded as to the exact locality:—

Analysis of Australian Iridosmine.

Iridium .	58·13
Rhodium	3·04
Platinum	...
Ruthenium	5·22
Osmium .	33·46
Copper .	·15
Iron	...
	100·00

(Deville and Debray, "Ann. Chem. et Phys.," III. lvi. p. 481.)

This also was probably a New South Wales specimen.

MERCURY.

NATIVE MERCURY.

The Rev. W. B. Clarke stated that he received his first sample of native quicksilver in 1841 from near Carwell Creek, on the Cudgegong River, in co. Phillip, where cinnabar occurs.

It has also been found in the Mookerawa Creek, and Great Water-hole at Ophir, in co. Wellington, mentioned by Stutchbury; and in his report he stated that mercury had never been used on that creek.

Native mercury is said to occur in the casing of the reef at the Clifton Mine, Boorook. Mercury is also said to have been detected at Wagonga,* co. Dampier, in a bed some 10 feet thick, and situated some 90 feet above the sea-level; and with cinnabar, in a clay lode, and the surface grit, at Calton Hill, Dungog.

CINNABAR.

Chem. comp.: Mercury, 86·2; sulphur, 13·8 = Hg. S. Found near Rylstone, on the Cudgegong River, some twenty-five miles from Mudgee, co. Phillip, in an argillaceous matrix, and in alluvial deposits associated with gold, gems, and other similarly occurring minerals, in the form of small rounded masses of a brilliant red colour. The Cudgegong Mine is no longer being worked. The cinnabar obtained at Rylstone, on the Cudgegong River, is regarded by Mr. Wilkinson, the Government Geological Surveyor, as having been deposited from thermal springs, and that the deposits are worthy of more attention. Reported also to occur at Moruya, co. Dampier. Cinnabar is reported with gold, silver, and copper on Grove Creek, Abercrombie Mountains.

* See "Mines and Mineral Statistics of New South Wales," 1875, p. 201.

COPPER.

NATIVE COPPER.

Cubical system. Crystallised native copper is by no means rare, but large and well-developed crystals, as elsewhere, are uncommon. It is met with massive, in plates, threads, wires, and arborescent forms, the latter being usually built up of elongated rhombic dodecahedra. I have been unable to find any analysis of New South Wales native copper, but it probably contains the usual small quantities of silver, lead, bismuth, and other metals.

In nearly all cases it is found in association with cuprite, malachite, and other oxidised copper ores, as at Carcoar and Bathurst, co. Bathurst. Native copper is found at Solferino, co. Drake; Girilambone, co. Canbelego; Milburn Creek and Cowra, co. Bathurst; near Oberon, co. Wellington; the Canoblas and Wellington, co. Wellington; Peelwood, with lead ores, Mitchell's Creek, Bell River, Pink's Creek, co. Roxburgh; Peel River, co. Inglis; the Belara Mine, twenty miles from Gulgong, co. Phillip; Manilla, co. Darling; Bingera, co. Murchison; Cobar, co. Robinson; the Peabody Mine, co. Ashburnham; Copper Hill, Pierce's Knob; and Mount Lyell, near the Stanley Ranges, co. Mootwingee. It occurs with smaragdite on Molong Creek, and with porphyry at Parkes, co. Ashburnham; and in the form of diffused grains in a dark grey phonolite near Kiama, co. Camden. It is found with alluvial gold near Nundle, co. Inglis; and in the Sara River, with gold, gems, &c.

CUPRITE.—Red Copper Ore.

Chem. comp.: Copper suboxide $= Cu_2O$; copper, 88·8; oxygen, $11·2 = 100$. Usually found massive, but occasionally well crystallised in cubes and octohedra, which, however, are seldom more than $\frac{1}{4}$ inch in diameter. The largest and best crystals I have seen have come from the Cobar Mine.

The variety crystallised in capillary crystals, known as chalcotrichite or plush copper, is met with at the Coombing Mine, near Carcoar, co. Bathurst; with chrysocolla at the Cadiangulong Mine, co. Bathurst; Parkes, co. Ashburnham; Boroell and Garryowen, co. Wellington; Dungowan Creek, co. Parry.

This mineral is usually associated with the other oxidised copper ores, such as malachite and chessylite.

It is abundant at Cobar, co. Robinson, both massive and crystallised; Clarence River, co. Clarence; Gordon Brook, co. Richmond; Cowra, Carcoar, Icely, Milburn Creek, Cow Flat, and the Bathurst District, co. Bathurst; Mitchell's Creek, and with tenorite and cerussite at Peelwood, co. Roxburgh ; Wiseman's Creek, co. Westmoreland ; Burrowa, co. King ; Molong, co. Ashburnham ; Mount Hope, co. Blaxland ; Copper Hill, West Bogan ; Courntoundra Range, co. Yunghulgra ; Apsley, co. Vernon ; Belara, co. Phillip ; Nymagee, co. Mouramba ; Thompson's Creek Mine, co. Georgiana ; Hurley and Wearne's Mine, near Wellington, co. Wellington ; and Frog's Hole, co. Auckland ; in the Armstrong Mine, where it contains both gold and silver ; on the Manilla Creek, co. Darling, with grey sulphide or redruthite ; Bungonia, co. Argyle ; Yass, co. King ; Bingera, co. Murchison ; at Temora, co. Bland, with iron pyrites, chalcopyrites, a little silver, and traces of gold.   Red oxide, with other copper ores, is found at Capertee, co. Hunter ; in the Barrier Range ; Condobolin, co. Gipps.

TENORITE.—Melaconite *or* Black Oxide of Copper.

Chem. comp. : Copper oxide $= CuO$; copper, 79·85 ; oxygen, 20·15 $= 100$.

Generally in the form of a black powder, massive, or sporadic, *i.e.*, disseminated in nests. Usually found associated with other oxidised copper ores, in the upper parts of veins, as at Carcoar, Icely, and Milburn Creek, co. Bathurst ; Wellington, co. Wellington ; Peel-wood, co. Roxburgh ; Burrowa, Gunning, Yass, and Bala, co. King ; Forbes, co. Ashburnham ; Nundle, co. Parry ; Nymagee, co. Mour-amba ; Belara, co. Phillip ; South Wiseman's Creek, co. Westmoreland ; between Monga and the Shoalhaven, co. St. Vincent ; Currawang, co. Argyle ; and Frog's Hole, co. Auckland ; at the Canoblas, co. Ash-burnham, with native gold ; at Apsley, co. Vernon ; Courntoundra Range, co. Yunghulgra ; Gordon Brook, co. Drake ; Clarence River ; and Cadiangulong, co. Bathurst.

MALACHITE.—Green Carbonate of Copper.

Chem. comp.: Hydrous copper carbonate $= Cu_2CO_3 + H_2O$. Copper oxide, 71·9 ; carbonic acid, 19·9 ; water, 8·2 $= 100$. Metallic copper, 57·5.

Oblique system. Colour from pale emerald to deep green. Occurs massive, also mammillated and botryoidal, with fibrous concentric struc-ture, the various layers often possessing different shades of colour, and forming a most beautiful and valuable stone for ornamental and inlaying purposes. Crystals are occasionally met with, and sometimes of large size ; those from the Cobar Mines are particularly beautiful. The silky lustre is often very remarkable, the capillary crystals being sometimes several inches long, and compacted together into fibrous bundles.

It is found in most of the upper workings of New South Wales copper-mines, as in the Bathurst district, with chlorite, vitreous, yellow, and other copper ores; at Cambalong, co. Wellesley, earthy and fibrous malachite is associated with barytes or heavy spar, and with yellow and peacock ore; at Cobar, co. Robinson, with steatite; Garryowen, Mitchell's Creek, and Wellington, co. Wellington, mixed with other surface ores, and often containing large quantities of gold and silver. Reedy Creek and Bingera, co. Murchison; Icely, co. Bathurst; Yass, co. King; Nymagee, co. Mouramba; Buckinbah, co. Gordon, in granite, with the sulphides of copper; at Lucknow, co. Wellington; Gundagai, co. Clarendon; Cow Flat and Milburn Creek, co. Bathurst; Belara, co. Phillip; Gordon Brook, co. Drake; Clarence River, co. Clarence; often containing beautiful specimens of coarse gold, Kaizer Mine, Mitchell's Creek, co. Bathurst; and Peelwood, co. Roxburgh; Wiseman's Creek and Oberon, co. Westmoreland; Courntoundra Range, co. Yunghulgra, sixty miles from Wilcannia; Condobolin, co. Gipps; between the Cotta and Queanbeyan Rivers, co. Cowley; Mount Hope, co. Blaxland. At Copabella, co. Goulburn; Barraba, co. Darling; Burraga, co. Wellington; Silver Dale, near Bowning, co. King; and Parkes, co. Ashburnham, with cuprite, redruthite, and other copper minerals.

Green carbonate of copper is also found near Goulburn, co. Argyle; Mount Gipps, Purnamoota, co. Yancowinna; Uralla, co. Sandon; Wyndham; Dungowan, co. Parry; Copper Hill, near Orange; Barrier Ranges; Braidwood, co. St. Vincent; Sewell Creek, Rockley, co. Georgiana; Abercrombie Ranges, co. Georgiana; Bermagui River, co. Dampier; Quedong, Bombala, co. Wellesley; Tomingley, on the Bogan River, co. Narromine; Brungle Hill; Pye's Creek, Bolivia, co. Clive; near Forbes, in garnet rock, co. Ashburnham; Illawarra District; near the Pinnacles, Tuena, co. Georgiana; near Silverton, co. Yancowinna; Costigan Mountain, co. Georgiana; Deepwater, co. Clive; Uralla, co. Sandon; Europambela Station, co. Vernon; containing both gold and silver, six miles east of Walcha; Gumble, near Molong, co. Ashburnham—this ore is remarkable on account of its containing some 25 per cent. of tin as well as some silver; Binalong, co. Harden; Glen Innes, co. Gough; Tuena, co. Georgiana; and Mount Trooper, Snowy River, co. Wallace. Many of the specimens analysed contained several ounces of silver to the ton, and some a little gold as well.

CHESSYLITE.—Azurite *or* Blue Carbonate of Copper.

Chem. comp.: Hydrous copper carbonate, $2CuCO_3 + CuH_2O_2$. Copper oxide, 69·2; carbonic acid, 25·6; water, 5·2 = 100.

Oblique system. Colour from azure blue to indigo, translucent to opaque. Found massive and crystallised. The best specimens of the latter come from the Cobar Mines, co. Robinson, where it has been very largely obtained. They often assume a radiated concretionary form, with the terminal planes of the crystals studding the surface of the balls, in the form of small projections. These concretions vary from almost imperceptible points up to balls several inches in diameter; in some cases they occur diffused through a pale grey or green-coloured steatitic clay, at other times the crystals are set off by a dazzling white felspathic clay; hence they often afford very attractive cabinet specimens. Well-developed crystals are also found lining vuggy cavities.

At Cobar chessylite is associated with atacamite, in addition to the other more commonly occurring minerals.

At Woolgarloo chessylite occurs with native copper, cuprite, and malachite, in pink and white fluorspar. This mixture has at times a very pretty effect from the manner in which the copper minerals are diffused through the cracks and reticulating cavities in the fluorspar. Something of the same sort of thing is to be seen in the fluorspar from South Wiseman's Creek, co. Westmoreland.

Amongst other localities for chessylite are Inverell, co. Gough, in quartz-veins; Bathurst and Icely, co. Bathurst; Ophir, co. Wellington; and Peelwood, co. Roxburgh. Also Abercrombie Ranges, co. Georgiana; Condobolin, co. Gipps; and Gumble Flat, near Molong, co. Ashburnham, where silver ore has also been met with.

ATACAMITE.

Chem. comp.: Hydrous oxychloride of copper $= 3CuH_2O_2 + CuCl$. Copper oxide, 53·6; copper chloride, 30·2; water, 16·2 = 100.

Crystallises in the rhombic system. Dark green in colour.

Occurs in the Cobar Mines, co. Robinson; Cowra and Icely, co. Bathurst.

Crystallised in radiated groups of small acicular crystals. A specimen, probably from Cobar, of a dark translucent olive green colour, with vitreous lustre and apple-green streak, yielded the following result:—

Analysis.

Water lost at 105°	·536
„ combined	13·955
Copper oxide	64·709
„ chloride	13·218
Silica and insoluble matter	7·599
	100·017

REMOLINITE, Hydrous Oxychloride of Copper, is also said to occur at Cobar.

BROCHANTITE.—Blue Vitriol *or* Copper Sulphate.

Crystallises in the doubly oblique or anorthic system, but most usually met with in the form of an efflorescence or incrustation.

Chem. comp.: $CuSO_4$, $3CuH_2O_2 =$ copper oxide, 70·34; sulphur tri-oxide, 17·71; water, 11·95 = 100.

A specimen from New South Wales gave Tschermak ("Berg-Ak. Wien." li, p. 131) the following results:—

Analysis.

Copper oxide	69·1
Sulphur tri-oxide	19·4
Water	11·5
	100·00

The late Mr. Stutchbury reported that at Kelloshiels the well water was found to be so impregnated with copper as to be unfit for domestic purposes. The copper was probably present as sulphate. Also found at Milburn Creek Mine, co. Bathurst.

DIOPTASE.

Chem. comp.: $CuSiO_3,H_2O =$ silica, 38·1; copper oxide, 50·4; water, 11·5 = 100.

Crystallises in the hexagonal system. Colour emerald green, with a vitreous lustre; sometimes mistaken for the emerald. This mineral is said to occur with chessylite at Cobar, in co. Robinson, and on the northern slope of Bowning Hill, near Yass.

CHRYSOCOLLA.

Chem. comp.: Hydrous copper silicate = $CuSiO_3$, $2H_2O$. Copper oxide, 45·3; silica, 34·2; water, 20·5 = 100.

Amorphous. In colour dark green. Reported to occur in a matrix of semi-opal at the Coombing Copper-mine, two miles from Carcoar, co. Bathurst; also at Cobar, co. Robinson, and at Essendon.

A massive piece, brought from Wheeo as a specimen of jasper, is of a bluish green colour, much darker outside than within. Breaks with a somewhat splintery and conchoidal fracture.

Hardness = 4. Specific gravity, varied in different parts from 2·37 to 2·43.

Analysis.

Water lost at 120° C.	11·92
,, ,, red heat	9·40
Copper oxide (CuO)	35·28
Iron oxide	trace.
Silica	43·11
Loss	·29
	100·00

As the above does not answer to the usual formula, it is probable that some of the silica exists in the free state.

PHOSPHOCHALCITE.—Pseudomalachite.

Chem. comp.: Hydrous copper phosphate = $Cu_3P_2O_8,3CuH_2O_2$. Copper oxide, 70·9; phosphoric acid, 21·1; water, 8·0 = 100.
Crystallises in the rhombic system. Colour, dark green.
Coombing Copper-mine, co. Bathurst.

ARSENIATE OF COPPER.

Mentioned as occurring in a quartz-vein on the Cox River, but it is not stated whether the mineral was condurrite, olivenite, or one of the other arseniates.

REDRUTHITE.—Vitreous Copper Ore—Copper Glance.

Chem. comp.: Copper disulphide = Cu_2S; copper, 79·8; sulphur, 20·2 = 100.
Crystallises in the rhombic system. It is of a lead-grey colour, soft, and leaves a shining streak something like galena.
I have only seen this mineral in the massive state, but it is found crystallised in South Australia.
Found at Cobar, co. Robinson; Mount Hope, co. Blaxland; Nymagee, co. Mouramba; South Wiseman's Creek, co. Westmoreland, with fahlerz; between the Lachlan and Bogan Rivers, 100 miles north-west of Forbes, co. Flinders; Parkes, co. Ashburnham; Mitchell's Creek, co. Roxburgh; Bocoble; Milburn Creek, co. Bathurst; Muswell, twelve miles from Goulburn, co. Argyle; Cullen Bullen, co. Roxburgh, with iron pyrites, copper pyrites, and calcite, containing both gold and silver; at Manilla Waters, near Bowral; near the Wellington Caves, co. Wellington, with blue and green carbonates in a quartzose vein-stuff; also at Wellbank, near Wellington; at Waterfall Creek, running into Cadiangulong Creek, co. Bathurst, with iron pyrites; at Bathurst, Icely, and Carcoar, co. Bathurst; Kroombit.
Argentiferous redruthite occurs at Spicer's Creek; Belara, co. Bligh; with fahlerz and galena, Webb's Mine, Emmaville, co. Gough; Bingera, co. Murchison; Severn River, Tenterfield, co. Clive; Barrier Ranges; with fahlerz, Pye's Creek, Bolivia, co. Clive; Cudgegong; with carbonate of copper, Wyndham; at Capertee, co. Hunter, with other copper ores and galena; Burrowa, co. King; Charlton, near Rockley; Ophir, co. Bathurst; Narragal, co. Gordon.
Siliceous Redruthite.—A peculiar copper ore was received from Coombing Copper-mine, about two miles from Carcoar, of a dark grey, almost black colour. In general appearance somewhat resembling redruthite, but of a duller lustre, and considerably harder, the hardness

being between 5 and 6. In parts a bronze tint and lustre is apparent. The specimen exhibits neither crystals nor crystalline structure; it breaks with a well-marked conchoidal fracture. Lustre somewhat resinous; streak shining.

Heated in a glass tube it gives off water, having a strongly acid reaction, from the sulphurous acid which is evolved. Before the blow-pipe it does not fuse, colours the flame green, and acquires a dull black colour. Treated with strong boiling nitric acid it is rapidly acted upon, a brown-coloured residue being left; the residue, when examined under the microscope, presents a honeycombed appearance; the walls of the irregular cellular cavities are pale brown and translucent, and apparently composed of quartz; when the powdered mineral is boiled with nitric acid a white residue of silica is left. Concentrated hydro-chloric acid also dissolves out the copper subsulphide, but much more slowly.

The mineral is intimately associated with quartz, both ordinary white vein quartz and a translucent variety of a greyish tint; this grey tint seems to be due to diffused, very finely divided copper sub-sulphide.

The specific gravity of a portion quite free from visible quartz was found to be 3·12 at 18° C.

The following analysis was made upon a portion which appeared to be perfectly homogeneous, even under a 1-inch objective; yet this yielded over 43 per cent. of silica.

Analysis.

Water, combined	2·354
Silica	43·420
Copper subsulphide (Cu₂S)	45·196
Iron sulphide (Fe S)	4·931
Iron sesquioxide	3·479
Undetermined and loss	·620
	100·000

The combined water was determined directly by collecting and weighing it in a chloride of calcium tube, a layer of lead oxide being placed in the front part of the combustion tube to arrest any sulphur or sulphur oxides.

The amount of silica soluble in a boiling solution of sodium car-bonate was also determined, and found to vary from 14·69 to 19·99 per cent.

The mineral, therefore, appears to be merely an intimate mixture of hydrated amorphous, and crystalline quartz, copper subsulphide with some iron oxide, and ferrous sulphide.

BORNITE.—Erubescite.

Purple Ore.   Buntkupfererz.

Chem. comp.: Varies considerably.   A double sulphide of copper and iron.   Copper, 56 to 70; iron, 6 to 17; sulphur, 21 to 26. Crystallises in the cubical system.   Colour, copper-red, purple to brown; fracture, even to small conchoidal; streak, blackish grey, shining.

Found at Cobar, co. Robinson; Bingera, co. Murchison; Wellbank and Louisa Creek, co. Wellington; and Cow Flat, co. Bathurst.

FAHLERZ.—Grey Copper Ore.   Tetrahedrite.

Chem. comp.: $4Cu_2S + Sb_2S_3$, but variable.   Part of the copper often replaced by iron, zinc, silver, mercury, or cobalt; and the antimony partly replaced by arsenic and occasionally by bismuth.   At times it is very rich in silver, even as much as 30 per cent.   Crystallises in the cubical system, usually in tetrahedral forms—hence one of its synonyms; colour, grey; soft, cuts with shining streak.

Occurs on the west side of Copper Hill, near Molong, co. Ashburnham.

Fahlerz is worked at Webb's Mine, Vegetable Creek, co. Gough; and some very fair crystals have been found.   The principal associated minerals are galena, copper pyrites, and green fluor spar, and much of the fahlerz is rich in silver.   An analysis gave the following results (*Report of the Mines Department*, 1885):—

Metallic copper	. . . .	31·500
„ antimony	. . . .	18·130
„ zinc	. . . .	6·140
„ iron	. . . .	6·440
„ lead	. . . .	·680
„ silver	. . . .	1·635
Gold	. . . .	traces
Sulphur	. . . .	26·180
Arsenic, &c.	. . . .	2·095
Insoluble	. . . .	7·200
		100·00

Some of the crystals yielded 3·5 per cent., or 1140 ozs., of silver per ton, and 30 per cent. of copper.

Argentiferous fahlerz is also obtained at South Wiseman's Creek, co. Westmoreland; at Tenterfield, co. Clive, with galena, in clay slate; on the Barrier Ranges; Pye's Creek, Bolivia, co. Clive; on the Mann River, co. Gough; Ten Mile, Park parish, co. Clive.

CHALCOPYRITES.—Copper Pyrites.

Chem. comp.: Copper-iron sulphide Cu_2S,Fe_2S_3, but variable. Copper, 34·6; iron, 30·5; sulphur, 34·9 = 100. Tetragonal system; hemihedral forms. A very abundant ore. Usually occurs massive; occasionally crystals are met with, but they are generally but imperfectly developed. Colour, usually brass yellow. Blister ore is more of a bronze colour, and occurs in mammillated and botryoidal forms. The tarnished variety of copper pyrites, known as peacock ore from the splendid colours which it acquires, is very common.

It occurs in nearly all the metalliferous districts in the colony—at Cobar, co. Robinson; Bingera, Elsmore, co. Murchison; Clarence, co. Clarence; Wiseman's Creek and Oberon, co. Westmoreland; Boroell, Ironbarks, and other places in the Wellington District, with zinc blende, steatite, quartz, and asbestos; Ophir, Carcoar, Cow Flat, and Mitchell's Creek, co. Bathurst; Wallabadah, co. Buckland; Cargo and Molong, co. Ashburnham; Peelwood, co. Roxburgh; Tuena, Charlton, Essington, co. Georgiana; Adelong, co. Wynyard, with gold; Lobb's Hole and Yarrangobilly, co. Buccleugh; Kiandra, co. Wallace; Snowball Mine, near Gundagai, co. Clarendon; Dundee, co. Gough; Gooderich and Narragal, co. Gordon; Cootalantra Mine and Belmore Mine, Monaro District; between Condobolin and Parkes; Frog's Hole, co. Auckland; Nymagee, co. Mouramba; Gordon Brook and Solferino, co. Drake; Apsley, co. Vernon; Bungonia and Currowang, Jacqua Mine and Nerrimunga, co. Argyle; and Mallone Creek, between Goulburn and Braidwood.

Copper pyrites also occur at Burraga, Crudine; at Bolivia with mispickel; with galena and zinc blende near Boro; Cooma; Nerriga; Ginderbyne; Moonbi; Holander's River, Oberon; six miles north of Tarana; Nowra; Deepwater; Pye's Creek, Bolivia; Blackman's Creek, Oberon; Trunkey; Ten Mile, co. Clive; Denisontown, Bermagui River; with calcite at Palmer's Oakley.

The copper pyrites from several of the New South Wales mines contain valuable amounts of gold and silver. *See* Gold and Silver.

Bell-metal Ore.—A peculiar variety of copper pyrites containing tin, Cobar, co. Robinson.

DOMEYKITE.

Chem. comp.: Copper arsenide, Cu_3As. Copper, 71·7; arsenic, 38·3 = 100.

Amorphous. Occurs in the Bathurst district with yellow sulphide of copper.

ANTIMONIAL COPPER ORE.

Said to occur at Eden, Twofold Bay, co. Auckland ; probably fahlerz (see p. 60).

ENARGITE.

A sulph-arsenite of copper. Crystallises in the rhombic system. Of a greyish iron-black colour. This mineral, which is a variety of Tennantite, is said to have been found in New South Wales.

Olivenite, liebethenite, bournonite, and other beautiful copper minerals have not apparently yet been found.

RETURN showing the Quantity and Value of Copper produced in the Colony of New South Wales.

Year.	Quantity.	Value.	Year.	Quantity.	Value.
	Tons.	£		Tons.	£
1858	58 ore	1,400	1872	1,452 ore	105,888
1859	150 ,,	2,250	1873	2,846 ,,	239,102
1860	43 ,,	1,535	1874	4,160 ,,	325,140
1861	144 ,,	3,390	1875	3,677 ,,	301,90
1862	2,200 ,,	12,000	1876	3,275 ,,	249,978
1863	125 copper	12,500	1877	4,513 ,,	324,226
1864	2,100 ore	22,100	1878	5,219 ,,	345,158
1865 {	295 copper 1,648 ore	} 37,845	1879	4,142 ,,	257,352
			1880	5,394 ,,	364,059
1866 {	304 copper 947 ore	} 28,135	1881	5,494 ,,	355,062
			1882	4,958 ,,	324,727
1867 {	296 copper 2,590 ore	} 35,316	1883	8,957·7 ore	577,201
			1884	7,305·4 ,,	416,179
1868 {	315 copper 5,151 ore	} 34,200	1885	5,746 ,,	264,920
			1886	4,027 ,,	167,665
1869	2,084 ,,	76,675			
1870	1,000 ,,	65,731	Total . . .		4,964,250
1871	1,444 ,,	88,886			

LEAD.

NATIVE LEAD.

The Rev. W. B. Clarke more than once mentions having found native lead on the Peel River, Hanging Rock, and elsewhere.

It has also been found by the miners on the gold-fields, in association with serpentine, on the spurs of the Curangora, near Bingera, co. Murchison. One specimen had a specific gravity of 11·04.

In 1880 I received an irregular piece of native lead from a miner, about 1½ inch long by 1 inch wide, and about ⅛ to 3/16 of an inch thick,

DIAGRAM SHOWING THE ANNUAL PRODUCTION OF TIN AND COPPER RAISED IN NEW SOUTH WALES TO 31st DECEMBER 1886.

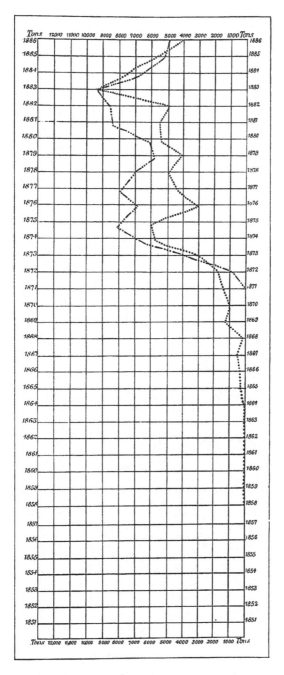

Copper ....
Tin .. ..

The Horizontal divisions indicate the Year and the Vertical divisions denote the Quantity of Metal

with rough surface, as if it had filled a jagged crevice, coated on the outside with impure oxide of lead of a brilliant red colour. The edges were slightly rounded, as if waterworn. It did not look at all as if it had been reduced artificially, or had been derived from bullets or sources of that kind. Weight = 32 grammes. Found near Gundagai.

The majority of the specimens of native lead which have been brought to me from time to time have evidently been derived from bullets which have found their way into the river deposits, and have been found by the miners when washing for gold.

Minium.—Native Red Lead.

Chem. comp.: Lead oxide = Pb_3O_4. Lead, 90·66; oxygen, 9·34 = 100. Occurs with cerussite at Peelwood, near Tuena; and near Gundagai. With ferruginous barytes at Captain's Flat, Molonglo; with ferruginous cerussite near Mount Trooper, Snowy River.

Cerussite.

Chem. comp.: Lead carbonate = $PbCO_3$. Lead oxide, 83·5; carbonic acid, 16·5 = 100. Occurs massive and in large prismatic crystals at Peelwood Mine; on the exterior they are often coloured red by a ferruginous clay.

Cerussite, both massive and crystallised, is found in the Silverton Mines, co. Yancowinna, in association with other lead minerals, native silver, silver chloride, &c. The amount of silver present in the mixed mineral is sometimes very large, but the cerussite itself often contains but little or none. Sometimes there is also gold present.

Also found in co. Argyle, at Tarrago and Slate Creek, Locksley; co. Bligh, at Belara; co. Clive, at Deepwater; co. Clyde, at Brewarrina; co. Drake, at Fairfield and at Solferino; co. Georgiana, at Abercrombie Ranges, Sewell's Creek (Rockley), and at Tuena in a red clay; co. Gipps, at Melrose Road, Condobolin; co. Gough, at Emmaville, and Pye's Creek, Bolivia; co. King, at Silverdale, near Bowning, with other lead ores and fluorspar; co. Mouramba, at Nymagee; co. Murray, at Captain's Flat, Molonglo; co. Roxburgh, on Mitchell's Creek; co. St. Vincent, at Nowra Ranges; co. Wellesley, at Bombala and Quedong; co. Wellington, at Crudine, with copper ores, at Lewis Ponds Creek, and at Sunny Corner; co. Yancowinna, at Mount Gipps. Also at Melrose, co. Wallace; Mount Trooper, Snowy River; and Weddin Mountain.

A ferruginous cerussite with quartz from Lewis Ponds Creek, near Orange, yielded 26 oz. 2½ dwts. silver per ton, and 2 oz. 9 dwts. gold per ton; and one from Fairfield gave, lead, 37·18 per cent.; gold, 4 oz. 1½ dwt. to 14 oz. 14 dwts. per ton, with a little silver.

ANGLESITE.

Chem. comp.: Lead sulphate = $PbSO_4$. Lead oxide, 73·6; sulphuric acid, 26·4 = 100. Said to have been found with galena on the Abercrombie River.

It is found fairly well crystallised at Silverton, also at South Wiseman's Creek, in association with the carbonate (cerussite) and other lead minerals; and at Campbell's Reef, Severn River.

PYROMORPHITE.

Chem. comp.: Lead phosphate = $3Pb_3P_2O_8,PbCl$. In round numbers, lead oxide, 75·0; phosphoric oxide, 15·0; lead chloride, 10 = 100. Small quantities of calcium fluoride and calcium phosphate are usually present, and part of the phosphoric acid is at times replaced by arsenic acid.

Hexagonal system; usually in six-sided prisms.

At Grenfell it is found as a bright green-coloured powder containing minute hexagonal prisms; it is also found of the same colour associated with galena and mimetite in a vein traversing clay slate near Bathurst. Another specimen from Bathurst was of a pale greyish-brown colour, with a waxy lustre and mammillated surface, upon which small crystals of chessylite were seated.

It occurs on the Sugar-loaf Hill, near Wellington; also on Mitchell's Creek; and at Silverdale, near Bowning, with galena.

MIMETITE.—Kampylite.

Chem. comp.: Lead arseniate = $3Pb_3As_2O_8,PbCl$. In this mineral the phosphoric is replaced by arsenic acid. Of a brown colour, and in much-curved or barrel-shaped hexagonal prisms. With pyromorphite at Sugar-loaf Hill, Wellington; Mitchell's Creek and Gulgong.

WULFENITE.

Chem. comp.: Lead molybdate = $PbMoO_4$ Lead oxide, 61·5; molybdic acid, 38·5 = 100. Mentioned as occurring on a spur of Mount Murulla, Kingdon's Ponds, and near Mount Wingen, co. Brisbane. The Rev. W. B. Clarke also records finding drifted molybdate of lead, waterworn and with a radiate structure, on the North Shore; at Molonglo, co. Murray; and at Munmurra, co. Bligh.—*Sydney Morning Herald*, August 17, 1850.

GALENA.

Chem. comp.: Lead sulphide = PbS. Lead, 86·6; sulphur, 13·4 = 100. This, as elsewhere, is the commonest ore of lead; it not only occurs in large deposits, but it is widely distributed over the colony.

It is usually found in the massive state, and with a granular structure which varies from fine to coarse. Occasionally it is met with fairly well crystallised, usually in cubes and in combinations of the cube and octahedron, as at Cambalong, but on the whole crystals are rare. In other respects it presents all the usual properties of the mineral as found in other countries.

Localities.—Argentiferous galena on the Chichester River; in co. Argyle, at Bungonia, near Goulburn, Goulburn District, and Slate Creek, Locksley; in co. Auckland, near Candelo with barytes, and Merimbula; co. Bathurst, at Waroo, Humewood, Cowflat Copper-mine, with carbonate and sulphide of copper; co. Beresford, at Cooma; co. Bligh, at Denisontown; co. Brisbane, on the Hunter, Isis, and Page Rivers; co. Buckland, at Wallabadah; co. Camden, at Burragorang; co. Clarendon, at Bethungra, and the Sebastopol Reef, Junee; co. Clive, at Deepwater; co. Clyde, at Brewarrina; co. Cook, near Hartley; co. Dampier, at Moruya; co. Drake, at Solferino; co. Dudley, at Kempsey; co. Georgiana, at Binda; co. Gipps, at Melrose Road, Condobolin; co. Gough, at Bolivia, Emmaville, Glen Innes, Inverell, and Strathbogie; co. Harden, at Binalong, Jugiong Creek, and Mylora Creek in a quartz porphyry, Murrumburrah and Bookham; co. Hardinge, at Sandy Swamp, and with zinc blende at Tingha; co. King, at Burrowa in quartz-veins, at Silverdale, Pudmore Creek, the Good Hope Mine, Mylora, and Yass; co. Menindie, at Menindie; co. Lincoln, at Dubbo; co. Monteagle, at Crookwell, and the Garibaldi Reef, near Young; co. Murchison, at Bingera and Reedy Creek; co. Murray, at Boro, Camberra Plains, and Gundaroo; co. Parry, on the Peel River, and at Mount Grosvenor; co. Phillip, with copper ores at Lawson's Creek, and Gulgong; co. Raleigh, Upper Bellinger River and Warrell Creek; co. Roxburgh, Mitchell's Creek, in quartz, with sulphides of copper and iron, and blue and green carbonates of copper, also at Bull Dog Range (Mitchell's Creek), Peelwood, and near Tarana; co. Sandon, at Armidale and Uralla; co. St. Vincent, on the Shoalhaven River, Talwal, Yalwal, and Major's Creeks, and near Braidwood; co. Wallace, at Kiandra, in quartz-veins; co. Wellesley, near Bombala, and at Quedong; co. Wellington, at Wellington and Ophir; co. Westmoreland, at Oberon with mispickel, on Holander's River, near Sewell's Creek, and at Wiseman's Creek; co. Woore, at Manaro; co. Wynyard, at Adelong, with gold in grey quartz, and at Eurongilly; co. Yancowinna, at Broken Hill, Leadville, Thackaringa, Umberumba Creek, and Umberumberka

E

with rich silver chlorides.   Also at Barrier Ranges; Everton (near Burrowa); Money Ranges (between Yass and Gundagai); Mount Trooper, Snowy River; Port Denison; Ravenswood; Woolgarloo; Bindor District.

In all cases the galena, when examined, has been found to contain more or less silver.   As examples, the following may be given :—

Galena from Manaro: lead, 69 per cent.; silver, 9 oz. 16 dwts. per ton.

Money Ranges, between Yass and Gundagai: lead, 80·9 per cent.; silver, 8 oz. 3 dwts. per ton, and a trace of gold.

Upper Bellinger River: lead, 42·04 per cent.; silver, 94 oz. 14½ dwts. per ton.

Bindon: lead, 60·6 per cent.; silver, 80 oz. per ton.

Goulburn District: lead, 52·5 per cent.; silver, 68 oz. 12 dwts. per ton; gold, 6 dwts.

Yass: lead, 83·31 per cent.; gold, a trace; silver, 52 oz. 5 dwts. per ton.

Brewarrina, lead, 75·8 per cent.; silver, 25 oz. 9½ dwts. per ton.

Fairfield: lead, 60·2 per cent.; gold, 9 oz. 16 dwts. per ton.

Additional assays can be consulted in the *Annual Reports of the Department of Mines.*

The lead ores containing galena from the Silverton District yield in some cases several thousand ounces of silver per ton, but it must be borne in mind that most of the silver is present as chloride, and cannot be regarded as being yielded by the galena.

STOLZITE.

Chem. comp.: Lead tungstate.   Has been obtained from near Peelwood.

LEAD VANADIATE.

Mr. Ranft reports the finding of a lead mineral resembling the vanadiate at Silverton.

RETURN showing the Quantity and Value of Lead produced in the Colony of New South Wales (*Annual Report of the Department of Mines,* Sydney) :—

Year.	Quantity.		Value.
	Tons.	Cwts.	£
1876	67	0	1,392*
1877	20	12	325
1878	5	0	258
1879	18	13	535
1880	27	14	890
1881	52	14	1,625
1882	11	19	360
1883	136	4	2,075
1884	9,167	11	241,940
1885	2,286	0	107,626
1886	4,802	2	294,485
Total	16,595	9	651,511

* These figures represent the quantity raised during 1876 and previous years.

CADMIUM.

A specimen of greenockite, the very rare cadmium sulphide, is said to have been found on Louisa Creek associated with zinc blende and quartz.

BISMUTH.

NATIVE BISMUTH.

Metallic bismuth in New South Wales usually occurs associated with carbonate of bismuth, and very often with oxide and sulphide of molybdenum. Many of the specimens which have been assayed have contained quite a large proportion of gold and silver; one specimen from near Glen Innes, probably Kingsgate, contained 123 oz. silver per ton; one from Kingsgate yielded 69·3 per cent. of bismuth, 4 oz. 1½ dwt. gold, and 57 oz. 3 dwts. of silver per ton; another from the same place yielded 79·5 per cent. metallic bismuth, 2 oz. 9 dwts. fine gold, and 10 oz. 12 dwts. of silver per ton; one from Pye's Creek gave 66·8 per cent. bismuth and 35 oz. 8½ dwts. silver per ton.

Occurs with copper ores at Cobar, co. Robinson, to the extent of 2 to 2·5 per cent. For analyses by Mr. W. A. Dixon see *Report of the Mining Department*, Sydney, 1880.

Localities.—In the New England District, at the Bruce Mine; co. Gough, near Glen Innes, at Redgate, on the Silent Grove Creek, where a vein averaging 8 inches wide was being worked in 1880; at the Elsmore Mine, also being worked, and Kingsgate, eighteen miles east of Glen Innes; in co. Sandon, Mount James, near Armidale; in the Vegetable Creek District, at the Gulf and on Duck Creek; a lode is stated to have been found near Kempsey, in the Macleay District. Also near Orange, co. Wellington; Mount Gipps, co. Yancowinna; Barrier Range; Pye's Creek, Bolivia, co. Clive; Gumble, near Molong, co. Ashburnham; Tingha, co. Hardinge, in rolled fragments.

Mr. Wilkinson, F.G.S., Geological Surveyor, gives the following account of the occurrence of this metal near Glen Innes:—

" *The Bismuth Lodes near the Yarrow Creek at Kingsgate, about sixteen miles east from Glen Innes.*—The formations here are granite and altered slate, forming rough broken country, with valleys about 500 feet deep. The line of junction of the two formations is well defined, and the bismuth lodes occur in the granite in proximity to this line or within about 400 yards from it. The mode of occurrence of these so-called " lodes " is very remarkable; they are *pipe-veins*, or oval masses of quartz of variable thickness, descending in a more or less vertical direction in the granite, and filled with quartz and metallic minerals.

Thus, in one lode in the Kingsgate Company's property two masses of quartz (which the manager, Mr. W. Yates, informed me were 30 feet apart at the surface), on being followed down, united and formed one large pipe-vein about 27 feet in diameter and of irregular shape, from portions of it protruding here and there into the granite. Nests or bunches of bismuth ore (native bismuth, sulphide, carbonate, and oxide of bismuth) were obtained about these protruding portions as well as through the mass of quartz; and in order to take out the vein-stone a large excavation about 60 feet by 40 feet has been made. The vein has only been sunk upon to a depth of 50 feet. The quartz is of a coarsely crystalline nature, and contains, in patches, a considerable quantity of molybdenite. The metallic bismuth and sulphide occur in the solid quartz, but the carbonate and oxide lie chiefly in the joint fissures in the quartz. Sometimes masses of native bismuth are found between crystals of quartz in the vein, and when removed the impress of the quartz crystals is well shown. Some splendid specimens, from 4 to 6 lbs. weight, from this mine were presented by the Company to the Mining and Geological Museum. The largest mass of native bismuth found here weighed nearly 30 lbs. Other similar veins have been proved, though only for a few feet in depth; one contains much arsenical pyrites and hexagonal plates of molybdenite. An average sample of these sulphides gave on assay:—Metallic bismuth, 2·6 per cent.; fine gold, 8 dwts. per ton; silver, 3 oz. 5 dwts. per ton.

"On Portion 25, about half a mile north-west from here, another large pipe-vein is being opened. Near the surface it consists of a very ferruginous mass of quartz, about 13 feet by 9 feet, containing bismuth, arsenical pyrites, wolfram, and molybdenite. The screened vein-stuff is said to yield about 50 lbs. of bismuth ore per ton, which will probably be equal to about 1 per cent. of ore for the whole vein-stuff. A sample of the ore, consisting of mixed particles of native bismuth, carbonate, sulphide, and oxide, yielded on assay:—Metallic bismuth, 69·3 per cent.; fine gold, 4 oz. 1½ dwt. per ton; fine silver, 57 oz. 3 dwts. per ton.

"The result of assay of the ironstone from this vein was 0·6 per cent. of bismuth and no gold; and the arsenical pyrites gave only a trace of gold and bismuth, with silver at the rate of 12 oz. 5 dwts. per ton. The gold, therefore, appears to be almost entirely contained in the bismuth ore, probably in the metallic portion of it.

"Several veins of a similar nature have been opened on the Glen Innes Company's property, which adjoins that of the Kingsgate Company. The Company is now sinking upon a vein which is said to have been 1 foot wide at the surface, but when I saw it at a depth of 40 feet, the lowest level then reached, it was 4 feet wide. This vein is in granite, and close to the boundary of the slate formation. The vein-stuff is thickly studded with large brilliant steel-grey plates of molybdenite, some of them being more than 3 inches in diameter. Nodules of native bismuth, larger than walnuts, with carbonate, sulphide, and oxide of bismuth, occur through the vein, and in greater quantity in places where the molybdenite becomes abundant. Another vein, situated about 100 yards from this, contains, besides bismuth ore and molybdenite, some arsenical pyrites, which latter yielded on assay 9·2 per cent. of metallic bismuth; fine silver, 92 oz. 14 dwts. per ton; and no gold.

"About three miles east from the Yarrow Creek Head Station, and about the same distance in a south-easterly direction from Kingsgate, is the Comstock Bismuth Company's mine. No work was being done at the time of our visit, but we saw three pipe-veins of hard white crystalline quartz, which had been opened for only a few feet from the surface. The shafts were partly filled with water, so that the exact size of the veins could not be measured; but the largest of them appeared to be about 6 feet by 15 feet near the surface. A sample of bismuth ore collected from the heaps gave on assay :—Metallic bismuth, 35·6 per cent.; fine gold, 2 oz. 9 dwts. per ton; fine silver, 9 oz. 16 dwts. per ton. Thus again we see that the bismuth ore contains gold. These veins are also in granite, and distant about 200 yards from the slate formation.

"It is a somewhat remarkable feature that all the bismuth veins (eighteen) as yet found occur in the granite within a short distance from the slate; and it is probable that, on further examination of the country along the line of junction of the two formations, other veins will be discovered. The bismuth lode in the Silent Grove mine occurs under the same conditions, viz., in granite close to its junction with altered slates, and it is of similar character to those above described.

"I may mention that about twelve miles north from Glen Innes, and about one mile east from the Tenterfield road, several bismuth and tin-bearing quartz-veins have been discovered. These occur in a different manner from those at Kingsgate. They form irregular veins and masses of quartz, traversing a fine-grained micaceous felsitic rock which is surrounded by altered sedimentary rocks. In one place this rock, for a length of about 100 yards and a width of 15 yards, is traversed by a network of quartz-veins. A small hole has been sunk here, and the stone taken from it contains bismuth ores, tin ore (cassiterite), molybdenite, arsenical pyrites, and wolfram. In another place, about 100 yards from that last named, a mass of hard crystalline quartz, in size at the surface about 40 feet by 20 feet, has been opened for a few feet in depth. It contains bismuth and tin ores, together with a large quantity of wolfram. Besides this, two other small veins of quartz, yielding bismuth and tin ores, crop out close by."—*Mines Report*, 1883.

BISMUTH GLANCE.

Bismuth sulphide. Bi_2S_3 or bismuth 81·25, sulphur 18·75 per cent. Rhombic system.

Readily recognised by its easy fusibility, and reduction on charcoal to a bead of bismuth surrounded by a yellow incrustation. Occurs with other bismuth minerals at Kingsgate, Glen Innes, and elsewhere.

BISMUTHITE.

Chem. comp.: A hydrated carbonate of bismuth. Bismuth oxide, 90·0; carbonic acid, 6·5; water, 3·5 = 100.

Found in the form of more or less rounded grains and pebbles with stream tin in the New England District.

Samples of bismuthite from Tingha, co. Harding, in the form of white and dark brown water-worn nodules, were found to contain 60·43 per cent. of bismuth; another with talc and sesquioxide of iron, 62·75 per cent. of bismuth. An earthy form of bismuthite from Tenterfield was also found by Mr. Dixon to yield: bismuth = 43·29 per cent., and molybdenum sulphide, 6·60.

Generally found in the form of dull grey or white earthy-looking rolled fragments, usually about the size of a pea, but sometimes larger pieces are found. Breaks with a dull earthy fracture. Found with the stream tin over most parts of the New England Tin District. A specimen from Ponds Creek gave the following results:—

Hardness = 3 to 4.

Analysis.

Silica	4·695
Bismuth trioxide (Bi_2O_3)	76·061
Alumina and traces of iron sesquioxide	1·983
Carbonic acid	5·426
Water, by difference	11·835
	100·000

The above does not agree with the usual formula given for this mineral. The specimen is more or less impure, as is shown by the presence of the silica, alumina, and iron.

RETURN showing the Quantity and Value of Bismuth produced in the Colony of New South Wales (*Annual Report of the Department of Mines*, Sydney).

Year.	Quantity. Tons.	Cwts.	Value. £
1881	12	10	2,728
1882	2	14	162
1883	3	14	650
1884	14	7½	2,770
1885	14	0	3,700
1886	20	18	3,870
Totals	68	3½	13,880

TELLURIUM.

Native Tellurium.

A rare element; reported to occur at Bingera, co. Murchison.

MOLYBDENUM.

Molybdenite.

Chem. comp.: Molybdenum sulphide MoS_2. Usually found massive, with a coarsely granular structure; also in grains, scales, plates, and rosette clusters of crystals. Sometimes the flat hexagonal plates or crystals are of large size; I have found some as large as a half-crown on the Elsmore Tin-mine, co. Gough.

The colour is usually bluish-white, with a strong metallic lustre.

Associations.—It is rather common in the New England Tin Districts, especially at the Elsmore and Newstead Tin-mines, where it occurs in the tin-veins traversing the granite. It is most usually associated with quartz. On the Hunter River it is found associated with gold, galena, pyrites, and other minerals; at Kingsgate, near Glen Innes, with bismuth.

Localities.—It also occurs at Bullin Flat, near Goulburn, co. Argyle; at Kiandra, co. Wallace, with quartz; and Cleveland Bay; Oban, co. Clarke; Gooderich Mine, co. Gordon; near Kempsey, co. Dudley; also at Kingsgate, near Glen Innes, on Glen Creek, co. Gough; at Capertee, co. Hunter, and Tarana, co. Roxburgh, in quartz with iron pyrites; and at Bolivia, co. Clive.

Molybdenum Ochre, MoO_3, often occurs with molybdenite as a yellowish amorphous incrustation.

ARSENIC.

Native Arsenic.

In massive pieces with mammillated surfaces, Lunatic Reef and Lion Reef, Solferino, co. Drake, with mispickel; Winterton's Mine, Mitchell's Creek, with gold and silver; Louisa Creek, co. Wellington. Occurs crystallised and massive at Lunatic, fifteen miles from Boorook, where it is sometimes very rich in gold, and is associated with anti-monite and cervantite.

MISPICKEL.—Arsenical Pyrites.

Chem. comp.: Sulph-arsenide of iron $FeS_2 + FeAs_2$. Arsenic, 46·0; sulphur, 19·6; iron, 34·4 = 100.

Rhombic system. Colour almost silver white. Streak dark-greyish black.

Localities.—Rather large crystals occur with quartz near Goulburn, co. Argyle, also on the Shoalhaven River associated with small hexagonal prisms of beryl which penetrate the mispickel; in New England, the Grampians and other places; ·in co. Argyle, at Marulan; co. Bathurst, at Carcoar; with marcasite and common pyrites at Icely, Mount Grosvenor, and Yetholme; co. Brisbane, at Scone; co. Camden, at Bundanoon; co. Clarendon, at South Gundagai; co. Clive, at Bolivia and Pye's Creek; co. Cook, near Lithgow; co. Drake, at Fairfield, bearing silver, 41 oz. 31 dwts. per ton; co. Dudley, at Kempsey, on the Macleay River; co. Gloucester, on Back Creek, Barrington, with gold; co. Gough, at Deepwater, Dundee, Elsmore, Emmaville or Vegetable Creek with fahlerz, Ferrucabad, Glen Creek with molybdenite, Glen Innes, and Kingsgate; co. Inglis, at Bendemeer and Moonbi; co. Hardinge, at Tingha, co. Murchison, at Bingera; co. Phillip, at Gulgong; co. Raleigh, on the Upper Bellinger River, and Warrell Creek; co. Richmond, at Gordon; co. Roxburgh, at Peelwood, auriferous mispickel with iron pyrites, in grey steatite and with talc, also at Tarana and Wattle Flat; co. Sandon, at Armidale and Uralla; co. St. Vincent, at Nerriga, at Moruya with blende and galena, and containing a fair proportion of gold and silver; co. Wellesley, at Bombala; co. Wellington, at Hargraves, large well-formed crystals on Louisa Creek, and very rich in gold near Orange; co. Westmoreland, at Native Dog Creek and Oberon; co. Wynyard, at Tarrabandra; co. Yancowinna, at Thackaringa. Also on the Mann River, the Moama or Mitchell River, near Cooradooral; at Mount Ovens, fourteen miles from Bathurst.

At Ournie, co. Selwyn, payable quantities of gold and silver occur in mispickel. Occasionally the mispickel is exceedingly rich in gold, as at Lucknow, amounting to many thousand ounces per ton.

LÖLLINGITE.—Leucopyrites.

Chem. comp.: Iron arsenide = $FeAs_2$; arsenic, 72·8; iron, 27·2.

Louisa Creek, co. Wellington, and near Gundagai, in small but well-formed crystals.

REALGAR.

Chem. comp.: Arsenic sulphide, AsS; As 70·1, S 29·9 = 100.
Oblique system. Orange red, translucent. Louisa Creek, co.
Wellington.

PHARMACOLITE.

Chem. comp.: A hydrated calcium arseniate, H_2CaAsO_4, $5H_2O$.
On Louisa Creek. In large imperfect crystals, dark-grey colour,
coated with white and yellow incrustations in part.

ANTIMONY.

NATIVE ANTIMONY.

Native antimony occurs at Gara, near Armidale, and in calcite at
the North Redfern Gold-mine, Lucknow, with gold and silver.

ANTIMONITE.—Stibnite—Antimony Glance.

Chem. comp.: Antimony sulphide = Sb_2S_3; Sb = 71·8, S = 28·2
= 100.

This ore is met with in the massive state in mineral veins, and
occasionally in rolled masses; well-formed crystals appear to be rare.

At times the cleavage planes are particularly large and brilliant,
at others the structure is more compact and granular.

It occurs on the Clarence and Paterson Rivers; the mineral is
found in masses of large size, and showing broad, well-defined, striated
cleavage planes, portions of the surface usually being incrusted with a
yellow coating of cervantite, an oxide of antimony = SbO_4.

A specimen of antimonite from Pyramul, co. Wellington, in
splendid massive blocks, showing well-developed, striated cleavage
planes. Exterior coated with the yellow-coloured oxide of antimony
(SbO_4) gave—metallic antimony, 67·74, with traces of gold and silver.

A minutely crystalline stibnite, with a little oxide, and traces only
of iron and lead, contained—antimony, 56·41 per cent.; gold and
silver, 19 grains per ton. This sample was from near Uralla, co.
Sandon, and the precious metals consisted chiefly of gold.

The antimony-bearing veins appear to occur for the most part in
rocks of Devonian age, and they vary in width from a few inches to
four or five feet. Quartz is the usual vein-stuff or gangue, and the
contents are often much broken or brecciated.

ASSAYS OF ANTIMONITE.

Locality.	County.	Antimony, per Cent.	Gold and Silver, &c.
Bendemeer	Inglis .	...	Silver, 12 oz. 8 dwts. per ton.
Bingera	Murchison	68·68	...
Carangula	Dudley .	61·55	Gold, 5 dwts. per ton.
Dungog (twenty miles north of)	Gloucester	52·64	Gold, 4 dwts. per ton. Silver, 3 oz. 5 dwts. per ton.
Hell's Hole, Mudgee Line .	...	40·26	Gold, 2 oz. 10 dwts. per ton. Silver, a trace.
Hillgrove, Eleanora Mine .	Sandon .	57·00 to 67·26	Gold, 2 oz. 12 dwts. per ton. Silver, 19½ dwts. per ton. Gold, a trace.
Hillgrove, Garibaldi Mine .	,,	39·90	Traces of gold and silver.
M'Donald River (head of) .	Inglis .	43·75	...
Oppenheimer Mine	Bathurst	61·50	Gold, 3 oz. 12 dwts. per ton. Silver, 1 oz. per ton.
Peckett Hill, Bellinger River	Raleigh .	69·97	
Sofala	Roxburgh .	46·30	Gold, 14½ dwts. per ton. Silver, 7½ dwts. per ton.
Temora.	Bland .	34·14	Gold, 3 oz. 16 dwts. per ton.

It is found associated in many parts of New England with tinstone, molybdenite, wolfram, and other minerals.

Localities.—In co. Argyle, at Bungonia and Marulan; co. Auckland, at Eden; co. Bathurst, at Crudine Creek, near Bathurst; co. Bland, at Temora; co. Clarence, at Grafton; co. Clarendon, at Gundagai; co. Clarke, at Aberfoil and near Mount Mitchell; co. Clive, at Pye's Creek, Bolivia, and Tenterfield; co. Cook, at Wallerawang; co. Drake, at Drake, Lunatic, Solferino, and Washpool Creek; co. Dudley, at Carangula and on Munga Creek, near Macleay River; co. Durham, at Gresford; co. Gloucester, at Coolongolook, and twenty miles north of Dungog; co. Hardinge, at Rocky River; co. Inglis, at Bendemeer, and at the head of the M'Donald River; co. King, at the Sharpeningstone Creek, near Yass; co. Murchison, at Bingera and Gineroi; co. Parry, at Nundle, and on the Peel River, twelve miles north of Hanging Rock; co. Phillip, in the Cudgegong District and at the Rockwell Mines; co. Raleigh, at Bowra (on the Nambucca River), Peckett Hill, and Warrell Creek, Nambucca River; co. Roxburgh, at Palmer's on the Upper Turon, Peelwood, Razorback, Rylstone, and Sofala; co. Sandon, at Boorolong, Dangar's Falls, Gara, Hillgrove (near Armidale), and Uralla; co. St. Vincent, at Shoalhaven; co. Vernon, at Walcha; co. Wellington, at Hargrave's Falls. Antimonite also occurs at Hell's Hole, Mudgee Line; Hermine; in the Macleay and Hastings Districts; at the Oppenheimer Mine, Bathurst District; twelve miles from Uralla; eighteen miles north-west of Gobandry and Wiseman's Creek; and at the Watershed, Macleay River.

The Rev. W. B. Clarke records finding a rolled mass of 3 lbs.

weight in the superficial ironstone gravel on an unfrequented hill on the north shore of Sydney Harbour.

CERVANTITE.

Chem. comp.: Antimony oxide $= SbO_4$; Sb 79·2, O 20·8 $= 100$.

Usually occurs massive, as an incrustation upon antimonite, sometimes as minute acicular crystals of a dull yellow colour.

Localities.—Almost the same as those for antimonite, as in co. Bligh, at Talbragar; co. Phillip, in the Cudgegong District, and at Ford's Creek, near Gulgong; co. Raleigh, on the Bellinger River; co. Sandon, at Gara, near Armidale; co. Wellington, Pyramul; and other places.

It is not unusual for it to contain small quantities of gold and silver.

The following assays of mixed antimonite, cervantite, and galena, as well as those on p. 74, are by the Government Analyst:—

Carangula, co. Dudley: 45 per cent. metallic antimony; gold, 5 dwts. per ton.

Marulan, co. Argyle: 16·75 per cent. antimony; gold, 2 oz. 17 dwts.; silver, 164 oz. 3 dwts. per ton; and 22·27 per cent. of lead.

Pye's Creek, Bolivia, co. Clive:—

Antimony sulphide	16·250
Lead „	63·280
Iron „	13·716
Silver „	1·124
Silica, &c.	5·630
	100·000

JAMESONITE.

Chem. comp. $= 2(PbFe)S + Sb_2S_3$; sulphur, 21·1; antimony, 32·2; lead, 43·7; iron, 3·0 $= 100$.

This mineral usually occurs in fibrous masses of a bluish lead-grey colour.

It occurs with cervantite in a soft quartz near to Campbell Creek and Nuggetty Gully, Bathurst District.

Return showing the Quantity and Value of Antimony produced in the Colony of New South Wales (*Annual Report of the Department of Mines*, Sydney):—

Year.	Quantity. Tons. Cwts.	Value. £
1871	31 0	560
1872	0 13	5
1873	27 12	210
1874	12 15	122
1875	142 0	5,000
1876	40 0	140
1877	69 12	1,131
1878	64 0	1,964
1879	76 16	1,046
1880	99 19	1,652
1881	539 4	17,346
1882	1,068 18	16,732
1883	375 11	5,555
1884	433 12	6,458
1885	293 0	4,296
1886	273 3	3,381
Total	3,541 35	65,598

There were no returns previous to 1871.

TIN.

NATIVE TIN.

The presence of native tin in New South Wales has recently been discovered.

"In the washings from Aberfoil River, about fifteen miles from the town of Oban, New South Wales, native tin exists in the form of irregular, somewhat globular grains, or aggregations of such grains; they are distinctly crystalline, from 0·1 to rarely over 1 mm. in size. When magnified 60 diameters they appear to be of an uneven surface, showing planes, which are too indistinct, however, for determining their form. They are greyish-white and of metallic lustre. It was impossible to select enough of the purest grains to determine their specific gravity or to make a quantitative analysis. A portion, treated with hydrochloric acid, dissolved readily with disengagement of hydrogen, leaving fine scales of iridosmine behind. Not a trace of any other metal but tin could be found in the solution."—*F. A. Genth, "Contributions to Mineralogy,"* October 2, 1885.

CASSITERITE.—Tinstone.

Chem. comp.: Tin binoxide $= SnO_2$. Tin, 78·67; oxygen, 21·33 $= 100$.

Tetragonal system. Occurs massive, crystallised, and as rolled pebbles and masses known as "stream tin." Well-developed crystals are by no means rare; the forms assumed are very similar to those found in other countries, viz., the prism, or a series of prisms combined with the pyramid, or pyramids, with and without the basal pinacoid plane. Sometimes the crystals are very large, especially those which are made up solely of the planes of the pyramid.

The lustre is usually bright metallic, and many of the specimens are very beautiful, especially some of the ruby and amber-coloured transparent specimens, which, however, have not, as a rule, so high a lustre as the black crystals. The colour varies from almost colourless and transparent, through shades of grey yellow, amber, red, brown, to black and opaque. Often more than one of these colours are to be seen in the same specimen, when the effect is very fine, especially the admixture of the ruby-red and translucent amber colours.

The hardness and specific gravity do not appear to materially differ from tinstone obtained elsewhere. Specimens having a specific gravity of only 4·463 and 5 413 have been met with, but they are probably very impure, although fairly well crystallised.

Discovery of Tinstone.—The probable presence of tin in Australia was mentioned as early as January 1799. Collins, in his account of the English colony of New South Wales, states that Mr. Bass, the surgeon of H.M.S. *Reliance,* found on the beach of Preservation Island (on the north coast of Tasmania, near the south coast of Barren Island) "a very considerable quantity of the black metallic particles, which appear in the granite as black shining specks, and are in all probability grains of tin." Tinstone, or cassiterite, is commonly known simply as "tin" by Cornish miners and others.

Mr. Bass is not likely to have mistaken grains of black mica, hornblende, or other minerals for tinstone, since he appears to have possessed considerable geological knowledge; and, moreover, had he done so he would probably have "discovered" tin in other places where granite exists.

Prior to about 1872 the alluvial tinstone met with by gold-diggers was thrown aside as a valueless black sand. After its value was recognised by some Cornish miners a rush for tin-mining set in.

A miner named James Daw claims to have discovered tin on the Broadwater, a tributary of the Severn River, then in the colony of New South Wales, now Queensland, in the beginning of 1849.

The first public mention made of the occurrence of tin in New South Wales was by the Rev. W. B. Clarke. In the *Sydney Morning Herald,* August 16, 1849, he records having found it in the Alps along part of the Murrumbidgee.

The Rev. W. B. Clarke also found small quantities of tin in the

New England District, and drew attention to the same in his report, dated 7th May 1853. He also mentions having found tinstone pseudomorphous after felspar crystals in New England corresponding to those from St. Agnes Mine, Cornwall.

In the "Papers Relating to Geological Surveys," published by the Government, I find that Mr. Hargraves makes the following mention of tin ore in New South Wales (pp. 71, 72):—

"GUNTAWANG, 18th July 1851.

" I have received information from Mr. Rouse of this place (Guntawang) that a shepherd of his found tin at Warranbungall Mountains some years ago, distant a hundred miles north of this place. I have therefore determined to visit the locality, and start for that place to-morrow," &c.   "E. H. HARGRAVES."

"MUDGEE, 3d August 1851.

" In travelling six miles north-west of the Cudgegong I found the gold region ceased; and on arriving at the Warranbungall Mountains, a hundred miles north-west, I found coal and iron in great abundance on every hill, but was not successful in finding the tin. The shepherd, who knows the locality, gave me a piece which he had smelted into bars, a sample of which I herewith enclose, which, I should suppose, contains 30 per cent. of silver (sic.); and in a short time the locality will be known to me. The man wants a large consideration for disclosing the whereabouts at present.   E. H. HARGRAVES."

A vein of tinstone an inch thick was being worked near Bungonia in 1870.

The principal tin-veins in New South Wales, which have yet been worked, occur in granite; at once seen to be similar to that of Cornwall. In some parts, as at Elsmore and Newstead, co. Gough, New England, the veins occur in greisen (mica and quartz), and in eurite (felspar and quartz). At Newstead Mine, and also at the Albion Tinmine, crystals of tinstone are seen disseminated through large and well-formed transparent quartz crystals. At the former place the quartz crystals in which it occurs often weigh nearly a hundredweight.

Lode or *Vein Tinstone* occurs in association with quartz, mica, orthoclase felspar, molybdenite, fluor spar, usually of pale shades of purple and green, a yellow steatitic mineral, garnet, beryl, topaz; the matrix of the tinstone is sometimes in places composed solely of topaz, as in Tasmania, at Mount Bischoff, where the tinstone occurs in a topaz rock, resembling the Schneckenstein of Saxony; this was formerly mistaken for a porphyritic trachyte until Baron von Groddeck pointed out its true nature. Tinstone has not, however, been met with in topaz rock in New South Wales; the topaz has only been found in the tin-veins like any other mineral, and not as a rock mass. Other associated minerals are gold, metallic bismuth, bismuthinite, bismuth carbonate, malachite, copper and iron pyrites, mispickel, tourmaline or schorl; at Giant's Den, Bendemeer, with garnets, wolfram, and radiated groups of schorl crystals. I have not seen wolfram in the same veins, but in other veins almost in juxtaposition. From the foregoing it is interesting to note that nearly all the minerals found associated with tinstone in Cornwall, Germany, France, America, and elsewhere have been met with this mineral in New South Wales.*

Tinstone occurs with azurite and redruthite at Gumble, near Molong, in veins running through slate at the junction of the slate with granite. One of the veins is rich in copper ore which contains about 60 oz. of silver to the ton; the other vein, which meets it at an angle of about 90°, is rich in tinstone, about 45 per cent., but is poor in copper and silver.

Messrs. Wilkinson, F.G.S., and David, F.G.S., Geological Surveyors, in their report on the Inverell District, state that the tin-lodes occur much in the same manner as do those of the Vegetable Creek District. They are chiefly narrow quartz and felspar veins containing tin ore irregularly distributed through them; but at Elsmore, Long Gully, and Stannifer-Bischoff it is met with in some quantity. At Elsmore it occurs in quartz-veins and in irregular patches of mica rock, somewhat similar to those of the Ding Dong and Pheasant Creek Mines; at Long Gully it is contained in quartz and felspar veins; and at Stannifer-Bischoff it is disseminated in separate coarse grains or crystals through porphyritic granite.

Large crystals of tinstone $1\frac{1}{2}$ inch in thickness have been found embedded in mispickel at the Jingellic Tin-mines, sixty-five miles above Albury, where quartz, mispickel, and scheelite form the principal associates of the tinstone.

Mr. Ranft states that at Paradise Creek, Mole Tableland, a large chlorite dyke occurs which carries about 5 per cent. of cassiterite and 7 per cent. of manganese garnet. The latter, on being roasted, yielded on washing from 4 per cent. to 5 per cent of tin ore.

Upwards of fifty tin-bearing veins or lodes have been opened in the Vegetable Creek Tin-field. The occurrence of the ore in the lodes is very variable; sometimes it is found in quartz, at others in felspar, greisen, chlorite, &c., and irregularly distributed through these, as bunches, veins, or as disseminated grains.

* See "Études synthétiques de Géologie Expérimentale," by Prof. A. Daubrée, p. 30 et seq. Paris, 1879.

The Ottery Lode, near Tent Hill, which has been traced a considerable distance, consists of a dyke mass of micaceous eurite from 40 to 100 feet wide, with numerous quartz-veins running irregularly through it, but generally along it, containing tin ore.

Some large lumps of oxide of tin ore have been discovered on the surface of the Ottery Lodes, five miles north-east of Vegetable Creek. One of these specimens weighed over 2 cwts.

The Folkestone Lode, about sixteen miles from Vegetable Creek, like one of the Ottery Lodes, contains iron pyrites and mispickel, blende, and probably silver.

The Dutchman's Lode, in the parish of Highland Home, has been traced for nearly half a mile, and strikes E. 30° N. to N.E.

It is noticeable that most of the tin-lodes, like the bismuth-lodes near Glen Innes, occur in the granite not far from its junction with the slate, which is highly metamorphosed near its junction with the granite.

Wood-tin occurs in veins at Glen Creek, co. Gough; at Tumberumba, ten or twelve miles from Kiandra, with gold in the granite.

Tin-lodes occur in co. Arrawatta, at Ashford; co. Ashburnham, at Gumble, near Molong; co. Blaxland, near Eremeran, in granite; co. Buller, twelve miles from Stanthorpe; co. Clive, at Grassy, Pheasant, and Pye's Creek, and Tenterfield; co. Gough, at Gilghi and other places in the Inverell District, at Elsmore, Newstead, Strathbogie, Bischoff-Stannifer, near Emmaville, The Gulf, the Grampian District, and at Rose Valley, Silent Grove, where the lodes are said to be rich in tin; co. Goulburn, at Jingellic, with quartz, mispickel, and scheelite; co. Hardinge, at Tingha; on the Mole Tableland; co. Wallace, at Mowembah; Meadow Flat. Vein tin is also reported to occur in quarries at Billabong, co. Clarendon.

Alluvial Tin Deposits.—There are two distinct sets of tin drifts—an older, or Tertiary, and newer, or Quaternary; the former are generally much more compact, and are often cemented together into a hard conglomerate, usually so hard as to require stamping. The tinstone is also much rounded and waterworn, whereas the tinstone in the newer drift is bright, and has undergone but little attrition. Some of the fragments or pebbles of rolled tinstone weigh many pounds, notably on the Butchart Tin-mine.

The minerals found associated with the stream tin are much the same as those found with it *in situ;* but, in addition, we find diamonds, sapphire, zircon, pleonaste, topaz (often of large size), metallic bismuth, bismuth carbonate, bismuthinite, rutile, and other minerals of high specific gravity. Gold in small quantities is usually present.

Rolled *wood-tin* of a grey and black colour at Abingdon; also at Grenfell and Lambing Flat, co. Monteagle, with extremely well-marked concentric and radiate structure, composed of red, brown, and black bands; other fragments are made up of alternate light and dark-grey bands; with diamonds near Mudgee and Bathurst. The variety known as toad's-eye tin is also met with here; in the Pliocene gold deposits at Grenfell, co. Monteagle; and on the Grampian Hills.

Nearly all the ore hitherto raised is stream tin obtained from the Tertiary and Quaternary drifts, where these are composed of the detritus from the stanniferous granites.

The neighbourhood of Deepsinker's Creek, Emmaville, is remarkable for containing nuggets of tinstone, ranging from several pounds in weight down to an ounce or two; one weighing 3 lbs. 5 oz. bears the appearance of having travelled a great distance. The ordinary tin is of a coarse kind.

Mr. Ranft states that on the Murray and Murrumbidgee tin-fields most of the cassiterite is grey, even to white; seldom any black. A small parcel of white, carefully selected out of a ton of stream tin, free from all earthy matter, gave only 70 per cent. of metallic tin, while the bulk gave close on 76 per cent. He thinks the difference may be due to the presence of tungstic oxide.

Tin is said to have been discovered at Little River, in Oberon District, and it has been obtained in ten gold claims on Sandy Creek, Tumut District.

Grey stream tin found at Manner's Creek, Tumberumba, near Kiandra, and at Attunga, near Albury; it is also said to occur in the belt of dry country between the Lachlan and Bogan Rivers, commencing at about a hundred miles north-west of Forbes. The same district is said to be rich in gold, copper, and iron; at Boona West, co. Blaxland, and Jumble Plains.

The cassiterite in the Aberfoil River is mostly in small rounded grains; some are of a deep aurora-red colour, others are hyacinth-red, reddish-brown, or variegated, black, red, and white.

Localities.—In co. Argyle, at Bungonia, Long Gully, and Spring Creek; co. Bathurst, at Wambrook, near Cowra; co. Buccleuch, at Dabarra and Tumut; co. Buckland, at Carroll's Creek and Quirindi; co. Buller, at Bookookoorara, Boonoo Boonoo Creek, Herding Creek, Maryland Creek, and the Undercliff; co. Burnett, at Warialda Creek; co. Clarence, at Tea-Tree Creek; co. Clarke, on the Aberfoil River, at Mount Mitchell, Oban, and on the Sara River; co. Clive, at Deepwater, the Mole River, Sandy Mount, and the Severn River; co. Darling, at Mangahra, Mount Lowry Creek, and Tiabundie Creek; co. Drake, at Fairfield, Gordon's Creek, and Lunatic; co. Farnell, eighteen miles north of Mount Gipps; co. Gough, at Blair Hill, Yarra Creek, near Glen Innes with pyrites, Glen Creek, The Gulf, near Emmaville, near Inverell, Kingsgate, M'Intyre River, Middle Creek, Paradise

F

Creek, Ranger's Valley, Stockyard Creek, Swan Creek, and Yarrow River; co. Goulburn, at Jingellic Creek; co. Gresham, on the Henry, Mitchell, and Ann Rivers; co. Hardinge, at Auburn Vale, Bundarra, Cope's Creek, Honey's Creek, Honeysuckle Creek, Kentucky Ponds, Long Arm Creek, Moredun Creek, Sandy Creek, and ruby tin ore at Tingha; co. Inglis, at Bendemeer, Carlyle, Watson's, and other Creeks; co. Mitchell, at Pullitop Creek; co. Murchison, at Bald Rock Creek, Bingera, the Gwydir River, Myall Creek, Reedy Creek, and Rocky River; co. Roxburgh, at Sheep Station Creek, and on the Turon River; co. Sandon, at Boorolong and Uralla; co. Selwyn, at Burra Creek; co. St. Vincent, on the coast south of Jervis Bay and at Shoalhaven; co. Tongowoko, at Granite Diggings; co. Wallace, at Adaminaby and at Mowembah, in quartz associated with chalcedony; co. Wellington, at Spring Creek; co. Wynyard, at Tumberumba and in the Wagga District; and at Broken Dam, Merool Creek. Alluvial tin deposits, covered by 60 or 70 feet of basalt, are worked in Swinton parish, co. Hardinge.

The stanniferous area in New South Wales is estimated at $5\frac{1}{2}$ million acres, or 8500 square miles.

Up to the present most of the tin has been obtained from the New England District.

Analyses of Tinstone.

A specimen of dark-coloured, almost black stream tinstone from the Jupiter Mine, Vegetable Creek, co. Gough, New England, New South Wales, gave the following results:—

Analysis.

Stannic oxide (SnO$_2$)	89·92
Titanic acid (TiO$_2$)	·69
Alumina	6·75
Silica	·80
Iron sesquioxide	2·30
	100·46

Specific gravity, 6·629.

A specimen crystallised in prisms from Carabuco, Bolivia, South America, was found to contain:—

Tin dioxide (SnO$_2$)	96·339
Iron sesquioxide (Fe$_2$O$_3$)	2·177
Silver	0·115
Wolfram trioxide (WO$_3$)	·020
Lead	·250
Water	1·737
	100·638

The following are added for comparison :—
1. Schlackenwald, Bohemia.
2. Alternon, Cornwall.
3. Finbo, Fahlun, Sweden.
4. Wicklow County, Ireland. Specific gravity, 6·75.
5. Heres, Mexico. Specific gravity, 6·86.
6 and 7. Tiperani River, Bolivia.

Description.	1.	2.	3.	4.	5.	6.	7.
Tin dioxide (SnO₂)	95·4	98·60	93·6	95·26	89·43	91·81	91·80
Tantalum dioxide	...	...	2·4	...	...	...	...
Silica (SiO₂)	...	·75	...	·84	2·21	...	...
Iron sesquioxide (Fe₂O₃) . .	·7	·36	1·4	2·41	6·63	} 1·02	2·69
Manganese dioxide (MnO₂) . .	...	...	0·8	...	...		
Alumina.	...	...	...	...	1·20	·73	...
Insoluble matter	...	...	...	...	...	6·48	5·51
Total . . .	96·1	99·71	98·2	98·51	99·47	100·04	100·00

Watt's " Dictionary of Chemistry," p. 817.

Small quantities of other substances are often present, such as copper, bismuth, calcium, magnesium, tungsten, titanium, &c.

ASSAYS OF TIN ORE.

Locality.	County.	Description of Ore.	Metallic Tin, per Cent.	Other Metals.
Ashford . .	Arrawatta .	Greisen, with crystals of tinstone .	30·8	
Bungonia . . .	Argyle . .	Tin ore	74·9	
Gilgai	Gough . .	Lode tin ore . .	52·2	
Mount Gipps (18 miles north of)	Farnell . .	Tin ore	34·10	
Glen Innes (near)	Gough . .	Tin ore and pyrites	49·4	
,, ,,	,,	Tin ore	34·0	
,, ,,	,,	,, ...	63·1	
,, ,,	,,	,, ...	58·4	
Gumble (near Molong)	Ashburnham	Brown earthy lodestuff, and earthy blue and green carbonates of copper . .	25·5	{ Copper, 14·85 per cent. ; silver, 6 oz. 10½ dwts. and 60 oz. per ton ; gold, a trace.
Inverell (near) .	Gough . .	Tin ore	64·9	
Jervis Bay (coast south of) . .	St. Vincent .	Black sand . . .	54·50	
Long Arm Creek, Bundarra . .	Hardinge .	,, . . .	65·2	
Wagga District .	Wynyard .	Tin ore	24·1	
Wambrook (near Cowra) . . .	Bathurst. .	,, . . .	68·0	
Watson's Creek .	Inglis . .	Black sand . . .	15· to 58·2	

RETURN showing the Quantity and Value of Tin produced in the Colony of New South Wales (*Annual Report of the Department of Mines,* Sydney):—

Year.								Quantity. Tons.	Value. £
1872	.	.	.	.	.	.	.	896	47,703
1873	.	.	.	.	.	.	.	4,571	334,436
1874	.	.	.	.	.	.	.	6,219	484,322
1875	.	.	.	.	.	.	.	8,080	561,311
1876	.	.	.	.	.	.	.	6,958	439,638
1877	.	.	.	.	.	.	.	8,054	508,540
1878	.	.	.	.	.	.	.	7,210	395,822
1879	.	.	.	.	.	.	.	5,921	372,349
1880	.	.	.	.	.	.	.	6,159	471,337
1881	.	.	.	.	.	.	.	8,200	724,003
1882	.	.	.	.	.	.	.	8,670	833,461
1883	.	.	.	.	.	.	.	9,125·5	824,552
1884	.	.	.	.	.	.	.	6,665·9	521,587·
1885	.	.	.	.	.	.	.	5,193	515,626
1886	.	.	.	.	.	.	.	4,968	467,653
				Total	.	.		96,890·4	7,502,340

TITANIUM.

RUTILE.

Chem. comp.: Titanic acid = TiO_2. Crystallises in the tetragonal system, usually in prisms. Up to the present time I have only found it in the form of fragments of crystals with striated surfaces, or in rounded grains of a hair-brown colour. It is found with the gem sand at Bald Hill, near Bathurst, and at Mount Walsh, near Uralla, and in quartz crystals at Cope's Creek.

Brookite, which is an allotropic form of titanic acid, crystallising in flattened forms belonging to the rhombic system, has also been found in New South Wales, at Burrandong, in waterworn, imperfectly crystallised, striated plates of a dark red-brown colour, with metallic lustre, but of a bright red colour by transmitted light.

In the diamond drift near Mudgee as flat, transparent, red and translucent reddish-white plates, with striated surfaces. H = 6, and specific gravity = 4·13. Chem. comp.: Pure titanic acid, except a minute trace of iron oxide.—*Dr. A. M. Thomson, Jour. Roy. Soc. N. S. W.*, 1870, p. 102.

Anatase.—A third allotropic form of titanic acid, crystallising in tetragonal pyramids. This has been found at the dry diggings of Burrandong. Some fairly good crystallised specimens have also been found in the Cudgegong River, co. Phillip.

For the account of titaniferous iron minerals see Iron Minerals, p. 102.

SPHENE.

A calcium silico-titanate. Crystallises in oblique system. I have met with but one well-crystallised specimen, of a green colour; the locality in New South Wales from which it came is uncertain.

TUNGSTEN.

WOLFRAM.

Chem. comp : Iron and manganese tungstate = $(FeMn)WO_4$. It is found in rolled masses in association with tinstone in many parts of New England. It is also found *in situ* in the quartz-veins on Elsmore and Newstead Mines, on Glen Creek, co. Gough; the Grampians; parish of Boyd, co. Gough; near Emmaville, on the Severn River, from a lode 7 feet wide, containing 74·41 per cent. tungstic acid. It has also been found at Kingsgate with bismuth; at Wilson's Downfall, on Hogue's Creek, near Glen Innes, and other places, in the usual form of imperfectly developed tabular crystals. It is commonly accompanied by iron pyrites.

A specimen found in quartz-veins with tinstone, Inverell, co. Gough, of the usual bronzy-black colour, sub-metallic lustre, opaque, lamellar structure, with only traces of crystal faces, had the following composition :—

Analysis.

Tungstic acid	77·640
Iron protoxide	18·760
Manganese	4·121
	100·521

SCHEELITE.

Calcium tungstate = $CaWO_4$. Crystallises in the pyramidal system. Occurs in New England and at Adelong.

A specimen from the Victoria Reef Gold-mine, Adelong, co. Wynyard, was massive, but with a portion of a crystal showing on one side, of an amber colour, translucent, resinous lustre, brittle, splintery fracture. Hardness, 4–5; specific gravity, 6·097. Associated with a dark-green chloritic vein-stuff.

The following analysis was kindly made for me by Dr. Helms :—

Analysis.

Loss at red heat	·25
Tungstic acid	79·53
Lime	19·14
Alumina	·58
Magnesia	·07
	99·57

The above results correspond to the formula $CaWO_4$.

It is found in massive lumps, in association with molybdenite and molybdenum ochre, at Hillgrove, near Armidale, co. Sandon.

Some from a lode near Peelwood gave :—

Tungsten trioxide	69·31	per cent.
Lime	19·35	„
Silica	4·88	„
Copper oxide	4·05	„
Iron oxide	2·01	„
	99·60	„

(*Government Analyst.*)

Crystallised scheelite has been found at Jingellic; also at The Gulf, Mole Tableland, in sandy deposits.

IRON.

NATIVE IRON.

Out of a large number of specimens of so-called native iron which have come before me from time to time not one was entitled to be so called; they had all without exception been derived from iron or steel tools.

Native iron, apart from that derived from meteorites, however, probably does occur in the colony, and it is most likely to be found in or near to igneous rocks; *c.g.*, melted globules of native iron have been met with at Ballarat, in Victoria, in connection with basalt.

Accounts of two nickel-iron meteorites are given at pp. 213, 214.

MAGNETITE.—Magnetic Iron Ore.

Chem. comp.: Iron oxide = Fe_3O_4. Iron, 72·4; oxygen, 27·6 = 100. Cubical system.

This is the richest of all the ores of iron, and when perfectly pure it only contains rather more than 72 per cent. of metallic iron; hence the absurdity of the statement so commonly made by the promoters of mining companies, that the iron ores on a certain property contain over 90 per cent. of metal, will be at once apparent; and, moreover, it is a very rare thing indeed for large masses of any ore to be quite pure; therefore, instead of the amount of metal in the vaunted mineral even approaching to the alleged richness, it falls far below it, and most probably it is much nearer to 40 than to 90 per cent.

It is found in the colony both massive and crystallised in octahedra, which are usually small. In structure it varies, being compact, granular, or lamellar.

Large deposits of magnetite exist at Wallerawang, co. Cook; Mount Lambie, co. Cook, with micaceous hæmatite, in a chloritic matrix; Mount Wingen, co. Brisbane; Solferino, co. Drake, in quartz-veins; Grafton, co. Clarence, with copper ores; on the Clarence and the Shoalhaven Rivers; Brown's Creek, co. Bathurst; with sulphide of copper from near Binalong, co. Harden.

The following extracts are from a paper read before the Royal Society of New South Wales:—

" The deposits of iron ore at present opened out are situated some six miles from Wallerawang, and near the junction of the coal-measures with the Upper Silurian or Devonian beds, which there crop out to the surface. These deposits contain two varieties of iron ore, viz., magnetite or the magnetic oxide of iron, and brown hæmatite or goethite—the hydrated oxide; then, in addition to these, there are deposits of the so-called " clay band," which are interstratified with the coal-measures. These clay bands are not what are usually known as clay iron ores in England. They are brown hæmatites, or limonite, while the English clay iron ores are impure carbonates of iron, which seldom contain much more than 30 per cent. metallic iron, against some 50 per cent. contained by these hæmatites.

" A highly ferruginous variety of garnet accompanies the veins of magnetite; this garnet is very rich in iron, and it will probably be found advantageous to smelt it with the other ores, not only on account of the large percentage of metal which it contains, but also on account of the increased fluidity which it would impart to the slag.

" *Magnetite.*—The vein of magnetic iron ore runs apparently N.E. by S.W. This can only be stated approximately, for, owing to the action exercised by it on the needle, the compass was found to be perfectly useless in the vicinity of the lode.

" The ore is scattered over the ground in blocks and nodules along its outcrop; but at a little depth it is in a solid and compact body, merely broken across here and there into large masses by joints and fissures.

" In one part the vein has a width of thirteen (13) feet; but at another spot, where a trench was cut across, it was there found to be not less than 24 feet in width.

" Two shafts have been sunk on this vein—one to a depth of 10 and the other to a depth of 23 feet. At these depths the quality of the ore is about the same as that at the surface; but certain portions of the vein are evidently richer than others.

"At present the average yield of metallic iron from the vein, as a whole, is not rich for a magnetite, which, when perfectly pure, contains 72·41 per cent. of iron, and under ordinary circumstances about 70 per cent., whereas the Wallerawang vein yields only 40·89 per cent.

"This average was obtained by taking samples from different parts, across the whole width of the trench cut across the vein, and then crushing them all up together. As I have before mentioned, picked portions yield a much larger percentage.

"On the whole, taking all the circumstances into consideration, we may come to the conclusion that the true capabilities of the deposit of magnetite have not yet been fully tested or proved.

"The vein-stuff or gangue accompanying the magnetic iron ore is siliceous. In some parts of the lode this appears to be replaced by the ferruginous garnet rock.

"A partial analysis of this ore yielded the following results:—

Silica and insoluble matter	18·70 per cent.
Metallic iron	40·89 „
Phosphorus	traces
Sulphur	traces

. "Both the phosphorus and the sulphur are present in such minute quantities that the ore may be regarded as virtually free from them; and these are the only really deleterious substances present, for although there is too large a quantity of silica and gangue present in this superficial portion of the vein to permit of malleable iron being made from it by a direct process, it is extremely well adapted for reduction in the blast furnace.

"*Garnet.*—The garnet occurs both crystallised, in the form of the rhombic dodecahedron, and in the massive state. The crystals are, as is usually the case, very uniform in size; they are nearly all of them either about ⅛ or ¼ of an inch in diameter.

"The faces of the crystals are smooth, free from pits and irregularities, and bounded by sharp and well-defined edges. The colour is brown without any red shade.

"Portions of the massive garnet and aggregations of crystals are hard and compact, whilst in other parts they are more or less disintegrated and friable.

"The average percentage of metallic iron is 21·05—an amount not much less than that contained by many commonly smelted ores." *

* See also "Iron and Coal Deposits, Wallerawang." A. Liversidge, *Jour. Roy. Soc. N. S. W.,* 1874.

The following analyses were made upon an intimate mixture of the two minerals as they occur in specimens collected by myself in 1874 * :—

Analysis.

Water lost at 100° C.	·30
„ combined	1·63
Silica	16·28
„ soluble	2·51
Alumina	1·35
Iron protoxide	3·67
„ sesquioxide	55·74
Manganese protoxide	2·99
Lime	14·28
Magnesia	·62
Sulphur	traces
Phosphoric acid	traces
Carbonic acid	·54
Loss	·09
	100·00

Iron protoxide 3·67 and sesquioxide 55·74 } = 41·87 per cent. metallic iron.

The finely divided ore was then separated by means of a magnet, the magnetic and non-magnetic parts being examined separately.

The portion removed by the magnet amounted to 56 per cent., but, as will be seen by the following analysis, it was found impossible by this means to obtain the magnetite quite free from the vein-stuff:—

Analysis.

	Magnetic.	Non-magnetic.
Water lost at 100° C.	·26	·21
„ combined	1·69	1·14
Silica	8·61	28·66
„ soluble	·65	3·88
Alumina	1·97	1·13
Iron protoxide	6·91	·56
„ sesquioxide	70·47	35·91
Manganese protoxide	2·39	1·62
Zinc-nickel (traces of)	·13	...
Lime	6·96	24·44
Magnesia	·20	1·00
Phosphoric acid	traces	...
Sulphur	traces	...
Carbonic acid	absent	1·66
	100·24	100·21

* "New South Wales Minerals." A. Liversidge, *Jour. Roy. Soc. N. S. W.*, 1880.

MAGNETITE.

Description.	No. 1. Brown's Creek, co. Bathurst.	No. 2. Wallera- wang, co. Cook.	No. 3. Wallera- wang, co. Cook.	No. 4. Wallera- wang, co. Cook.	No. 5. Wallera- wang, co. Cook.
			Magnetic.	Non- magnetic.	
Water lost at 100° C. . .	} 0·13	0·30	0·26	0·21	} 2·16
„ combined . . .		1·63	1·69	1·14	
Iron peroxide* . . .	60·48	55·74	70·47	35·91	64·01
„ protoxide * . . .	18·67	3·67	6·91	0·56	8·99
Manganese protoxide .	...	2·99	2·39	1·62	traces
Zinc and nickel . . .	...	...	0·13	...	...
Alumina	14·22	1·35	1·97	1·13	2·75
Silica	6·50	16·28	8·61	28·66	6·70
„ soluble . .	...	2·51	0·65	3·88	...
Magnesia	...	0·62	0·20	1·00	0·41
Lime	...	14·28	6·96	24·44	3·75
Fluoride of calcium . .	...	...	...	...	10·68
Carbonic acid . . .	...	0·54	...	1·66	...
Phosphoric oxide . .	traces	traces	traces	...	traces
Sulphur	...	traces	traces	...	...
Total . .	100·000	99·91	100·24	100·21	99·45
* Equal to metallic iron .	56·85	41·87	54·72	25·57	51·79
Analyst . . .	{ Gov. Analyst. }	Liversidge.	Liversidge.	Liversidge.	Dixon.

The non-magnetic part thus answers to the general formula for the iron-lime-garnet, $3CaO,2SiO_2 + Fe_2O_3,SiO_2$.

A lamellar magnetite of good quality occurs in quartz at Carcoar associated with iridescent botryoidal brown hæmatite, and at Combullanarang with copper ores.

It is also found at Inverary Quarry, co. Argyle, where Stutchbury mentions that it occurs in the pisolitic form, associated with a black non-magnetic ore in rounded particles the size of peas, and cemented together by a variety of crystallised minerals. Crystallised and compact magnetite occurs near the limestone quarries on Belubula Creek, co. Bathurst. Rounded and polished nodules of magnetic iron ore occur in the Lachlan River with ilmenite. It is also found in nearly all the gold and gem bearing drifts and deposits.

Deposits of magnetite are said to exist between the Bogan and Lachlan Rivers, about 100 miles north-west of Forbes; on Jugiong Creek, near Wellington, and Binalong with copper ores, co. Harden, associated with malachite; between Cooyal and Warrabil Springs, co. Phillip, associated with brown hæmatite; Rocky River; Barraba, co. Darling, with chrome iron; massive magnetite with a granular structure at Bogolong; Errol; on Clear Creek, Peel River, co. Parry, containing both gold and silver—one sample yielded 2 dwts. 5 grains per ton of the two metals (*Annual Report of the Department of Mines,* Sydney, 1878, p. 11). Also found ten miles from Cowra, on the Grenfell road, co. Forbes; at Burra Burra, Parkes District, co. Ashburnham; Mitchell's Creek, co. Roxburgh; and at Brown's Creek, near Carcoar, co. Bathurst; with zircons at Talbragar, co. Bligh. Magnetite in the form of small grains and crystals is common in the creeks in basaltic districts.

HÆMATITE.—Red Hæmatite, Specular Iron.

Chem. comp.: Iron oxide, Fe_2O_3. Iron, 70; oxygen, $30 = 100$.

Hexagonal system, in rhombohedral forms. Usually massive, platy, or micaceous. Well-formed crystals are at present almost unknown here. Specular iron ore occurs in a coarse-grained granite at Summer's Hill, near Bathurst, and at Mount Lambie, co. Cook; also at Bookham, and with carbonate of copper from Binalong, co. Harden, and Yass, co. King, with micaceous and massive red hæmatite; micaceous hæmatite also occurs at Pine Bone Creek, with titaniferous iron; at Dungowan's Creek, co. Parry; at Bibbenluke, co. Wellesley.

"*Specular hæmatite* was found at Carwary, in the Shoalhaven District, in abundance; near the spot was a vein of ironstone of a fused appearance; a quartzose ferruginous conglomerate and a calcareous tuff containing fragments of these rocks."—*Mitchell's* "*Eastern Australia,*" vol. ii. p. 321. Also found at Carwell; micaceous hæmatite at Boro, co. Murray; parish of Ponsonby, near Bathurst; between Mylora and Bookham, in the Yass District; O'Connell Plains, co. Westmoreland, and in the New England District. Specular iron also occurs at Tumut, co. Buccleuch.

Of the hæmatite near Carcoar the late Mr. Stutchbury speaks as follows:—" In a gully or creek called the Waterfall Creek, running into the Cadiangulong Creek, and at the extremity of a mountain spur known as the Rocky Ridge, there is an immense mass of oxydulous iron (hæmatite) forming in one solid mass a precipitous waterfall of about sixty feet in height; in this mass of iron, especially in the joints, there are brilliant crystals of iron pyrites, with a small quantity of yellow copper ore and traces of blue and green carbonate of copper. Here also is found iron sulphate, from the decomposition of the pyrites."

In the cliffs at Shepherd's Hill, Newcastle, there are trunks of trees converted into red hæmatite.

Large deposits of massive and somewhat ochry red hæmatite occur at Brisbane Water, also over large areas in the county of Argyle. This same mineral enters also largely into the composition of the so-called " red hills " occurring in the New England tin districts and other parts. A siliceous red hæmatite is also common in the Hawkesbury sandstone, about Sydney, and elsewhere, in irregular deposits, filling veins, crevices, and joints; also as concretionary masses and nodules. Is often more or less mixed with sand and other impurities.

The following analysis was made upon a specimen collected in the neighbourhood of Sydney :—

Specific gravity, 4·49.

Analysis.

Water lost at 104° C.	·646
Silica	4·210
Alumina	·713
Iron sesquioxide	90·555
Iron protoxide	3·632
Manganese	trace
Lime	...
Magnesia	...
Sulphur	...
Phosphoric acid	absent
Loss	·244
	100·000

The above results show the specimen to be an extremely good iron ore.

One of the nodules used for gravelling garden walks about Sydney contained 28·0 per cent. of metallic iron, and one of the compact red hæmatite from Nattai gave 45 per cent.

GOETHITE.—Brown Hæmatite.

Chem. comp.: Hydrated sesquioxide of Iron = $Fe_2O_3,3H_2O$. Iron sesquioxide, 89·9; water, 10·1 = 100. Crystallises in the rhombic system.

Generally massive, or with fibrous radiate structure; minute velvety crystals are sometimes met with; also scaly, mammillated, pisolitic, reniform, and stalactitic.

Externally the colour is often jet black with high lustre; within yellow, yellowish-brown, and full-brown. Streak, brown.

Many of the nodules of brown hæmatite contain cavities and hollows holding a soft black substance like manganese dioxide, which hardens on exposure.

Many specimens of iron oxide contain gold, especially those derived from the "gossan" or upper decomposed portions of lodes; at Gundagai cubes of brown hæmatite pseudomorphous after iron pyrites have been found showing free gold on the faces of the crystals.

Very large and extensive irregular deposits and pockets of brown hæmatite occur at Wallerawang, Mount Lambie, Piper's Flat, Clarence Tunnel, Mount Clarence, Blackheath, Newbridge, and Lithgow Valley, co. Cook; Jamberoo, Nattai, Berrima, Mount Keira, Mittagong, and Broughton Vale, co. Camden; Port Hacking, co. Cumberland; near Gundagai, co. Clarendon; Mount Tellulla; Newbridge or Back Creek, near Blayney, co. Bathurst—deposits of this ore are being worked and smelted at Lithgow; at Goulburn, Boro, Camberra, Joppa, Kingsdale, and Norwood, co. Argyle; at Bingera, co. Murchison; Uralla, co. Sandon; Ballimore, near Dubbo, co. Lincoln; eight miles from Jervis Bay, co. St. Vincent; Burra Burra, co. Ashburnham; Narrandera, co. Cooper; fifty miles west of Forbes, Lachlan River; Narellan Creek, co. Monteagle; Scone, co. Brisbane; near West Maitland; in the Coal Ranges, Clarence River; at Tamworth, co. Inglis; between the Lachlan and Bogan Rivers; and in many other places, such as between Mount Tomah and Mount King George, co. Cook. In fact, this mineral is one of the most widely diffused. Between Cooyal and Warrigal Springs, co. Phillip, a wide vein of brown hæmatite is reported with magnetite. Pseudomorphous crystals of iron pyrites changed into brown hæmatite occur at Carwell, co. St. Vincent.

The principal deposits of iron ore consist of brown hæmatite, and occur either with the coal-measures or near to them, as at Mittagong and Berrima, on the Southern Railway Line, and at Piper's Flat, Wallerawang, Lithgow, in the Coolah Valley, Mudgee District, and other places on the Western Line. This is a matter of the highest importance, and adds very greatly to the value of these iron ores, since it is absolutely necessary to have the two close together for smelting purposes—if either the coal or iron have to be carried any distance, the cost of freight renders it impossible to make cheap iron.

The brown hæmatite from Manly Beach, near Sydney, often possesses a somewhat laminated and concentric structure, with small vesicular cavities, many of which were filled with white and yellow clay-like substances. See analysis No. 14, Brown Hæmatites, p. 95.

The 60·720 per cent. of sesquioxide of iron is equal to 42·504 of metallic iron. The undetermined constituents were chiefly alumina, lime, &c. The amounts of sulphur and phosphorus are small, so that the mineral is adapted for use as an ore of iron.

Hæmatite such as that from Manly Beach is very common throughout the Hawkesbury sandstone. It usually occurs in small veins, pockets, and nodules, and is very often used as a road-metal and for gravelling paths. It would probably be useful to mix with other ores for iron-smelting.

Similar nodular brown hæmitite occurs at Wallerawang. See analysis No. 28, p. 96.

A massive specimen, but somewhat vesicular in places, from the neighbourhood of Jamberoo, dark brown to pitchy black colour, brown streak, on analysis gave the results entered under No. 9, p. 94.

Auriferous hæmatite is reported to occur in the Conoblas, near Orange.

BROWN HÆMATITE.

Description	No. 1. Berrima, co. Camden.	No. 2. Berrima, co. Camden.	No. 3. Berrima, Atkinson's Property.	No. 4. Bingera, co. Murchison.	No 5. Clarence Tunnel, co. Cook.‡	No. 6. Gosford, co. Northumberland.	No. 7. Gosford, co. Northumberland.	No. 8. Goulburn, co. Argyle.	No. 9. Jamberoo(near), co. Camden.	No.10. Jamberoo Mountains, co. Camden.
Hygroscopic moisture . .	} 15·47	{ 2 50	1·94	3·173	0·48	} 10·73	7·39	{ 1·56	1·335	1·45
Loss on ignition .		...			10·80			...	...	...
Combined water .	...	10·57	11·31	7·304	...	...	...	11·43	11·872	9·75
Iron peroxide * .	82·54	67·42	83·10	81·877	79·20	60·99	41·24	80·60	77·155	76·03
Iron protoxide * .	...	...	...	...	...	...	...	traces	...	...
Manganese protoxide	...	...	...	0·561	...	...	...	traces	0·428	traces
Alumina . .	0·84	3·43	1·05	0·634	traces	0·82	1·71	3·35	1·232	3·33
Silica . . .	...	15·52	...	5·819	...	25·10	46·43	...	8·507	8·90
Magnesia . .	...	traces	traces	traces	traces	0·41	...	...	trac e	traces
Lime . . .	...	traces	traces	0·503	...	traces	traces	...	0·257	traces
Phosphoric oxide .	0·63	traces	traces	...	traces	0·03	0·02	1·35	...	0·44
Sulphur trioxide .	...	0·13	traces	...	...	...	...	traces	...	0·12
Sulphur . .	...	...	...	...	...	traces	traces	...	...	...
Insoluble in acids .	0·52	...	2·50	...	8·80	1·84 (alumina)	3·65 (alumina)	1·90	...	...
Loss, undetermined, &c. . . .	...	0·53	...	0·129	0·72	...	...	...	...	...
Total .	100·00	100·10	99·90	100·000	100·00	99·92	100·44	100·19	100·786	100·02
Specific gravity .	...	...	...	3·52	...	...	...	...	3·52	...
* Equal to metallic iron . .	57·8	47·2	58·17	57·31	55·44	42·69	28·86	56·42	54·00	53·22
Analyst . .	Gov. Analyst.	Gov. Analyst.	Gov. Analyst.	Liver-sidge.	Gov. Analyst.	Dixon.	Dixon.	Gov. Analyst.	Liver-sidge.	Gov. Analyst.

Stalactites of hæmatite are often formed by the ferruginous springs found over the coal-measures, as at Berrima and Nattai, co. Camden, and elsewhere, and the deposits of brown iron from these often contain beautiful impressions of leaves and other objects; also in botryoidal and mammillated forms, with a well-marked concentric structure.

A deposit of brown iron, some fifty feet in diameter, from a ferruginous spring rising from trachyte rocks, occurs in "The Valley," about one mile from Springwood on the Western Line. An ochreous deposit is also found at Temora.

Brown hæmatite is common on the Bingera Diamond-fields in the form of small concretionary nodules, some of which are as spherical as

† Traces of cobalt protoxide.
‡ Another sample from the same locality yielded 56 per cent of metallic iron.

BROWN HÆMATITE.

Description.	No. 11. Joppa, co. Argyle.	No. 12. Kingsdale, co. Argyle.	No. 13. Lithgow (Eskbank Iron Co.), co. Cook.	No. 14. Manly Beach, near Sydney.	No. 15. Mittagong, co. Camden.	No. 16. Mittagong, Butler's Property.	No. 17. Mittagong, Fitzroy Mine. Raw.	No. 18. Mittagong, Fitzroy Mine. Calcined.	No. 19. Mittagong, Fitzroy Mine.	No. 20. Mittagong, Waite's Property.	No. 21. Mount Clarence, co. Cook.
Hygroscopic moisture . . .	1·48	2·00	...	1·600	} 10·43	} 1·66	...	...	...	1·82	1·76
Loss on ignition .	...	...	8·47	...			...	...	...	...	8·64
Combined water .	11·40	10·84	...	13·770	...	11·60	10·80	...	12·00	12·20	...
Iron peroxide * .	69·44	71·37	46·11	60·720	73·67	84·36	80·00	72·00	81·25	79·64	81·40
Iron protoxide * .	traces	...	...	...	...	...	...	14·14	...	traces	...
Manganese protoxide .	traces	1·06	...	...	6·83	0·09	3·40	1·00	...	traces	...
Alumina .	6·16	2·04	5·89	...	9·07	...	} 4·40	11·20	} 3·45	1·17	} 8·20
Silica . .	...	11·65	37·80	...	...	...			3·20	4·70	
Magnesia .	1·05	traces	traces	...	...	traces	...	...	...	traces	...
Lime . . .	traces	traces	...	...	...	traces	1·00	1·12	traces	traces	...
Phosphoric oxide .	1·30	0·74	traces	traces	traces	traces	...	...	...	traces	...
Sulphur trioxide .	traces	traces	...	...	...	traces	...	...	...	traces	...
Sulphur . .	...	...	...	0·075	...	...	...	...	} minute traces	} ...	...
Insoluble in acids .	7·50	...	...	12·660 (with silica)	...	...	...	...	...	...	...
Loss undetermined, &c. .	†2·07	...	†1·73	‡11·175	...	2·30	...	...	0·10	...	...
Total .	100·40	99·70	100·00	100·000	100·00	100·01	99·60	99·46	100·00	99·53	100·00
* Equal to metallic iron . .	48·60	49·96	32·27	42·504	51·57	59·05	56·00	61·39	56·75	55·74	56·98
Analyst .	Gov. Analyst.	Gov. Analyst.	Gov. Analyst.	Liversidge.	Gov. Analyst.	Gov. Analyst.	Noad.	Noad.	Gov. Analyst.	Gov. Analyst.	Gov. Analyst.

marbles; in other cases they are more or less elongated; or two or three of the globular forms may be joined together. Some possess a curiously wrinkled or corrugated surface, but most are quite smooth, but not polished, the material being rather soft. On breaking them open they are seen to have traces of a concentric structure; the outer portions occasionally present indications of a radiate fibrous structure also. The hydrated oxide of iron seems to have been originally diffused through an impure carbonate of lime and magnesia, and afterwards to have segregated together into these concretionary forms. Occasionally the nodules are met with enclosed in the matrix of impure magnesite.

† With alkalies.
‡ The undetermined constitutents in No. 14 were chiefly alumina, lime, &c.

BROWN HÆMATITE.

Description	No. 22. Newbridge, co. Bathurst.	No. 23. Norwood, co. Argyle.	No. 24. Norwood, co. Argyle.	No. 25. Norwood, co. Argyle.	No. 26. Norwood, co. Argyle.	No. 27. Uralla, co. Sandon.	No. 28. Wallerawang, co. Cook.	No. 29. Wallerawang, co. Cook.	No. 30. Westbrook, Singleton Graham's Property.	No. 31. Westbrook, Singleton Graham's Property.	No. 32. Willerot Station, Lake George, co. Murray.
Hygroscopic moisture	{12·35	{0·50	1·57	1·50	0·35	1·787	{1·28	{15·25	5·60	{2·80	0·57
Loss on ignition		...	...	...	...	...	...			...	...
Combined water	...	2·76	11·50	10·82	2·12	10·652	12·04	...	†12·03	†10·09	10·55
Iron peroxide *	68·04	76·32	68·47	74·20	85·91	77·132	73·60	75·52	60·11	59·11	84·55
Iron protoxide *	...	11·23	...	...	5·59	3·526	...	1·01	traces	traces	...
Manganese protoxide	...	traces	traces	traces	traces	0·940	...	traces	traces	traces	traces
Alumina	10·38	3·04	5·54	4·31	2·18	0·159	...	‡3·08	9·36	5·74	1·14
Silica	6·95	6·25	11·85	8·70	3·80	3·782	...	4·25	10·90	19·25	2·47
Magnesia	...	traces	traces	traces	traces	traces	...	0·21	0·55	0·75	traces
Lime	...	traces	traces	traces	traces	2·022	...	0·19	0·72	0·35	traces
Phosphoric oxide	2·28	traces	0·89	0·51	traces	...	0·12	0·38	0·73	0·70	0.66
Sulphur trioxide	...	traces	traces	traces	traces	...	0·06	0·03	traces	0·92	traces
Insoluble in acids	...	...	...	...	...	...	12·19 (with silica)	...	...	...	...
Loss, undetermined, &c.	...	...	...	...	...	...	0·71	...	...	...	...
Total	100·00	100·10	99·82	100·04	99·95	100·000	100·00	99·92	100·00	99 71	99 94
Specific gravity	...	...	...	...	...	3·611	...	...	...	...	...
* Equal to metallic iron	47·62	62·15	47·93	51·94	64·48	56·734	51·52	53·64	42·07	41·37	59·18
Analyst	Gov. Analyst.	Gov. Analyst.	Gov. Analyst.	Gov. Analyst.	Gov. Analyst.	Liversidge.	Liversidge.	Dixon.	Gov. Analyst.	Gov. Analyst.	Gov. Analyst.

Hardness, 3–4; specific gravity, 3·52. The streak or powder is yellow. See analysis No. 4, p. 94.

Similar concretions have been found on the Cudgegong with the diamond drift; also near Maryborough, Queensland.

Hollow nodules of ironstone are found in the bed of the Macquarie River, near Dubbo, where they apparently are not uncommon.

The outer shell consists for the most part of brown hydrated oxide of iron, and when first found they are quite soft, and can be cut with a knife. I am informed that the interior is usually filled with sand, which can be shaken out, leaving a hollow cavity. Although hard and compact, they are evidently of quite recent origin. The first of these were collected by Mr. Murdoch, of the Railway Department.

† 2·11 per cent. of the alumina in No. 29 is insoluble.
‡ And organic matter.

Limonite.—A variety of brown hæmatite. Extensive deposits of what are termed *clay band* iron ores occur interbedded with the coal-measures. (See also "Iron and Coal Deposits at Wallerawang," by A. Liversidge, *Jour. Roy. Soc. N. S. W.*, 1874.) These are an earthy variety of brown hæmatite; yet they are often very rich, and as they occur in immense quantities in close association with coal, they form a most valuable source of iron. Analyses are given—Nos. 4, 5, 6, and 7.

LIMONITE.

Description.	No. 1. Eskbank, co. Cook.	No. 2. Jamberoo, co. Camden.	No. 3. Lithgow, co. Cook.	No. 4. Wallerawang, co. Cook.	No. 5. Wallerawang, co. Cook.	No. 6. Wallerawang, co. Cook.	No. 7. Wallerawang, co. Cook.†	No. 8. Wallerawang, co. Cook.
Hygroscopic moisture	1·730	1·452	} 8·47	} 1·28	1·31	1·35	0·97	} 12·00
Loss on ignition	...	...			...	...	...	
Combined water	13·560	11·000	...	3·54	4·17	10·29	10·07	
Iron peroxide *	66·320	13·019	} 46·11	80·00	85·32	78·96	77·29	59·87
Iron protoxide *	...	1·255		...	0·52	0·67	0·46	2·26
Manganese protoxide	...	0·257	...	...	1·60	2·43	0·76	traces
Alumina	...	15·070	5·89	...	2·13	1·38	1·20	‡7·96
Silica	...	...	37·80	...	§4·14	§3·73	§8·61	17·21
Magnesia	...	traces		...	0·29	0·14	0·28	0·17
Lime	...	0·158		...	0·35	0·65	0·19	0·16
Phosphoric oxide	traces	traces		0·49	traces	traces	traces	0·44
Sulphur trioxide	...	...	} 1·73	0·11	0·04	0·04	traces	0·04
Sulphur	0·192	...		...	...	...	...	
Insoluble in acids	13·520	57·528 (with silica)		4·60 (with silica)	...	...	...	
Loss, undetermined, &c.	4·678	0·531		9·98	...	...	...	
Total	100·000	100·000	100·00	100·00	99·87	99·60	99·83	100·11
Specific gravity	...	2·73	...	...	...	...	...	3·255
* Equal to metallic iron	46·424	10·089	...	56·00	60·13	55·80	54·46	43·50
Analyst	Liversidge.	Liversidge.	Gov. Analyst.	Liversidge.	Liversidge.	Liversidge.	Liversidge.	Dixon.

Limonite, or " clay band ore," occurs at Eskbank, interbedded with the coal-measures; in masses of an irregular cuboidal form, containing cavities, closely answering in shape to the external form; in some instances these cavities are more or less completely filled with yellow ochre. See analysis No. 1.

† Other specimens from these seams in the same locality yielded 49·28 and 53·31 per cent. of metallic iron respectively.

‡ 7·45 per cent. of alumina in No. 7 is insoluble.

§ 3·63, 3·66, and 8·34 per cent. of silica in Nos. 5, 6, and 7, respectively, is insoluble.

A "clay band" iron ore occurs in Jamberoo; of a dark reddish-brown colour. Analysis No. 2 shows how very much some of these "clay band" ores vary. It has a somewhat laminated structure; breaks with a flat conchoidal fracture, and dull earthy surfaces.

The specimen yielding the results given under No. 5, p. 97, was taken from the outcrop of the uppermost seam at Wallerawang, and had probably been subjected to bush fires, since the proportion of water is far less than is required; and, moreover, the mineral contains a trace of magnetic iron, and yields a dark chocolate powder instead of the usual yellow-coloured one.

Nos. 6 and 7 are specimens from two other similar deposits in the same locality.

No. 8. This is the analysis of a "clay band" from Wallerawang. It had a curious concretionary structure, containing numerous cavities filled with yellow ochre in some cases; in others with a dark grey matter scarcely soluble in acid.

Similar clay bands exist in the Buttar Ranges, near to East Maitland; at Mount Wingen, co. Brisbane; at Mount Lambie, in the coal-measures, where both magnetite and micaceous hæmatite also occur; and elsewhere.

Large outcrops of limonite occur at Lithgow, Piper's Flat, and Bowenfels, co. Cook; also in the Illawarra District, at Bulli, where it is said to have a thickness of 20 feet. Assays of this, made at the Royal Mint, Sydney Branch, yielded 32·9, 38·9, 44·3, and 55·7 per cent. metallic iron.

Pisolitic Iron Ore is another of the less pure forms of hæmatite.

Large superficial deposits of pisolitic and brecciated iron ore, red and brown, occur near Bungonia and Windellama Creek, co. Argyle, and overlie the slate more or less continuously between Bungonia, Jacqua Creek (with limestone), Dog Trap, and Spring Creeks, forming what are known as the "Made Hills;" also at Windsor. Concretions of ironstone more or less diffused throughout the shales of Cumberland. A pea-iron ore occurs in the coal at Nattai, co. Camden; and near Bungonia there is an auriferous argillaceous iron ore. At the Boro Creek, co. Argyle, there is a botryoidal pisolitic ore. The same variety occurs at Brisbane Water, co. Cumberland.

The "Made Hills," which lie between the Macintyre River and Cope's Creek, are composed of the same material.

Also found with limestone gems, titanite, &c., in the drift deposits of New England, and elsewhere.

BOG IRON ORE.

These are impure limonites, usually of a more porous and ochry character.

Description.	No. 1. Jamberoo Mountains, co. Camden.	No. 2. Mittagong, co. Camden.	No. 3. Mittagong, Butler's Property.	No. 4. Mittagong, Waite's Property.	No. 5. Mittagong, Waite's Property.
Hygroscopic moisture . .	2·40	3·00	1·20	2·20	2·30
Combined water . . .	11·07	9·72	10·38	9·70	10·86
Iron peroxide * . . .	60·26	68·37	57·61	74·71	65·84
Iron protoxide * . . .	...	traces	...	...	...
Manganese protoxide . .	traces	traces	6·41	traces	1·40
Alumina	7·61	4·63	24·30	3·04	4·49
Silica	17·70	14·10	...	10·10	14·27
Magnesia	0·28	traces	traces	0·43	0·48
Lime	traces	traces	traces	traces	traces
Phosphoric oxide. . .	0·38	traces	traces	traces	0·25
Sulphur trioxide . . .	0·30	traces	traces	traces	0·11
Total . .	100·00	99·82	99·90	100·18	100·00
* Equal to metallic iron .	42·18	47·86	40·32	52·29	46·08
Analyst . . .	Gov. Analyst.	Gov. Analyst.	Gov. Analyst.	Gov. Analyst.	Gov. Analyst.

Red and Yellow Ochres are closely allied to the above hæmatite iron ores, and are usually found associated with them, but they generally contain more earthy matter.

SPATHIC IRON ORE.—Chalybite.

Siderite, Sphærosiderite.

Chem. comp.: Iron carbonate $= FeCO_3$. Iron oxide, 62·1; carbonic acid, 37·9 = 100.

Crystallises in the hexagonal system in rhombohedral forms.

Occurs in minute crystals at Gulgong, co. Phillip. It is also found at Newstead Mine, New England, with arragonite; and in amygdaloidal cavities in basalt at Inverell; in basalt and conglomerate at Rocky Ridge; at Jordan's Hill, Cudgegong River, co. Wellington; O'Connell Plains and Essendon, co. Westmoreland; Mount Gipps, co. Yancowinna; also in the Hawkesbury sandstone.

Thick bands of grey-coloured impure carbonate of iron, some of which contain about 10 per cent. of metallic iron, occur in the coal-measures at Jamberoo, co. Camden; the siderite is in the form of small particles diffused through a compact grey-coloured argillaceous lime-stone.

The following analysis was made of the whole to ascertain its value as an ore of iron, as it was found impossible to separate the particles of siderite.

Specific gravity, 2·79.

Analysis.

Water lost at 105° C.	·932
„ combined	11·922
Silica and insoluble matter	42·292
Alumina	22·837
Iron protoxide	12·870
Manganese protoxide	1·048
Magnesia	traces
Potash soda	traces
Phosphoric acid	traces
Carbonic „	7·816
Titanic „	·716
	100·433

Siderite is also said to occur in the neighbourhood of Wentworth.

SCORODITE.

Chem. comp.: An arseniate of iron, $Fe_2As_2O_8, 4H_2O$. Arsenic acid, 49·8; sesquioxide of iron, 34·7; water, 15·5 = 100.

Rhombic system. With iron pyrites, Cadell's Reef, Mudgee Road, nine miles south-east of Mudgee; also at Louisa Creek, co. Wellington.

PHARMACOSIDERITE.

Chem. comp.: An arseniate of iron. Arsenic acid, 43·13; sesqui-oxide of iron, 40·0; water, 16·87 = 100.

Cubical system. Found crystallised in small olive green cubes. Sub-translucent.

Localities.—To the east of Bungonia, co. Argyle; also six miles south-east of Emmaville, with massive galena and mispickel.

VIVIANITE.

A hydrated phosphate of iron. $Fe_3P_2O_8, 8H_2O$. Phosphoric oxide, 28·3; iron protoxide, 43·0; water, 28·7 = 100.

Dr. Leibius, of the Royal Mint, Sydney Branch, forwarded for identification a specimen of this mineral to me in March 1882, which, I believe, is the first found in the colony.

The specimen came from the Nymagee Copper-mine, where it was found associated with copper pyrites. Externally the fragment is partially surrounded by layers of càrbonate of iron and iron pyrites; it looks as if the vivianite had crystallised within a kind of geode. The mineral is translucent, and shows the usual changing green and deep blue tints when viewed from different positions. No complete crystals were present.

Mr. Ranft states that it is also found in fissures in sandstone at Boorook, and in auriferous granite at Timbarra.

TITANIFEROUS IRON.

Chem. comp.: Iron and titanic acid.

There are several different kinds of titaniferous iron, distinguished by their physical properties and by the amounts of titanic acid which they contain, such as ilmenite, iserine, menaccanite, &c. Until those found in New South Wales have been examined it will be as well, perhaps, to class them all under the general head of titaniferous iron.

Occurs in a quartz-vein near Wellington.

Found usually with alluvial gold, as at Ophir, Mudgee, and Wellington, in co. Wellington; Bathurst; Bingera, co. Murchison; and Uralla, co. Sandon, in the diamond drift. Large rolled masses occur at Uralla. Ilmenite, menaccanite, nigrine, and iserine are said to occur with gold, garnets, and chrysolites in the Five-mile Flat Creek, Cudgegong River; in the Lachlan and at Talbragar, with magnetite; also near Wagga Wagga, co. Wynyard, and the Rocky River, co. Hardinge. With tinstone, zircons, spinel, &c., at the Y. Waterholes.

Nigrine.—Burrandong, co. Wellington. In small grains, for the most part rounded, but with traces of crystal faces on some of the fragments.

A variety of titaniferous iron ore found in the river deposits near Uralla, by miners working for gold, in the form of black pebbles, with a submetallic lustre, was composed as follows:—

Analysis.

Silica .	9·491
Alumina	14·799
Titanic acid	44·506
Metallic iron	23·019
Oxygen	8·185
Lime .	traces
Magnesia .	traces
	100·000

Specific gravity, 4·44.

The iron exists in the form of both protoxide and sesquioxide, the former being present in the larger quantity. As it is difficult to determine accurately the amount of protoxide in a difficultly soluble mineral such as this, the total iron has been stated as metallic iron, and the oxygen estimated by difference. The alumina and silica doubtless exist in combination as silicate.

CHLOROPAL.

Found in veins in the basalt at Two-mile Flat, near Mudgee. Of a pistachio-green colour—earthy, somewhat fibrous in parts, looks like a decomposition product. Friable; the fracture is splintery to earthy. H. 2–3.

Specific gravity, 1·94. Yields a green powder. Emits an argillaceous odour when breathed upon. Before the blowpipe blackens, does not fuse, becomes magnetic. With hydrochloric acid is decomposed, silica being left. Does not gelatinise.

Analysis.

Water lost at 105° C. . .	12·313
,, combined	5·224
Silica	49·657
Iron sesquioxide	29·108
Manganese.	traces
Lime	2·606
Magnesia	·508
Soda	·599
Potash	·170
	100·185

It has also been found in a shale four miles from Burraga Coppermine.

IRON PYRITES.

Chem. comp.: FeS_2. Sulphur, 53·3; iron, 46·7 = 100. Crystallises in the cubical system. Occurs massive and crystallised, the most common forms being the cube and the pentagonal dodecahedron. Well-formed cubes, partially decomposed into brown hæmatite, are common in many deposits with gold, and are known to the miners by the name of "devil's dice," especially in the gem sand at Walker's Crossing, on the Cox River, about $1\frac{1}{2}$ mile below Wallerawang. All specimens of pyrites which I have examined have, without exception, contained traces of gold, and in some cases large amounts.

As is found to be the case in other parts of the world, this mineral is almost universally diffused throughout the metalliferous districts of the colony, and is found in rocks of all ages.

Well-formed crystals are found in the Manilla and Namoi Rivers, co. Darling. In the tin district of New England it is very common, also in the Bathurst District; at Gulgong well-formed pentagonal dodecahedra are common in the auriferous quartz-veins. Very abundant in the Adelong reefs, co. Wynyard; the Carcoar District; at Kiandra, co. Wallace, crystallised in cubes with molybdenite. Well-crystallised specimens are said to occur in a chlorite schist near Grenfell. Masses of iron pyrites, or even large crystals, which are superficially changed into brown hæmatite, break with deep conchoidal fractures; and these fresh surfaces possess a very remarkable lustre; two such specimens had a specific gravity of 4·975 and 4·990.

Marcasite.—Rhombic pyrites. Chem. comp.: Iron sulphide = FeS_2. The same as the former, of which it is an allotropic form. Very readily decomposes.

Fluted rhombic crystals occur with arsenical and common pyrites (auriferous) to the south of Reedy Creek, Shoalhaven River; also at Carcoar, co. Bathurst, with galena and other minerals. It is found near Cadia, also at Great Extended Gold-mine, Forest Reefs, investing fossil wood.

Some very interesting concretions of iron pyrites occur at the Sunny Corner Silver and Gold Mine, which is situated on Mitchell's Creek, some sixteen miles from Rydal, on the Western line. The rocks in which the Sunny Corner deposits occur are altered Devonian or Silurian shales and sandstones, penetrated by a porphyry dyke. The portion of the lode worked for silver, which bears nearly north and south with westerly dip, is mainly composed of a loose earthy ferruginous material, and is rather cavernous in places. The vuggs or cavities vary much in size, but are usually small, and are lined with stalactites of brown hæmatite, externally of a deep brown or black colour.

The vein-stuff is very variously coloured, yellow, brown, green, red, black, &c., and contains but little mineral matter of a definite and readily recognisable character except galena and pyrites; occasionally small crystals of barytes are found, and some black oxide of copper. In places the vein is as much as fifty feet across, but usually much less.

Formerly this mine, when owned by Messrs. Winter & Morgan, was worked for gold only, and yielded some very rich returns.

FIG. 2.

Concretion of Iron Pyrites, showing the radial lines.

In some respects these concretions of pyrites resemble the calcareous concretions of the London clay, known as septaria, and used for the preparation of hydraulic cement, *i.e.*, as far as general form and structure, both are more or less rounded and both are fissured, but the fissures or cracks in those from Sunny Corner are filled in either with pyrites or with quartz.   I am indebted to Mr. J. M. Smith, of Sydney, the chemist and superintendent of the mine, for specimens of these, as well as for many others which he has been good enough to obtain for me from time to time.

The concretions occur in a pale-coloured shale of a greyish tint, abutting against the vein, full of cavities, which can be seen to have formerly contained crystals of iron pyrites.   This gradually passes into a slaty shale of a dark bluish-grey colour, studded with cubical crystals of pyrites, most of which are twinned.

As will be seen from fig. 2, the concretions of pyrites have a somewhat concentric structure, and are fissured in a more or less regular radiate manner.

They vary in size ; some are an inch or less in diameter, and others are several inches through.

Some of the concretions (fig. 2) consist wholly of iron pyrites, with the fissures or cracks also filled in with the same material, but of a

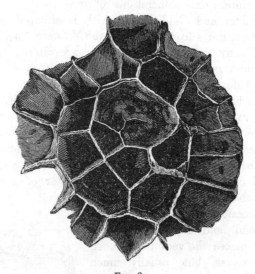

FIG. 3.

Siliceous Septa set free from Concretions of Iron Pyrites.

more compact character.   Others consist of soft friable pyrites with the fissures filled in with hard white quartz, thus forming septa.   As the rock weathers and exposes the concretions the granular pyrites falls out and the septa are left in the form of irregular, exaggerated honeycomb structures (fig. 3).

The changes which appear to have gone on are as follows:—

1. The iron pyrites crystals, as more or less well-developed cubes, were formed in the slaty shale, probably while it was in a soft and clay-like condition.
2. The pyrites crystals were gradually removed by solution.
3. The pyrites was gradually redeposited from solution, not in the form of cubical crystals, but in the form of nodules of marcasite, i.e., the rhombic and less durable form of iron pyrites.
4. The pyrites nodules (marcasite) became cracked or fissured. These cracks are probably due to the outer portions of such nodules having become hardened first; then, as the inner portions hardened and contracted, fissures would necessarily form within, since the hard outer portions would not give way so readily as the softer and weaker inner portions.
5. The fissures in the pyrites nodules were next filled in, in some cases with pyrites, in others with quartz, and the septa formed. It may have been that the latter were also filled in first with pyrites, which was afterwards dissolved out and replaced by quartz.
6. Finally the marcasite was removed and the siliceous septa set free (fig. 3).

The pyrites of the nodules oxidises with great rapidity; specimens kept for only a few months rapidly fall to powder, and become incrusted with crystals of iron sulphate.

PYRRHOTINE.—Magnetic Pyrites.

Chem. comp.: Fe_7S_8. Sulphur, 39·5; iron, 60·5 = 100.
Hexagonal system.

More of a copper-colour than the other pyrites, slightly magnetic, and crystallises in six-sided forms. Is found at Emmaville, with zinc blende at Ottery's, Vegetable Creek; on the Barrier Ranges; at Lunatic; with galena on Adelong Creek; with galena and blende on Pye's Creek; with copper pyrites and garnets in quartz six miles north of Tarana.

It occurs with gold and calcite at Hawkins' Hill, co. Wellington.

YENITE.—Ilvaite.

Chem. comp.: Double silicate of iron and calcium. Crystallises in the form of rhombic prisms.

The late Rev. W. B. Clarke reported that he had found drifted pieces on the Tuggerah Lake beach. As it appeared to be a new variety, he named it, provisionally, Baddeleyite, after the finder.

RETURN showing the Quantity and Value of Iron produced in the Colony of New South Wales (*Annual Report of the Department of Mines*, Sydney):—

	Quantity. Tons.	Cwts.	Value. £
1874	...		15,434*
1875	40	0	502
1876	2,680	0	13,399
1877	2,600	0	7,600
1878	900	0	6,666
1879	1,118	0	10,550
1880	2,322	0	15,335
1881	6,560	0	47,891
1882	7,476	0	37,224
1883	3,434	3	26,908
1884	3,759	2	24,572
1885	4,176	0	25,793
1886	3,685	17	19,068
Total	38,751	2	250,942

CHROMIUM.

CHROME IRON.—Chromite.

Chem. comp. : Iron chromate $= FeCrO_4$. Iron oxide, 32·0 ; chromic acid, 68·0 = 100.

Cubical system. Usually occurs massive, with a granular or lamellar structure, and as small crystals and waterworn grains in gold and gem bearing sands. Black in colour.

A specimen of the massive variety of chromate of iron from Woolomi, Tamworth, had a black colour, and sub-metallic lustre. On certain portions the specimen exhibits curved, somewhat fluted, polished surfaces, closely resembling the smooth and lustrous surface of a slickenside.[†] It may not be out of place to mention that this resemblance to a slickenside is not at all uncommon in many compact clay deposits, in steatite, serpentine, and other rocks; it is also often well shown in many specimens of the mineral noumeaite.

To distinguish this structure from the true slickenside I have pro-

* These figures represent the value of iron raised prior to the year 1875 ; quantity not known.
† A slickenside is the smooth polished and striated surface occasionally exhibited by the walls of faults and slides ; in such cases, however, the peculiar structure has doubtless been induced by friction, accompanied by intense pressure.

posed the term *petaloidal*, from the resemblance which the typical examples of such surfaces often roughly bear to the curved and fluted petals of an unopened flower-bud.

This specimen contained 64·72 per cent. of chromium sesquioxide and 21·11 per cent. of iron protoxide.

The outcrop from which this was taken is about 700 feet above Bowling Alley Point, and the apparent thickness of the vein is in one part some 40 odd feet; one huge block of the mineral lying loose on the surface measures about 12 feet long by 6 feet high and 5 feet wide.

The chrome iron vein is in association with serpentine, diallage rock, and black slates. This deposit ought to be easily and cheaply worked.

Chrome iron is found in the Gwydir River and many of its tributaries; in co. Clarence, near Grafton, in the Oaky Creek, and in Parish Pucka; co. Clarendon, in Jones' Creek, Gundagai; co. Darling, at Barraba; co. Drake, at Gordon Brook; co. Hardinge, at Stony Batta, with serpentine; co. Inglis, at Tamworth, with hyalite, and at Bendemeer; co. King, near Yass; co. Murchison, in the Angular,. Bingera, Gundalmulda, Kennedy, and Reedy Creeks, and in the Horton River; co. Parry, at Hanging Rock and Nundle Creek, near Nundle with serpentine, diorite, diallage rock, hyalite, and black slates; co. Phillip, at Gulgong; co. Roxburgh, at Two-mile Creek; co. Sandon, near Armidale, and with chrome ochre at Uralla; co. Wellington, at Ironbarks and Mudgee; in the Murrumbidgee River; and with serpentine beyond Young, in the Bland District. Chrome iron is usually to be expected where serpentine exists.

MANGANESE.

The ores of manganese do not appear to have been discovered in any great abundance in New South Wales.

PYROLUSITE.—Wad—Asbolite. Black Oxide of Manganese.

Chem. comp.: MnO_2. Crystallises in the rhombic system, but more usually found massive. Louisa Creek, co. Wellington; said to occur in large quantities near Caloola, co. Bathurst.

Wad is an impure oxide of manganese.

At Long Gully, near Bungonia, co. Argyle, it is met with having a more or less botryoidal form and platy structure; of a black colour; soft, with a black shining streak; in association with quartz, both as small veins running through the quartz and as an external coating or incrustation. A specimen from this locality was found by Dr. A. M. Thomson to contain 1·57 per cent. of cobalt and 0·36 per cent. of nickel.

Wad from Trunkey gave on analysis:—

Silica	25·84
Oxide of iron and traces of alumina	24·72
Oxide of manganese	34·93
Oxide of cobalt and traces nickel	2·11
Magnesia	1·00
Water	11·15
Alkalies and loss	·25
	100·00

(*W. A. Dixon*, 1879.)

Samples from Boro, Goulburn District, consisting of oxide of manganese mixed with quartz, contained:—

Available oxide of manganese (MnO$_2$)	23·27	37·84
Other substances soluble in acid, chiefly oxide of iron	29·33	22·76
Quartz	47·40	39·40
	100·00	100·00

(*W. A. Dixon*, 1879.)

Another specimen yielded 77·2 per cent. of available dioxide and traces of cobalt.

PYROLUSITE.—Manganese Dioxide.

Locality.	County.	Manganese Peroxide.	Cobalt, &c.
		Per cent.	
Bendemeer	Inglis	44·1 to 88·4	...
„ (near)	„	86·2	...
Bowning	King	91·0	...
Fontana Reef, near Bathurst	Bathurst	91·55	...
Glanmire	Roxburgh	71·97	...
„ (near)	„	43·4	...
Gulgong (near)	Phillip	62·05	Sesquioxide of cobalt, 0·63.
Mihi Creek	Sandon	75·86	Cobalt, a strong trace.
Mudgee	Wellington	23·6	Cobalt, 0·49.
Never Never	...	14·97	...
Newbridge (four miles from)	Bathurst	{ 54·35 / 61·17 }	...
Newcastle District	Northumberland	53·0	...
New England	...	86·85	...
Pentecost Island	...	88·4	...
Rockley (five miles from)	...	42·60	...
Sutton Forest	Camden	46·64	Cobalt, 1·05.
Tamworth District	Inglis	86·2	...
Walcha	Vernon	82·85	...

Some of the above are poor, but the specimens found may perhaps lead to the discovery of other richer deposits.

Description.	Stony Batta.	Bathurst.	Bendemeer.	
			I.	II.
Moisture	4·28	4·75	1·60	3·13
Manganese dioxide . . .	85·25	78·72	89·60	73·00
„ monoxide . .	...	3·66	...	...
Iron peroxide . . }	3·33	6·50	· 3·86	1·93
Alumina. . . . }			4·74	3·57
Silica	2·75	5·80	·20	18·37
Lime	1·50	...	...	...
Magnesia	1·06	...	...	trace
Phosphorus trioxide . .	·47	...	traces	...
Sulphur trioxide . . .	·32	...	...	...
Undetermined . . .	1·04	...	...	...
Total . . .	100·00	99·43*	100·00	100·00
Analyst	...	M. P. Muir.	Government Analyst.	

It is abundant in the diamond drift near Mudgee, both as a cement and incrustation; often dendritic in outline. The incrustation on many of the pebbles is evidently quite recent.

It is very common as dendritic markings on rocks in many parts of the colony.

A peculiar form of wad is found in cavities in the basalt at Hill End. This variety is very soft and porous, being composed of minute scales arranged loosely together in a concentric manner—in fact, having a structure similar to that of wood; externally it has somewhat a frothy appearance, with a metallic lustre, so soft that it blackens the fingers, and will hardly bear handling without crushing.

A massive variety of wad has been sent down in large blocks from the Wellington District from time to time; it occurs at Caloola, and on the Ellenborough River, in the Walcha District. A large and well-defined lode is said to exist at Fairy Meadows, and samples yielded 70 per cent. of available dioxide of manganese. It is found in co. Argyle, at Bungonia; co. Ashburnham, near Parkes; co. Bathurst, near Bathurst, at Caloola, four miles from Newbridge, and at Orange; co. Buckland, at Quirindi; co. Camden, at Sutton Forest; co. Cook, to the north of Katoomba, Govett's Leap, and other places on the Blue Mountains—in fact, it occurs in the Hawkesbury sandstone under similar conditions to the hæmatite, as embedded nodules and loose on the ground; in co. Georgiana, at Back Creek, near Rockley; co. Gordon, at Tomingley; co. Inglis, at Bendemeer and near Tamworth; co. Harden, at Cootamundra; co. Hardinge, at Stony Batta; co. King, at Bowning, Silverdale, and Yass; co. Northumberland, in the New-

* Of a greyish black colour, crystalline in structure; on the dried mineral the calculated amount of manganese dioxide is 82·21 per cent.

castle District; co. Phillip, near Gulgong; co. Roxburgh, near Glanmire
and on Mitchell's Creek; co. Sandon, on Mihi Creek; co. Vernon, at
Walcha; co. Wellington, on Curramudgee Creek and at Lucknow.
It is also found twenty miles from Armidale; at Bogenbung (two miles
east of Wollongelong Run); Never Never; New England; Pentecost
Island; five miles from Rockley; Sturt's Meadows, Barrier Range;
Tintin Hull; and the Warrumbungle Mountains.

Psilomelane.—This is an impure oxide of manganese, assuming
stalactitic and botryoidal forms.

Occurs in the drift, Three-mile Diggings, Kiandra; at King's
Creek, parish of Jocelyn, co. Westmoreland.

A mineral which looks like psilomelane occurs at the Broken Hill
Mine, and was met with at a depth of six feet in cutting a trench on
the outcrop of the lode. It is smooth, black, and shining in mammillated stalactitic form, and upon its surface crystals of silver chloride
have been deposited in many cases. Dr. G. S. Mackenzie tells me the
mineral is locally used as a flux in smelting the silver ores; on analysis
he found it to contain 19·00 per cent. lead, probably present as carbonate; 20 per cent. silica; 37:88 per cent. manganese (metallic), probably present as dioxide; 3·02 per cent. alumina; 2·97 per cent. iron.
Another specimen, which he also partially analysed, from the South
Broken Hill Mine, about half a mile south of the outcrop whence the
previous one was found, yielded 6·20 per cent. silica; 28·85 per. cent.
iron; 21·10 per cent. manganese, probably present as dioxide; 14·00
per. cent. lead, probably present as carbonate. They both contain
lime, &c., but the other elements have not been estimated.

KUPFERMANGANERZ.—Cuprous Manganese.

Chem. comp.: An impure oxide of manganese, containing a small
percentage of black oxide of copper and oxide of cobalt.

Found in the Coombing Copper-mine, with native copper, cuprite,
copper carbonates, and sulphides; also on Wiseman's Creek, co. Westmoreland, near Soldier's Hill, with wad.

BRAUNITE.

Chem. comp.: Manganese oxides, and manganese silicate. Crystallises in the pyramidal system, also massive. At Rylstone; Port
Macquarie; Bungendore, co. Murray; at Caloola; near Gundagai; and
in the Wellington District.

A hard compact specimen, with a very minute crystalline structure;

strikes fire with steel; fracture conchoidal; of a dark iron-grey colour. From near Wellington.

Specific gravity, 6·465; hardness, 6·5.

Soluble in hot strong hydrochloric acid, with evolution of chlorine, a residue of white silica being left.

Analysis.

Silica	11·778
Alumina	4·061
Iron sesquioxide	·3·153
Manganese protoxide	31·516
„ dioxide	50·125
Lime	traces
Magnesia	traces
	100·633

This mineral is one of unusual hardness and specific gravity for one consisting essentially of the oxides of manganese. The silica is probably present merely as an impurity in combination with the iron and alumina.

MANGANITE.—Grey Manganese Ore.

Chem. comp.: $H_2Mn_2O_4$ or Mn_2O_3,H_2O. Rhombic system. Reported from Back Creek, Rockley, with rhodonite and diallogite.

DIALLOGITE.—Rhodochroite—Manganese Spar.

Manganese carbonate = $MnCO_3$.

This mineral is usually of a pretty pink tint, and often associated with silver ores, together with other manganese minerals.

It is said to occur at Back Creek, Rockley.

RHODONITE.

Chem. comp.: Manganese silicate. Has been found at Back Creek, Rockley; Glanmire and Bendemeer.

MANGANBLENDE.—Alabandite.

Manganese sulphide. MnS. It is said to have been found at Rylstone.

Iron black in colour, with brown tarnish; green streak.

Mr. Ranft has also found it in Wiseman's Creek, co. Westmoreland, with wad and kupfermanganerz.

COBALT.

Minerals containing cobalt, except wad and pyrites, do not yet appear to have been found in New South Wales.

The following assays of cobalt and nickel bearing wad, &c., have been published by the Department of Mines at different times:—

Locality.	County.	Description of Mineral.	Sesquioxide of Cobalt.	Metallic Nickel.
			Per cent.	Per cent.
Bungonia . . .	Argyle . .	Wad	3·5	...
,, ,, . .	,, . .	,, ...	3·85 to 3·92	...
,, (13 miles S. of)	,, . .	,, ...	1·5	1·15
,, . . .	,, . .	,, ...	1·51	0·40
Goulburn (4 miles from)	,, . .	,, in sandstone .	1·21	1·32
Windellama (near) .	,, . .	,, ,, .	1·19 to 2·27	...
Sutton Forest . .	Camden . .	,, ...	1·05	...
Inverell District . .	Gough . .	Cement (51·2 per cent.)	0·5	...
Tamworth . . .	Inglis . .	Stone	0·4	...
Port Macquarie . .	Macquarie.	Wad	3·21 to 5·8	...
Boro	Murray .	Wad in claystone .	0·15	...
Lake George . . .	,, . .	Wad, concretionary .	5·04	...
Between Trial Bay and Nambucca River .	Raleigh .	Wad	2·9	...
Lismore . . .	Rous . .	Wad from joints in decomposed felsite .	2·46	0·20
10 miles from Capertee, on Mudgee Road .	Roxburgh .	Wad from ironstone grit	1·66	0·35
Mudgee	Wellington	Wad	0·49	...
Barrier Range . .	...	Decomposed ferruginous rock, with joints filled with wad . . .	1·95	3·05
Bombala (31 miles from)	Wellesley .	Wad	3·95	0·80

NICKEL.

MILLERITE.—Nickel Sulphide.

Chem. comp. : NiS.   Nickel = 64·45; sulphur, 35·55 per cent. Hexagonal system.   Usually in yellow capillary crystals.

Mr. Ranft has found this mineral, in the usual form of slender crystals, in a quartz-reef running in the bed of a creek above Bony Gully, about three miles from Elsmore.

KUPFERNICKEL.—Copper-nickel.

Chem. comp.: Nickel arsenide = NiAs, Ni = 44·1 ; As = 55·9 = 100. Hexagonal system.   A massive variety, of a copper-red colour, in parts incrusted with pale green nickel hydrate, is reported from near Bathurst.

Found by the Rev. W. B. Clarke (near Bowling Alley Point ?), on the Peel River, and to the south-west of Weare's Creek. Yellowish white in colour, highly magnetic. Sp. gr. = 8; H = 5·5; and dissolving readily in nitric acid.

ZINC.

ZINCITE.—Red Oxide of Zinc—Spartalite.

Chem. comp.: Oxide of zinc. Zinc = 80·25; oxygen = 19·75 = 100. Crystallises in the hexagonal system.

Occurs in the Vegetable Creek District, as reddish-brown fragments showing fairly well-marked cleavage planes, where it was mistaken by the miners for tinstone.

CALAMINE.—Zinc Carbonate.

Chem. comp.: $ZnCO_3$. Zinc = 52, CO_3 = 48 = 100.

Has been found at Collington, Bredbo, Cooma District, containing 51·28 per cent. of zinc (*Mining Department Report*, 1886).

ZINC BLENDE.

Chem. comp.: Zinc sulphide = ZnS. Zinc, 67·0; sulphur, 33·0 = 100. Found massive, and crystallised in small hemihedral forms belonging to the cubical system. Many of the crystals have beautiful bronze and purple metallic tints.

With tin, gold, manganese, copper pyrites, galena, and other minerals, on Major's Creek, near Bungonia, co. Argyle. A specimen from the Braidwood District was found to contain 15 dwts. 16 grs. of gold and 11 oz. 15 dwts. 4 grs. of silver per ton.

Zinc blende occurs at Mount M'Donald, co. Bathurst; Burragorang, co. Camden; eight miles west of Gundagai, co. Clarendon; Silverdale, co. King; Yalwal, co. St. Vincent; Orange, Louisa Creek, co. Wellington. With gold, iron, copper pyrites, black oxide of copper, galena, and asbestos in a quartz-vein, Wiseman's Creek, co. Westmoreland. With copper ores at Cow Flat, co. Bathurst. With galena thirty miles north of Orange, co. Wellington; ten miles south-west of Tingha, co. Hardinge. With argentiferous galena and copper pyrites, Sunny Corner, co. Roxburgh. With galena and copper pyrites, Pye's Creek, co. Clive. With galena, copper, and iron pyrites in quartz from Denisontown, co. Bligh. With galena and iron

H

pyrites from Nambucca, co. Raleigh. With iron pyrites from Fairfield, co. Drake; Folkstone Lode, co. Gough; Bolivia and Pye's Creek, co. Clive; Yalwal Creek, co. St. Vincent; Sunny Corner, co. Roxburgh; Adelong, co. Wynyard; Binda District. With magnetic iron pyrites near Emmaville, co. Gough. With copper and iron pyrites, Deepwater, co. Clive; at Bluff Rock, south of Tenterfield.

Analysed specimens from Fairfield gave 50·2 and 69·98 per cent. of zinc; a zinc blende with pyrites from the Grampians yielded 20 per cent. zinc, 5 per cent. copper, and 61 oz. 13 dwts. of silver per ton; a mixed ore from Mitchell's Creek gave 19·67 per cent. zinc, 16·70 per cent. lead, 3 dwts. 10 grains silver, and 1 oz. 1 dwt. per ton of gold; and one from Sunny Corner Mine gave 29·7 per cent. zinc, 33·5 per cent. lead, 2 per cent. copper, 5 oz. 4 dwts. silver, and a little gold.

CERIUM, LANTHANUM, AND DIDYMIUM.

MONAZITE.

The following description and analysis of a specimen of monazite from Vegetable Creek, co. Gough, are by Mr. W. A. Dixon, F.I.C., of the School of Arts, Sydney :—

Analysis.

		Duplicate.
Phosphoric acid	25·09	24·61
Oxide of cerium	36·64	
„ lanthanum	} 30·21	68·08
„ didymium		
„ thorinum	1·23	
„ manganese	traces	
„ magnesium	traces	
„ aluminium	3·11	
Silica	3·21	
	99·49	

Specific gravity, 5·001.

"The mineral was crystalline, but the crystals were broken and ill defined. One piece, however, appeared to be a monoclinic prism. Colour, yellowish red, in thin pieces semi-transparent; it gave a white streak, showing a hardness about 5; it was rather brittle, and gave a yellowish powder infusible before the blowpipe."

The accompanying table is extracted from the *Mining and Mineral Statistics of the United Kingdom* for 1886, for the purpose of showing the relative positions of the various colonies in respect to the

Year.	British Colonies and Possessions.	Coal.		Copper.		Gold.	
		Quantity.	Value.	Quantity.	Value.	Quantity.	Value.
	AFRICA and MEDITERRANEAN—	Tons.	£	Tons.	£	Ounces.	£
1885	Cape of Good Hope . .	16,480	15,050	Ore 20,113	395,675	...	...
1884	Gold Coast . . .	...	...	...	...	24,994	89,981
1885	Natal	No information.		...	...	...	...
1885	Cyprus	...	...	...	...	...	...
	ASIA—						
1885	British North Borneo .	No information.		...	...	No information.	
1885	Ceylon	...	...	...	...	...	...
1884	India	1,397,818	629,018	...	...	...	...
1885	Straits Settlements . .	...	...	...	...	...	...
	AUSTRALASIA—						
1885	New South Wales . .	2,878,863	1,340,213	Metal 5,746	264,920	103,736	378,665
1885	Queensland . . .	209,698	87,228	Ore 1,340	18,920	310,941	1,088,293
1885	South Australia . .	...	...	{ Ore 18,639 } { Metal 3,517 }	{ 128,893 } { 194,090 }	4,692	18,295
1885	Victoria	Lignite	62	...	...	735,218	2,940,872
1885	Western Australia . .	...	...	Ore 120	1,792	...	...
1885	New Zealand . . .	511,063	400,000	...	...	237,371	948,615
1885	Tasmania	5,334	5,215	...	...	41,241	155,310
	NORTH AMERICA—						
1886	Dominion of Canada .	1,868,247	808,923	Metal 1,000	46,571	76,779	265,748
1885	Newfoundland . .	...	...	{ Ore 4,401 } { Regulus 300 }	{ 17,604 } { 2,880 }	...	...
	WEST INDIES —						
1886	Redonda	...	...	...	...	...	...
1885	Sombrero	...	...	...	...	...	...
1875	Trinidad	...	...	...	...	...	...
	Totals . .	6,887,626	3,285,709	{ Metal 10,263 } { Ore 45,013 }	1,071,345	1,534,972	5,885,779

* Export value. A Government monopoly. † These figures relate only to salt which is liable to the British Sa

MINERAL SUBSTANCES PRODUCED.

The black figures are taken from the Export Returns of the Colony.

IRON.		LEAD.		SILVER.		TIN.	
Quantity.	Value.	Quantity.	Value.	Quantity.	Value.	Quantity.	Value.
Tons.	£	Tons.	£	Ounces.	£	Tons.	£
...	...	...	...	...	...	...	...
...	...	...	...	...	...	...	...
...	...	...	...	...	...	...	...
...	...	...	...	...	...	...	...
...	...	...	...	...	...	...	...
...	...	...	...	...	...	...	...
No information.		...	...	...	...	...	...
...	...	...	...	...	...	Metal 7	527
Metal 4,176	25,793	Ore 2,096§ Metal 190	107,626	794,174	159,187	Ore 535 Metal 4,658	415,626
...	...	Ore 7,124§	64,235	...	...	Ore 13,992‖	151,871
...	...	Ore 37§ Metal 8	1,496 137	...	...	Ore 3	100
...	...	Ore 50	404	28,951¶	5,790	...	...
...	...	Ore 465	3,255	...	...	...	...
Ore 50	208	...	...	16,624	3,169	...	...
...	...	...	...	...	...	Ore 5,461	395,098
Ore 63,250	27,136	Metal 27	492	25,000	5,000	...	...
...	...	...	...	...	...	...	...
...	...	...	...	...	...	...	...
...	...	...	...	...	...	...	...
...	...	...	...	...	...	...	...
Metal 4,176 Ore 63,300	53,137	Metal 225 Ore 9,772	177,645	864,749	173,146	Metal 4,665 Ore 19,991	963,222

Salt-Tax, and do not include a quantity of salt made in certain Native States.   ‡ Estimated cost of production.

MINERAL SUBSTANCES PRODUCED.

The black figures are taken from the Export Returns of the Colony.

IRON.		LEAD.		SILVER.		TIN.	
Quantity.	Value.	Quantity.	Value.	Quantity.	Value.	Quantity.	Value.
Tons.	£	Tons.	£	Ounces.	£	Tons.	£
...	...	...	...	...	...	...	...
...	...	...	...	...	...	...	...
...	...	...	...	...	...	...	...
...	...	...	...	...	...	...	...
...	...	...	...	...	...	...	...
...	...	...	...	...	...	...	...
No information.		...	...	...	...	...	...
...	...	...	...	...	...	Metal 7	527
Metal 4,176	25,793	Ore 2,096§ / Metal 190	107,626	794,174	159,187	Ore 535 / Metal 4,658	415,626
...	...	Ore 7,124§	64,235	...	...	Ore 13,992‖	151,871
...	...	Ore 37§ / Metal 8	1,496 / 137	...	...	Ore 3	100
...	...	Ore 50	404	28,951¶	5,790	...	...
...	...	Ore 465	3,255	...	...	...	...
Ore 50	208	...	...	16,624	3,169	...	...
...	...	...	...	...	...	Ore 5,461	395,098
Ore 63,250	27,136	Metal 27	492	25,000	5,000	...	...
...	...	...	...	...	...	...	...
...	...	...	...	...	...	...	...
...	...	...	...	...	...	...	...
...	...	...	...	...	...	...	...
Metal 4,176 / Ore 63,300	53,137	Metal 225 / Ore 9,772	177,645	864,749	173,146	Metal 4,665 / Ore 19,991	963,222

alt-Tax, and do not include a quantity of salt made in certain Native States.　　　　‡ Estimated cost of production.

COMPILED FROM OFFICIAL REPORTS.

MISCELLANEOUS MINERALS.			TOTAL VALUES.	SOURCE OF INFORMATION.	BRITISH COLONIES AND POSSESSIONS.
Kind of Mineral.	Quantity.	Value.			
	Tons.	£	£		AFRICA—
monds . (carats)	2,440,788	2,489,659 ⎰	2,900,890	Agent-General, London . .	Cape of Good Hope.
s	365	506 ⎱			
...	...	...	89,981	Blue-Book for 1884 . . .	Gold Coast.
...	...	...	...	Blue-Book for 1885 . . .	Natal.
sum	1,393	888 ⎰	1,445	Blue-Book for 1885 . . .	Cyprus.
ber	908	557 ⎱			
					ASIA—
No information.	...	...	...	...	British North Borneo.
as (no information)					
mbago . . .	9,820	163,667 ⎰	164,776	Blue-Book for 1885 . . .	Ceylon.
	1,774	*1,109 ⎱			
as (no information) .	...	...	1,021,919	India Office, London . . .	India.
. . . . .	†1,187,245	‡392,901 ⎰			
...	...	...	527	Blue-Book for 1885 . . .	Straits Settlements.
imony Ore . .	288 ⎰	4,296			AUSTRALASIA—
)o. Metal . .	5 ⎱				
estos . . .	6	90			
nuth	14	3,700	2,775,175 ⎰	Annual Report of the Department of Mines, 1885 . .	⎰ New South Wales.
as (no information).	...	...			
Shale . . .	27,462	67,239			
dry Minerals . .	457	7,820			
imony Ore . .	70	300 ⎰	1,429,872	Blue-Book for 1885 . . .	Queensland.
ie . . .	86,367	19,025 ⎱			
ganese Ore . .	130	893			
fing Slates (No.)	107,090	907			
s	7,336	2,417	350,456 ⎰	Statistical Register of South Australia for 1885 . . .	⎰ South Australia.
iter	31	547			
ite Salt . . .	1,890	2,681			
es	471	942 ⎰	2,950,527 ⎰	Mineral Statistics of Victoria for 1885	⎰ Victoria.
ie (Flagging) . .	1,996	2,457 ⎱			
...	...	...	5,047	Blue-Book for 1885 . . .	Western Australia.
imony Ore .	666	5,289 ⎰			
ri Gum . . .	5,876	299,762 ⎱	1,659,752	Blue-Book for 1885 . . .	New Zealand.
ganese Ore . .	602	1,716			
ed Minerals . .	114	993			
ding Stone, &c. .	228,359	9,437 ⎰			
estone . . .	1,330	532 ⎱	570,972	Blue-Book for 1885 . . .	Tasmania.
fing-Slates (No.)	538,000	5,380			
imony Ore . .	721	6,594			
estos . . .	3,156	42,198			
iite . . .	3,395	7,702			NORTH AMERICA—
ohite . . .	407	717			
idstones (incomete return) . .	⎱ 4,020	11,099		Return furnished by Dr. Alfred R. C. Selwyn, C.M.G., F.R.S., Director of the Geological and Natural History Survey of Canada	
sum . . .	161,222	39,093	1,494,436 ⎰		⎰ Dominion of Canada.
ganese Ore . .	1,535	8,095			
. . . (lbs.)	5,333	3,200			
oleum . (galls.)	20,384,000	81,536			
phate of Lime .	19,252	57,689			
tes . . .	40,000	40,000			
ing-Slates . .	5,618	13,600			
. . (barrels)	459,273	29,043			
...	...	...	20,484	Blue-Book for 1885 . . .	Newfoundland.
					WEST INDIES—
phate of Lime .	2,200	7,700	7,700	The Dee Oil Company . .	Redonda.
phate of Lime .	4,635	16,222	16,222	Messrs. Pickford & Winkfield	Sombrero.
alt, purified .	6,731	13,456 ⎰	41,961	Blue-Book for 1885 . . .	Trinidad.
., raw . .	28,505	28,505 ⎱			
...	...	3,892,159	15,502,142		

§ Silver Lead Ore. ‖ Mainly undressed Tin Ore. ¶ Silver extracted from Gold at the Melbourne Mint.

value of their mineral productions of all kinds. From it will be seen that the principal colonies stood in the following order in 1885 :—

Victoria	£2,950,527
Cape of Good Hope	2,900,890
New South Wales	2,775,175
New Zealand	1,659,752
Canada	1,494,436
Queensland	1,429,872
India	1,021,919
Tasmania	570,972
South Australia	350,456
Ceylon (export returns)	164,776
Gold Coast	89,981
Trinidad	41,961
Newfoundland	20,484

The value of the mineral productions of the other colonies are all below £20,000 annually.

PART II.

NON-METALLIC MINERALS.

Class I.

CARBON AND CARBONACEOUS MINERALS.

Diamond.

Chem. comp. : Carbon, usually accompanied by a small percentage of ash or mineral matter. Cubical system.

The diamond was first found in New South Wales by Mr. Stutchbury in the Turon River in 1851, and by Mr. E. H. Hargraves, who, in his report, dated from the Wellington Inn, Guyong, on the 2d July 1851, refers to some enclosed specimens of gold, gems, and " a small one of the diamond kind," from Reedy Creek, sixteen miles from Bathurst. The next record of the occurrence of the diamond in New South Wales appears to have been made by the Rev. W. B. Clarke, in an appendix to his " Southern Gold-fields," published in 1860. He records that four were brought to him on September 21, 1859, which were obtained from Macquarie River, near Suttor's Bar; the crystalline form which they exhibited was that of the triakisoctahedron, or three-faced octahedron, and one of them had a specific gravity of 3·40. Another which was received from Burrendong on December 29, 1859, had a specific gravity of 3·50. One from Pyramul Creek, crystallised in the hexakis or six-faced octahedron, weighed 9·44 grains, and had a specific gravity of 3·49. Another was sent to him in August 1860, which had been found in the Calabash Creek by a digger as far back as 1852.

Diamonds were found by the gold-diggers on the Cudgegong Diamond-diggings, about nineteen miles from Mudgee, in 1867, but were not especially worked until 1869. Although not at present worked, diamonds are still occasionally found.

The diamonds were obtained from outliers of an old river-drift which had in parts been protected from denudation by a capping of hard compact basalt. This drift is made up mostly of boulders and pebbles of quartz, jasper, agate, quartzite, flinty slate, silicified wood, shale, sandstone, and abundance of coarse sand mixed with more or less clay.

Many of the pebbles and boulders are remarkable for the peculiar brilliant polish which they possess. The principal minerals found with the diamond are gold, garnets, wood-tin, brookite, magnetite, ilmenite, tourmaline, zircon, sapphire, ruby, adamantine spar, barklyite, common corundum, and a peculiar lavender-coloured variety; quartz, topaz, magnesite and nodules of limonite which had been set free from an impure magnesite; the chemical composition of similar limonite nodules from Bingera is given on p. 94; black vesicular pleonaste, spinel ruby, and osmo-iridium.

The character of the "wash dirt" or alluvial drifts in which the diamond is found in New South Wales differs altogether from the matrix of the diamond in South Africa. These drifts cover large areas, and are for the most part the same as or similar to those worked for gold and tin, whereas the Cape deposits are quite local or limited, filling pipe-like oval or irregular cavities, which descend vertically to untried depths. In places some of the rock fragments and minerals are rounded, and give the appearance of a conglomerate, but in other parts the fragments are more or less angular, and the structure is that of a breccia, or there may be a mixture of the two. Although some of the rocks have the appearance of being waterworn, the rounding may have been due to attrition caused by the working up and down of the contents of a volcanic pipe or vent, as is known to occur in some cases prior to a volcanic eruption, i.e., before the eruptive forces become sufficiently strong to force the contents out of the pipe or crater.

On comparing a list of the rocks and minerals found in the Cape diggings with those found in Australia many differences are noticeable. The most common rocks at the Cape diggings are amphibolite, augite rock, basalt, dolerite, eclogite, granite, gneiss, amygdaloidal neelaphyre containing agates, mica schist, pegmatite, serpentine, and talc schist.

Amongst the minerals are bronzite, calcite, enstatite, garnets of various colours—green, red, and brown—hornblende, ilmenite, iron pyrites, opal, smaragdite, staurolite, vaalite, zeolites, and zircon.

It is noticeable that many important minerals found with the diamond in New South Wales are absent from the above list, including gold, platinum, osmo-iridium, cassiterite, topaz, sapphire, ruby, barklyite, brookite, rutile, tourmaline, and different varieties of quartz; while, on the other hand, the diamond-yielding deposits of Brazil, Borneo, United States, and India contain most of the minerals which are characteristic of the Australian diggings, but not those peculiar to the Cape.

The diamond is found in Australia, Brazil, Borneo, India, &c.,

as a "derived" mineral, to use a palæontological term, whereas at the Cape it probably occurs *in situ* also, *i.e.*, it has probably been formed in the old volcanic pipes or vents now being worked. The occurrence of "cleavage" or fractured diamonds at the Cape is not an argument in favour of the deposit or matrix being of alluvial origin, since the working up and down or heaving of the solid upper portions of volcanic vent prior to an abortive eruption might cause the fracture of some of the diamonds, as well as produce both the brecciated and conglomeratic structure of the matrix; moreover, I think that many of the fractured or cleavage diamonds are probably due to blasting operations in sinking.

The largest diamond found weighed 16·2 grains, or about 5⅝ carats. Was found in the river-bed between Two Mile Flat and Rocky Ridge.

The average specific gravity was 3·44, and the average weight of a large number of those obtained was but 0·23 carat. (For further particulars see paper on the Mudgee Diamond-fields, by Professor Thomson and Mr. Norman Taylor, in the *Transactions of the Royal Society of New South Wales*, 1870, and *Geological Magazine*, London, 1879.) The total number found has been stated roughly at about 6000; the number, also, from Bingera must be nearly as many—in all 12,000 at least.

In colour they vary from colourless and transparent to various shades of straw-yellow, brown, light green, and black. One of a rich dark emerald green was found in the form of a flattened hemitrope octahedron.

The most common crystalline forms which have been met with are the octahedron, the hemitrope octahedron, the rhombic dodecahedron, the triakis and hexakis octahedron, but they are all usually more or less rounded. The flattened triangular hemitrope crystals are very common; one specimen of the deltoidal dodecahedron was met with. A twin crystal of two imperfect three-faced octahedrons, united by a plane parallel to a face of the octahedron, was obtained from the Sydney Diamond-Mining Company, near Inverell. The colour is somewhat yellow; one of the two crystals is nearly twice the size of the other; the weight is 3·5 grains troy.

The lustre is usually brilliant or adamantine, but occasionally they have a dull appearance. This want of lustre is not due to any coating of foreign matter, or to abrasion or attrition, as is often the case with softer minerals, but it is due to the surface being covered with innumerable edges and angles belonging to the structure of the crystal; these reflect the light irregularly at all angles and give the stone its frosted appearance.

The diamonds at Bingera occur under almost exactly the same circumstances as at Mudgee, and with the same minerals, except that I did not come across either the black vesicular pleonaste or barklyite.

From a series of determinations made on nineteen of the Bingera diamonds I obtained a mean specific gravity of 3·42.*

Some other uncut diamonds from unknown localities, but found in New South Wales, yielded the following specific gravities:—

	Weight.	Specific Gravity.	Temperature.
1 diamond, off colour . . . =	·2920	3·4762	at 20° C.
5 small dark diamonds . . =	1·3220	3·5633	,, 18·5° C.
6 ,, light-coloured diamonds . =	2·2790	3·5278	,, 18·5° C.
12 ,, ,, ,, ,, . =	2·7390	3·5233	,, 17·5° C.
8 ,, dark ,, ,, · =	1·4376	3·5166	,, 17·5° C.

Diamonds have also been found at Bald Hill, Hill End, with the same gems as at the above-mentioned places; one octahedral crystal, rather flattened, which I examined, weighed 9·6 troy grains, and had a specific gravity of 3·58.

A specimen of "bort" or black diamond was obtained near Bathurst. It is of about the same size as a large pea, black in colour, with a graphitic or black-lead lustre; it is very nearly spherical in form, but has a few slight irregular processes, which seem to be due to an attempt to assume the form of the hexakis octahedron.

In weight it is 7·352 troy grains, and at 70° Fahr. the specific gravity is 3·56.

Mr. Wilkinson mentions that from the Bengonover Tin-mine, near the Borah Tin-mine, several diamonds were obtained, the largest being 7·5 grains. From the Borah Tin-mine, situated at the junction of Cope's Creek with the Gwydir, 200 were obtained in a few months; out of a batch of eighty-six, averaging 1 carat grain each, the largest weighed 5·5 grains. Diamonds have been found on most of the alluvial tin-workings at Cope's, Newstead, Vegetable, and Middle Creeks, also in the Stannifer, Ruby, and the Britannia Tin-mines, and elsewhere in the district.

Amongst other places, the diamond has been found in the gravels along the course of the Gwydir, Turon, the Abercrombie, the Cudgegong, Macquarie, and Shoalhaven Rivers. Many are reported as having been found at Auburn Vale, Tingha Division. One was found in August 1874, in Brook's Creek, Gundaroo, near Goulburn, valued at £3. At Uralla, Oberon, and Trunkey they are by no means uncommon. In 1881 I purchased a small hemitrope octahedron, obtained from the Lachlan River, weighing 1·5 grains; since then several have been found, including one of 1½ carats. They have also been obtained from diggings on the sea-shore near to Ballina.

Diamonds are found in the gravels under the basalt at Monkey

---

* "Bingera Diamond-fields," A. Liversidge, *Trans. Royal Soc N S W.*, 1873, and *Journal of the Geological Society of London*, 1873. See also Appendix.

Hill and Sally's Flat, co. Wellington, just as is the case on the Cudgegong River and at Bingera. A few diamonds and other gems have been
found near Mittagong, seventy-seven miles from Sydney (see *Jour. Roy.
Soc. N. S. W.*, 1886), on a small abandoned gold-field which exists there.

During the last two or three years the Bingera deposits have been
again worked with more or less success; the chief difficulty has been
the want of water. There are also several companies at work in the
Inverell District.

A drift having almost exactly the same characters as those at
Bingera and Mudgee occurs in other districts, as at Wallerawang
and on the Mary River, Queensland, including masses of conglomerate
of jasper, quartz, and other pebbles agglutinated together by a ferruginous and manganiferous cement, as on the Mudgee and Bingera
workings. These masses of hard conglomerate are probably derived
from the coal-measures.

With respect to the quality of the New South Wales diamonds
Messrs. Etheridge, F.G.S., and Davies, F.G.S, in a report upon the
diamonds at the Colonial and Indian Exhibition in 1886, state that
out of 275 diamonds from Inverell they find—

100 colourless.
126 straw yellow, " off colour."
39 slightly tinted, " bye waters."
6 cinnamon yellow.
4 dirty grey " rejections."

GRAPHITE.—Plumbago.

Chem. comp.: Carbon. Hexagonal system. Occurs with quartz,
iron pyrites, and pyromorphite at the head of the Abercrombie River;
possesses a curved lamellar structure. Occurs in small radiating
masses in the granite at Dundee, in New Valley, and near Tenterfield.

Reported also from Bungonia, but its existence there is doubtful;
also from Pambula, near Eden, in quartz; the Cordeaux River, near
Mount Keira; and Plumbago Creek, near the junction of Timbarra
Creek, co. Drake.

Small particles are not uncommon in the Hawkesbury sandstone
about Sydney and other places.

Any black clay or other substance which can be made to leave a
mark on paper is brought into Sydney as a sample of a valuable
deposit of graphite; but I have not yet seen, out of many highly
extolled specimens, one fit for even the commonest purpose.

The following analysis of a black clay shows the composition of
one of these reputed graphites *:—

* A. Liversidge, *Report of the Department of Mines*, Sydney, 1876, p. 183.

CARBONACEOUS EARTH.

" A black, earthy, friable material from near Mudgee; soils the fingers readily. In parts it is grey in colour, and here and there an occasional white streak is seen; falls to powder when immersed in water. " Specific gravity, 2·88.

Analysis.

Hygroscopic moisture	1·60
Combined water (by difference)	13·38
Silica	46·00
Alumina	32·32
Lime	absent
Magnesia	absent
Potash	·17
Soda	·13
Carbon	6·40
	100·00

" The mineral, as shown by the above analysis, is essentially a hydrous silicate of aluminium mixed with a small proportion of carbonaceous matter. The carbonaceous matter is easily burnt off.

" As a fireclay this material would not be of any great value, since it only possesses average refractory qualities. It should be remarked that it is totally distinct from graphite, the mineral for which it is often mistaken by miners."

COAL.

As far as is at present known, the coal-fields of New South Wales cover an area of about 24,000 square miles. Mr. Wilkinson, F.G.S., is of opinion (*Journal of the Linnæan Society of New South Wales*, 1885) that the seams which are now being worked underlie, within a readily accessible depth, 3328 square miles (nearly half the area of the coal-fields of Great Britain), and that they contain, after deducting one-half of the total contents as waste, &c., 14,370,000,000 tons of coal. This estimate does not include the other seams of good coal within the same area not at present worked; neither does it include the coal in the remaining 20,000 and odd square miles of the coal-measures in which coal-seams are known to exist, but which have not yet been proved.

The thickness of the seams worked varies from 3 to 25 feet. Most of them are nearly horizontal, or have but a slight dip, and none of the shafts are at all deep as compared with those in Great Britain and elsewhere; many are, in fact, worked by drives or adits.

The existence of coal in New South Wales appears to have been discovered in the month of August 1797. The following reference is

made to its occurrence by Collins in his account of the English colony
in New South Wales * :—

"Mr. Clark, the supercargo of the *Sydney Cove*,† having mentioned
that, two days before he had been met by the people in the fishing-
boat, he had fallen in with a great quantity of coal, with which he and
his companions had made a large fire, and had slept by it during the
night, a whale-boat was sent off to the southward, with Mr. Bass, the
surgeon of the *Reliance*, to discover where an article so valuable was to
be met with. He proceeded about seven leagues to the southward of
Point Solander, where he found, in the face of a steep cliff, washed by
the sea, a stratum of coal in breadth about 6 feet, and extending eight
or nine miles to the southward. Upon the summit of the high land,
and lying on the surface, he observed many patches of coal, from some
of which it must have been that Mr. Clark was so conveniently supplied
with fuel. He also found in the skeletons of the mate and carpenter
of the *Sydney Cove* an unequivocal proof of their having unfortunately
perished, as was conjectured.

"By the specimens of the coal which were brought in by Mr. Bass,
the quality appeared to be good; but, from its almost inaccessible
situation, no great advantage could ever be expected from it; and,
indeed, were it even less difficult to be procured, unless some small
harbour should be near it, it could not be of much utility to the
settlement."

During the following month of the same year—*i.e.*, in September
1797—coal was found to the north of Sydney, at the place now known
as Newcastle. On page 48 Collins states:—

"Lieutenant Shortland proceeded with a whale-boat as far as Port
Stephens. On his return he entered a river, wnich he named Hunter
River, about ten leagues to the southward of Port Stephens, into which
he carried three fathoms water in the shoalest part of its entrance,
finding deep water and good anchorage within.

"The entrance of this river was but narrow, and covered by a high
rocky island lying right off it, so as to leave a good passage round the
north end of the island between that and the shore. A reef connects
the south part of the island with the south of the entrance of the
river. In this harbour was found a very considerable quantity of coal
of a very good sort, and lying so near the water-side as to be conveniently shipped, which gave it in this particular a manifest advantage
over that discovered to the southward. Some specimens of this coal
were brought up in the boat."

In 1799 it seems to have become customary to send regularly to

* "An Account of the English Colony in New South Wales," by David Collins, Esq.,
late Judge Advocate and Secretary to the Colony, vol. ii. p. 45, London, 1798, 1802.

† The *Sydney Cove* was wrecked on the coast of Tasmania when on a voyage to Sydney
from Bengal.

the Hunter River for supplies of coal, and under the heading of April 1799 Collins has the following entry in his journal:—

"The discovery of vast strata of coal must be reckoned among the new lights thrown upon the resources of the colony. The facility that this presents in working the iron ore (some of this iron ore, which has been smelted in England, has been reported to be equal, if not superior, to Swedish iron) with which the settlement abounded must prove of infinite utility whenever a dockyard shall be established here; and the time may come when the productions of the country may not be confined within its own sphere."

In the early days of the colony the Hunter was for some time known as the Coal River.

In September 1800 another entry records the discovery of coal, although in this case the seam appears to have been valueless:—

"It having been reported that coal had been found upon the banks of George's River, the Governor visited the place, and on examination found many indications of the existence of coal, that useful fossil, of which, shortly after, a vein was discovered on the west side of Garden Island Cove."

The Australian Agricultural Company, formed in 1826, with a capital of £1,000,000 and a free grant of 1,000,000 acres of land, gave the first impetus to the great coal trade now carried on in the colony. The charter possessed by the company conceded to them the sole right to work the Newcastle coal-beds. This monopoly expired in 1847.

The following account of the coals of New South Wales contains the results of an examination into the chemical composition of certain samples of coal and "kerosene shale;" included with these are one or two carbonaceous minerals, which, although they cannot properly be classed with the coals, yet can conveniently be included with them.

I may mention that most of my own analyses were made upon samples of the coals which were collected by the officers of the Mining Department, and were reported upon by me to that Department in 1875.* The proportions of moisture, volatile matter, fixed carbon, ash, coke, and sulphur only were then determined, as information upon these points is quite sufficient for all ordinary purposes. Shortly afterwards, as I had the remains of the specimens, I thought it would be desirable to determine the ultimate composition, and to ascertain the chemical composition of the ashes of these coals; the results of these further examinations were published in a paper read before the Royal Society of New South Wales in December 1880, and published in its *Journal* for that year. Together with the above are incorporated the analyses made by Mr. W. A. Dixon, F.C.S., published in the *Annual Reports of the Department of Mines* for 1878, 1879, and 1880. The

* *Annual Report of the Department of Mines,* 1875, p. 127.

samples examined by Mr. Dixon * were collected five or six years after those analysed by me.

I particularly wished to see how the New South Wales coals compared with those of Europe, and especially with English coals, and to do so ultimate analyses had to be made, *i.e.*, the amount of carbon, hydrogen, nitrogen, &c., had to be determined. This of course necessitated the expenditure of considerable time and trouble, but it enabled me to ascertain how the calorific intensity of the fuels, calculated from the percentage amounts of carbon and hydrogen, correspond with their evaporative powers as determined experimentally by means of Thompson's calorimeter.

The ashes were analysed because it was thought that a knowledge of their chemical composition would be of service to the metallurgist, as well as of some general scientific interest; it is, of course, of great importance to the metallurgist to know the composition of the ashes of the coal which he uses, since some of the constituents may have a bad effect upon the products of his furnaces, and in some cases even render the metal useless for certain purposes.

Methods of Analysis.—I may perhaps mention the methods of analysis followed, since it is sometimes of interest to any one going over similar ground to know what processes were employed; and when it is wished to compare results, it is often a great advantage to be able to use the same methods. The proximate analyses were made according to the well-known process described in Crooke's " Methods in Chemical Analysis," p. 368; in each case upon about two grammes of the freshly powdered coal.

The sulphur was estimated by heating about two grammes of the coals with chlorate of potash and strong nitric acid, and then adding strong hydrochloric acid; the solution being largely diluted, filtered, and precipitated in the ordinary way. The reagents were rendered sulphur-free before use.

The specific gravity was determined upon the coal in the form of a coarse powder; the powder was allowed to soak in the specific-gravity bottle, and kept in a warm place, until air-bubbles ceased to be evolved; when cold the second weighing was proceeded with.

The carbon and hydrogen were determined by combustion with lead chromate in a current of oxygen; it was found that when copper oxide and a current of oxygen were employed the carbon was liable to be understated. The nitrogen was determined in the ordinary way by the soda-lime process.

All the determinations were made in duplicate.

Calculated Calorific Intensity and Evaporative Power.—The theoretical evaporative power was determined experimentally by means of

Thompson's calorimeter, for a description of which see Dr. Percy's "Metallurgy," vol. i. p. 541. The results given are the means of several experiments. The calorific intensity was calculated according to the formula given by the same author, p. 537.

On examining the two sets of results, i.e., the calculated calorific intensity and the calculated evaporative power as determined by the calorimeter, it will be at once apparent that they do not in all cases place the coals in the same order; there is no doubt that other things besides the absolute quantities of carbon, hydrogen, oxygen, and ash influence the production of heat and help to determine the value of a coal. We as yet really know very little as to how the combustible elements are combined in coals, or whether there are differences in the mode of such combination in different coals—it is most probable that there are—but we do know that there are considerable variations in the mechanical structure of coals, which must necessarily influence the rate of combustion and the amount of heat generated.

It is a well-known fact that many commanders of steam-vessels belonging to the Royal Navy, the great mail companies, and to the intercolonial lines prefer southern to northern coal, although the former contains more ash. The disadvantage of the greater proportion of ash is considered to be counterbalanced by the fact that the southern coal burns uniformly and does not form a clinker; but when it is desired to get up steam rapidly, then the rich, so-called bituminous, northern coal is preferred.

In the report * to the Mining Department upon the theoretical evaporative power of certain coals I pointed out that these results represent the theoretical calorific or evaporative power of the samples, i.e., the weight of water which would be converted into steam by the complete combustion of one pound of each of the various coals respectively.

It must, however, be clearly understood that the actual heat-producing or evaporative power of a coal obtained in practice depends very greatly upon the size, construction, and form of both furnace and boiler, as well as upon the method of firing or burning, and upon many other equally obvious circumstances; it will, therefore, be apparent that the results can only be *rigidly* compared when the conditions under which the fuels are burnt are alike, as was the case in the experimental trials.

Analysis of the Ash.—The ash was prepared for analysis by incinerating the powdered coal in a muffle furnace at a dull red heat; in order to obtain the ash as expeditiously as possible from a fairly large quantity of coal a tray $10 \times 6 \times 1$ inch deep, made out of stout platinum foil, was used for the incineration.

The ash was rendered soluble by direct fusion with the mixed alkaline carbonates, and proceeded with in the usual manner for silica,

* *Annual Report of the Department of Mines*, Sydney, 1877, p. 207.

alumina, iron, lime, &c.; the alkalies were determined in separate portions by Dr. J. Lawrence Smith's process, *i.e.*, by fusion with calcium carbonate and ammonium chloride.

The phosphoric and sulphuric acids were also determined in separate portions of the ash; as the proportion of phosphoric acid, where present, was shown by the qualitative tests to be small, the molybdic acid process was employed, about two grammes weight of ash being taken in duplicate in each case.

COMPOSITION OF COALS.—Proximate Analyses.
Northern District.

Name of Colliery, &c.	Water.	Volatile Hydro-carbons.	Fixed Carbon.	Ash.	Sulphur.	Specific Gravity.	Coke.	Analyst.
1. Anvil Creek	1·74	41·10	47·90	7·80	1·46	1·323	55·70	Liversidge.
2. Ashford (Toose's)	1·40	20·50	71·94	6·16	·52	1·350	78·10	Gov. Analyst.
3. Branxton (Wyndham's)	1·80	39·29	50·63	8·28	2·35	1·297	58·91	„
4. „ „	1·45	50·91	38·58	9·06	...	1·240	47·64	„
5. Burnett, Gwydir District	3·35	51·61	34·71	10·33	0·50	1·230	45·04	„
6. Greta	2·25	39·21	54·41	2·72	1·41	1·287	57·13	Liversidge.
7. Gunnedah (Pye's)	3·00	29·50	58·80	8·70	·61	1·291	67·50	Gov. Analyst.
8. „ „	1·75	14·75	76·76	6·74	·64	1·255	83·50	„
9. „ „	2·80	30·47	56·83	9·90	·52	1·278	66·73	„
10. „ Darcy's Well	3·10	39·60	48·23	9·07	·78	1·281	57·30	„
11. Hexham, Minmi Colliery	2·59	33·87	56·49	5·61	1·44	1·280	62·10	Dixon.
12. Lake Macquarie,	3·65	31·93	54·66	8·82	·94	1·340	63·48	„
13. „ „	·98	37·72	55·14	6·16	·48	1·228	61·30	Gov. Analyst.
14. „ (near)	3·80	33·90	55·50	6·80	...	...	62·30	„
15. „ Cardiff Mine	1·85	43·35	49·49	4·94	·34	1·286	54·43	Liversidge.
16. „ G. N. R. C. Co.	2·98	34·02	53·57	9·43	·39	1·380	63·00	Gov. Analyst.
17. „	·49	31·15	59·46	8·90	·80	1·330	68·36	„
18. „ Quigley Estate	3·15	33·05	55·11	8·43	·55	1·330	63·54	„
19. „ „	3·30	30·00	57·72	8·43	·55	1·350	66·15	„
20. Leconfield	1·60	44·82	45·18	8·40	...	...	53·58	„
21. Maitland, Deep Creek	2·35	40·79	52·76	4·10	·839	1·243	56·86	„
22. Maitland, W., Homeville Coll.	2·21	36·45	51·51	9·83	·68	1·318	61·34	„
23. „ „	2·10	40·27	51·80	5·83	·89	1·270	57·63	„
24. „ „	2·27	35·39	53·91	8·43	·63	1·300	62·34	„
25. Narrabri	4·56	49·14	38·03	8·27	·55	1·252	46·30	„
26. Newcastle, A. A. Co.	1·65	35·45	57·84	4·44	·62	1·286	63·28	Dixon.
27. „ „	2·20	33·60	57·52	5·35	1·33	1·297	62·87	Liversidge.
28. „ Ferndale Colliery	2·10	36·22	57·24	3·84	·60	1·296	61·08	Dixon.
29. Newcastle C. M. Co., Glebe	2·14	33·36	59·16	4·76	·58	1·283	63·92	„
30. „ Wallsend Co.	2·75	34·17	57·22	4·64	1·22	1·333	61·86	Liversidge.
31. „	2·29	34·21	58·60	4·28	·62	1·347	62·88	Dixon.
32. New Lambton Colliery	2·61	30·62	59·56	6·72	·49	1·291	66·28	„
33. Northumb. C. Co., Cockle Ck.	2·12	31·24	55·13	11·51	...	...	66·54	Gov. Analyst.
34. Plattsburg Co-operative Coll.	2·45	34·38	58·24	4·20	·73	1·310	62·44	Dixon.
35. Redhead Coal Co.	2·09	33·48	57·04	6·84	·55	1·325	63·88	„
36. Rix Creek, Singleton	2·80	37·00	54·00	5·06	·51	...	59·06	Latta.
37. Russell's Mine	1·85	44·09	49·95	2·70	1·41	1·274	52·65	Liversidge.
38. Springfield, near Gunnedah	2·30	39·73	52·47	5·50	·52	1·308	57·97	Gov. Analyst.
39. Teralba, near Newcastle	4·65	32·84	52·68	8·16	1·67	1·350	60·84	Dixon.
40. „ „	3·81	30·22	54·44	8·52	3·01	1·290	62·96	„
41. Waratah Colliery	2·21	36·70	55·82	4·15	1·12	1·303	59·97	Liversidge.
42. „	2·45	38·16	54·12	4·64	·63	1·293	58·76	Dixon.
Mean	2·45	35·51	54·29	6·60	·88	1·298	61·14	

No. 1. ANVIL CREEK.—Structure laminated, but compact; not so much mother-of-coal present as in that from the Waratah Mine. Breaks into cuboidal masses. Does not readily soil the fingers. *Coke.*—Good, firm, bright silvery lustre, not much swollen up. *Ash.*—White.

No. 2. ASHFORD (Toose's).—The total thickness of clean coal is over 30 feet, the greatest thickness of coal without bands being 7 feet. The analysis is taken from one of the loose blocks found in the river-bed, and shows the sample to belong to the class of bituminous caking coal, suitable for smelting and steam-coal purposes.

No. 3. BRANXTON (Wyndham's Tunnel).—Bituminous.

No. 4.   ,,   ,,   ,,   Described as a cannel coal.

No. 5. Coal from GRAGIN STATION, Burnett, Gwydir District.

No. 6. GRETA.—In appearance very similar to the Waratah coal, but with less mother-of-coal. Does not soil the fingers; streaky appearance. Fracture conchoidal across the layers.

This seam is 26 feet thick, with an occasional band or layer of torbanite or cannel coal.

Coke.—Good, firm, not quite so bright as the former, but rougher in the grain and more swollen up. *Ash.*—Loose, buff-coloured.

No. 7. GUNNEDAH (Pye's).—The analysis is taken from a mixed sample from the upper part of James Pye's 6-feet seam, and from 72 feet 4 inches to 74 feet 6 inches from the surface. A caking, bituminous coal, quick lighting, and useful for gas, steam, &c. No true coke. *Ash.*—Yellowish grey.

No. 8. GUNNEDAH (Pye's).—Soft coal taken from 2 feet below No. 7. A caking, bituminous coal, quick lighting, and useful for gas, steam, &c. No true coke. *Ash.*—Yellowish grey.

No. 9. GUNNEDAH (Pye's).—The analysis is taken from a mixed sample from the lower part of James Pye's 6-feet seam, and from 75 to 78 feet from the surface.

Coke.—Of fairly good quality, though the percentage of ash is rather high. *Ash.*—Yellowish grey.

No. 10. GUNNEDAH (Darcy's Well).—Coal from 95 feet deep. A caking, bituminous coal, quick lighting, and useful for gas, steam, &c. *Coke.*—Of a fair quality. *Ash.*—Yellowish grey.

No. 11. MINMI COLLIERY (Hexham).—Bituminous, bright, with a few narrow dull streaks. *Coke.*—62·10. Coke bright, dense, with fused appearance, little swollen. *Ash.*—Reddish, somewhat fusible.

No. 12. LAKE MACQUARIE.—Bright and semi-bituminous. In steaming power it would lie between the ordinary Newcastle coal and those of the Illawarra District.

Coke.—63·48 per cent.; dense, hard, and fairly bright. *Ash.*—White and loose.

No. 13. LAKE MACQUARIE. Coal from an unworked seam. (*Mines Report*, 1886.)

No. 14. LAKE MACQUARIE. (*Mines Report*, 1883.)

No. 15. LAKE MACQUARIE (Cardiff Mine).—A bright, firm, and compact-looking anthracitic coal; when struck emits a clear ringing sound, very unlike the dull sound given out by soft and friable varieties of coal. This specimen came from a depth of 434 feet.

Across the joints and planes of stratification it breaks with a somewhat splintery and conchoidal fracture.

Tough, and does not yield readily to pressure.

Does not soil the fingers; no mother-of-coal or mineral charcoal observed. When ignited, decrepitates somewhat, and burns with but a small amount of flame.

A few scattered grains of pyrites were observed in the sample, but the total amount of sulphur present is below the average.

Coke.—54·430 per cent.; bright in lustre, and fairly well swollen up. *Ash.*—Grey, loose; contains traces of copper.

The presence of copper is rather an unusual occurrence in coal ashes; the copper probably existed as copper pyrites. An examination for gold was made upon this ash, but without success; the ash from some 30 or 40 lbs. weight of coal was tested.

No. 16. LAKE MACQUARIE (G. N. R. Coal Co.)—Bituminous coal.

No. 17. LAKE MACQUARIE (Parbury, Lamb, & Saddington's).—Bituminous coal.

Nos. 18 and 19. LAKE MACQUARIE (Quigley Estate).—Coal from a new seam.

No. 20. LECONFIELD.—Described as cannel coal.

No. 21. DEEP CREEK, Maitland.—A good coal for gasmaking. *Coke.*—Of a good quality.

No. 22. MAITLAND WEST (Homeville Colliery).—Coal obtained by jumping-rods from a seam 50 feet below the Homeville seam.

No. 23. MAITLAND WEST (Homeville Colliery).—Bituminous. A band of clay shale, from ¼ inch to 1 inch in thickness, occurs intermittently at about 4 inches below the roof of the seam. The coal is of an excellent quality for steam, gas, and household purposes, and, from its very hard nature, is well suited for shipment.

No. 24. MAITLAND WEST (Homeville Colliery).—Splint coal.

No. 25. Coal from NARRABRI.

No. 26. AUSTRALIAN AGRICULTURAL COMPANY'S MINE, Newcastle. *Coke.*—63·28 per cent. *Ash.*—Reddish.

No. 27. AUSTRALIAN AGRICULTURAL COMPANY'S MINE, Newcastle. —Very similar to the Waratah coal, but a shade less bright. Breaks into irregular cuboidal fragments. Does not soil the fingers. Contains films of mineral charcoal.

Coke.—A good firm coke; very large cauliflower-like excrescences. No. 27-A.—From the same mine. Examined by W. Skey, *Trans. N. Z. Inst.*, 1871, p. 150. His results were:—Moisture, 1·42; vol. hydrocarbons, 7·25; fixed carbon, 61·21; sulphur, 1·02; ash, 8·80 = 99·70.

No. 28. FERNDALE COLLIERY, Tighe's Hill.

Coke.—61·08 per cent. *Ash.*—Buff-coloured.

No. 29. NEWCASTLE COAL COMPANY, Glebe, Newcastle.

Coke.—63·92 per cent. *Ash.*—Buff-coloured.

No. 30. Another sample from the same mine. A bright coal; laminated structure well marked; breaks into irregular cuboidal fragments. Does not soil the fingers readily. Contains a little fibrous mineral charcoal, or mother-of-coal.

Coke.—Much the same as from the Greta coal, but with large cauliflower-like excrescences. *Ash.*—Of a pinkish shade, being white mixed with reddish particles.

No. 31. WALLSEND, Newcastle.

Coke.—62·88 per cent. *Ash.*—Red.

No. 32. NEW LAMBTON MINE.

Coke.—66·28 per cent. *Ash.*—Reddish-coloured.

No. 33. Coal from the NORTHUMBERLAND COAL COMPANY, Cockle Creek.

No. 34. PLATTSBURG.—Coal from the Co-operative Mine.

Coke.—62·44 per cent. *Ash.*—Reddish.

No. 35. REDHEAD COAL COMPANY.

Coke.—63·88 per cent. *Ash.*—Grey-coloured.

No. 36. RIX CREEK, Singleton.—Coal bright, but rather tender; slightly coking. Analysed by Mr. Latta.

Coke.—59·06.

No. 37. RUSSELL'S MINE.—Made up of alternate bright and dull laminæ, which merge one into the other irregularly, giving the coal a streaky appearance quite distinct from the laminated appearance of a coal made up of well-defined bright and dull layers. The bright layers have a very brilliant pitchy lustre. Fracture somewhat conchoidal. Does not soil the fingers.

Coke.—Good, firm, bright silvery lustre, with cauliflower-like excrescences. *Ash.*—Loose, colour red, but paler than the Waratah coal ash.

No. 38. SPRINGFIELD, near Gunnedah.—Bituminous coal from a seam between 6 and 7 feet thick, 78 feet from the surface.

No. 39. TERALBA, near Newcastle.—Semi-bituminous. Bright, with small conchoidal fracture, stained with oxide of iron.

Coke.—60·84. Coke swollen, fairly bright, with small excrescences, showed distinct prismatic fracture. *Ash.*—Reddish and somewhat friable.

No. 40.—A second specimen from the same place was for the most part bituminous, bright, with a few narrow dull layers.

Coke.—62·96. Coke bright and lustrous, very little swollen, dense, splits readily. *Ash.*—Grey, not easily friable.

No. 41. WARATAH COLLIERY.—A good, firm, bright coal, with well-marked lines of lamination ; bright layers preponderate. Fracture fairly even, breaking into cuboidal masses. Layers of fibrous "mineral charcoal" or mother-of-coal in between the bright layers; these are also to be observed in nearly all the other coals. The coal from this mine is sometimes beautifully iridescent.

Coke.—Good, firm, bright and silvery lustre, well swollen up, with small cauliflower-like excrescences. *Ash.*—Loose and flocculent, reddish colour.

No. 42. The WARATAH COAL COMPANY's old tunnel at WARATAH.

Coke.—58·76 per cent. *Ash.*—Buff colour.

Nodular Coal.—A smooth, rounded nodule of anthracitic coal from the Waratah Mine, about 2 inches in diameter, harder than the ordinary coal, in which I understand it was found embedded; the rounded form is apparently not due to attrition or the action of running water, but appears to be of a concretionary nature. Similar anthracitic nodules occur in the Australian Agricultural Company's mine.

On being struck with a hammer the mass flew to pieces, as if it had been in a state of strain or tension ; the fragments were small and showed conchoidal fracture surface. I believe that these nodules are sometimes met with of much larger size.

It will be noticed that the amount of ash is much less than in the ordinary coal from this mine.

Specific gravity, 1·294.

Dried at 100° C.

Proximate Analysis.

Loss at 100° C.	3·32
Volatile hydrocarbons	32·41
Fixed carbon	62·35
Ash	1·72
Sulphur	·19
	99·99

Ultimate Analysis.

Carbon	83·828
Hydrogen	5·437
Oxygen	8·236
Sulphur	·190
Nitrogen	·530
Ash	1·779
	100·000

ANTHRACITE.

A splintery anthracite is said to occur at Gordon Brook, in the county of Richmond. As far as I have seen at present, only one of the so-called New South Wales anthracites are really deserving of that name ;

the others are merely very poor or else baked coals, *i.e.*, coal which has been more or less destroyed by the intrusion of a dyke of some igneous rock. See p. 135.

WESTERN DISTRICT.

Name of Colliery, &c.	Water.	Volatile Hydro-carbons.	Fixed Carbon.	Ash.	Sulphur.	Specific Gravity.	Coke.	Analyst.
1. Blackheath . . .	2·75	27·15	63·86	6·24	·70	1·320	70·10	Gov. Analyst.
2. Bowenfels . . .	2·36	28·35	56·54	11·40	1·35	1·399	none	Liversidge.
3. Capertee . . .	3·25	25·41	60·24	11·10	·39	1·390	71·34	Gov. Analyst.
4. Clarence Siding . .	3·60	32·74	54·18	8·48	...	...	62·66	,,
5. ,, . .	2·12	8·16	79·42	10·30	·71	1·400	89·72	,,
6. ,, . .	3·60	27·15	61·00	8·25	1·44	1·298	69·25	,,
7. Eskbank . . .	2·70	28·78	57·88	9·88	·76	1·329	none	Dixon.
8. ,, . .	2·00	33·55	49·97	12·91	1·57	1·335	62·88	Liversidge.
9. Grose Valley . . .	3·60	29·70	59·74	6·96	·05	1·340	66·70	Gov. Analyst.
10. ,, . .	3·66	30·38	54·46	11·50	·062	1·380	65·96	,,
11. Katoomba . . .	2·90	25·82	61·34	9·26	·68	1·326	70·60	Dixon.
12. ,, . .	2·25	26·28	60·84	10·04	·57	1·400	none	,,
13. ,, . .	2·71	25·31	60·90	10·84	·24	1·343	71·74	,,
14. ,, . .	3·02	29·84	59·51	7·63	·61	1·320	67·14	Gov. Analyst.
15. Lawson (near) . .	2·45	25·31	59·41	12·83	·74	1·440	72·24	,,
16. Lithgow Valley . .	2·24	28·48	58·80	9·68	·80	1·340	none	Dixon.
17. ,, . .	1·95	34·18	52·34	10·12	1·41	1·329	62·46	Liversidge.
18. Rylstone . . .	1·70	36·42	51·48	9·76	·64	1·300	61·24	Gov. Analyst.
19. Vale of Clwydd . .	2·15	35·02	52·36	9·72	·75	1·328	none	Dixon.
20. ,, . .	2·10	33·35	53·38	9·80	1·37	1·323	63·18	Liversidge.
21. Wallerawang . .	3·85	27·69	61·56	6·88	·02	1·326	none	Dixon.
22. ,, . . .	1·95	27·25	61·86	8·94	...	1·398	70·10	Liversidge.
23. ,, . . .	1·51	33·24	55·74	9·50	...	1·333	65·24	,,
24. ,, . . .	3·10	34·73	51·28	10·36	·53	1·327	61·64	Dixon.
25. No locality . .	3·40	30·60	55·17	10·83	·76	1·316	66·00	Gov. Analyst.
Mean . .	2·67	28·99	58·13	9·73	·73	1·347	67·90	

No. 1. Coal from BLACKHEATH.

No. 2. BOWENFELS.—Dull lustre, rather strongly laminated; laminæ of bright coal, very thin. Does not soil the fingers. Fracture is in parts large conchoidal.

Coke.—Does not cake; only a loose and incoherent black powder left. *Ash.*—Heavy, white.

No. 3. Coal from CAPERTEE.

No. 4. CLARENCE SIDING.

No. 5. CLARENCE SIDING.—Anthracitic.

No. 6. CLARENCE SIDING.—Bituminous coal.

No. 7. ESKBANK.

Ash.—Grey.

No. 8. A good compact coal; soils the fingers; lustre dull; laminæ not well defined.

Coke.—Fair, but rather tender. *Ash.*—Brilliant white colour.

Nos. 9 and 10. Coal from GROSE VALLEY.

No. 11. KATOOMBA.—A fairly bright and tolerably hard coal, from

Katoomba, from 106 feet in the tunnel.   Did not soil the fingers, and showed layers of "mother-of-coal" in places.

Coke.—70·60 per cent.; only slightly fritted together, dull-coloured, with a few bright specks.   *Ash.*—A greyish white.

No. 12. KATOOMBA.

Ash.—Greyish white.

No. 13. A sample of the whole thickness of a 4-feet seam at KATOOMBA.   It consists of a mixture of a bituminous and splint coal, with bright and dull coloured pieces.

The coke is dense, scarcely swollen, but fairly lustrous.   The ash is white.

This is a fairly good coal, the low percentage of sulphur being particularly noteworthy.

No. 14. KATOOMBA.

No. 15. LAWSON (near).—Coal from an unworked seam.

No. 16. LITHGOW VALLEY COLLIERY.

Ash.—Greyish white.

No. 17. LITHGOW VALLEY.—Has much the appearance of the Vale of Clwydd coal.   Does not soil the fingers.

Coke.—Hard, compact, and fairly lustrous.   *Ash.*—White in colour.

No. 18. RYLSTONE.—Dull, with bright fracture.

Coke.—Strong and fairly bright.   *Ash.*—Greyish-white and bulky.

No. 19. VALE OF CLWYDD.—A compact coal, rather bright on the whole, the bright layers being fairly numerous; fracture irregular; a fresh surface; does not soil the fingers.

Coke.—Hard, compact, and fairly lustrous.   *Ash.*—Of a very feeble grey tint.

No. 20. VALE OF CLWYDD.

Ash. Grey.

No. 21. WALLERAWANG.—"A sample of true splint coal.   It was very firm, and contained some layers of mineral charcoal; colour of a dull brownish black.

"The powdered coal did not form a coke, an incoherent black powder being left.   On heating the coal in lumps it left a hard coke, slightly lustrous, the pieces having the same shape as the original coal, and showing no signs of fusion.

"The ash was grey-coloured and bulky, and contained 0·19 per cent. of phosphoric oxide.

"This is a good coal for the purpose to which it is intended to be applied, namely, iron-smelting, as it could be used raw with hot blasts, or coked with either hot or cold; and it is sufficiently firm to carry a heavy burden of ore and fluxes.   On heating, it decrepitates slightly; but this does not interfere with the firmness of the coke obtained from lumps, but the small, from its character, would be useless for coking."—*W. A. Dixon.*

No. 22. A sample from another seam 6 feet 6 inches thick.

No. 23. A specimen of the WALLERAWANG coal, from a seam 17 feet 6 inches thick.

No. 24. " Another sample of coal from the same locality contained dull and bright layers in about equal proportions, the bright parts breaking with a rather large conchoidal fracture, considering their thickness. This coal is a bituminous moderately coking coal, and the specimen was of only moderate firmness.

" The coke produced from the powdered coal was scarcely swollen, of moderate brightness, and not very firm. As I have learned, however, that the sample had been exposed to the air for about a year, it is probable that the freshly dug coal would yield a much superior coke to that produced by the specimen analysed.

" The ash was white, very dense, and contained 0·29 per cent. of phosphoric oxide.

" This coal is not so good for iron-smelting as the last, as it contains more sulphur; and as a large quantity of it would be required to produce a ton of iron, as it would have to be used coked, more of the obnoxious ingredients would be introduced into the furnace."—*W. A. Dixon, F.C.S., Annual Report of the Department of Mines, Sydney,* 1880.

No. 25. No LOCALITY.—Semi-bituminous coal.

SOUTHERN DISTRICT.

Name of Colliery, &c.	Water.	Volatile Hydro-carbons.	Fixed Carbon.	Ash.	Sulphur.	Specific Gravity.	Coke.	Analyst.
1. Berrima	1·70	32·78	53·84	10·40	1·28	1·364	64·24	Liversidge.
2. „ (Atkinson's)	1·26	26·61	62·28	9·40	·45	1·408	71·68	Dixon.
3. Bulli Colliery	·65	21·65	65·86	11·28	·74	1·369	76·96	„
4. „ „	1·03	23·65	61·61	13·17	·54	1·471	74·78	R. Smith.
5. Coal Cliff Colliery	1·61	19·68	68·08	10·28	·35	1·372	78·36	Dixon.
6. „ „	·86	18·22	69·84	10·80	·28	1·378	80·64	„
7. Goulburn, near	2·18	30·98	58·04	8·80	·228	1·350	66·84	Gov. Analyst.
8. Heathcote	·59	16·83	69·34	13·24	·37	1·372	82·28	„
9. „	1·34	16·16	70·87	11·63	·352	1·360	82·50	„
10. Holt-Sutherland	·24	14·03	73·96	11·16	·61	...	85·12	Dixon.
11. Illawarra C. Co.'s Coll.	·70	22·04	68·08	8·76	·42	1·354	76·84	„
12. Marulan	2·13	28·75	59·00	9·55	·57	1·404	68·55	„
13. „	1·97	31·77	55·64	9·94	·68	1·398	65·58	„
14. „	2·25	26·14	57·68	13·52	·41	1·341	none	„
15. Mittagong (near)	2·12	8·16	79·42	10·30	·71	1·400	89·72	Gov. Analyst.
16. „ Fitzroy Mine	2·80	33·52	53·35	9·15	1·18	...	62·50	Elouis.
17. Nattai	3·28	4·34	87·96	4·41	trace	...	93·37	Liversidge.
18. Mount Keira	1·15	23·51	64·65	9·70	·99	1·379	74·35	„
19. Mount Kembla	1·50	19·74	67·18	10·72	·86	1·363	none	„
20. Ringwood	1·62	29·40	56·82	12·16	·398	1·500	68·98	Gov. Analyst.
21. Shoalhaven	3·20	28·98	59·88	7·94	1·43	1·313	67·82	„
22. „	·85	32·15	56·18	10·82	1·63	1·302	67·00	„
23. „	1·60	32·30	59·22	6·88	1·21	1·210	66·10	„
24. Ulladulla	1·50	33·04	57·40	8·06	1·11	1·306	65·46	„
25. „	1·76	32·06	56·38	9·80	1·24	1·351	66·18	„
26. Wollongong (Osborne's)	1·19	21·07	66·92	10·20	·62	1·404	77·12	Dixon.
27. No locality	·75	23·37	65·81	8·19	1·88	1·307	74·00	„
Mean	1·55	24·11	63·53	10·01	·76	1·365	73·88	

No. 1. BERRIMA.—A good firm coal, but more tender than the others. The bright layers present in fair proportion.

Coke.—Bright and lustrous; very much swollen up. *Ash.*—White.

No. 2. BERRIMA.

Coke.—71·68 per cent. *Ash.*—Greyish white.

No. 3.—BULLI.

Coke.—76·96 per cent. *Ash.*—Grey.

No. 4. BULLI.—This analysis was made by Mr. Richard Smith, of the Metallurgical Laboratory in the Royal School of Mines, London; to compare it with the others, its proximate composition has been calculated from the ultimate analysis:—

"The theoretical calorific or evaporative power, that is, the weight of water converted into steam by 1 lb. of the coal, as determined by experiment with the calorimeter, is 12·21 lbs. A second experiment gave a like result.

"The colour of the ash is reddish white.

"When a portion of the powdered coal is heated in a closed vessel the gases evolved burn with a yellow, luminous, somewhat smoky flame, and a slightly lustrous coherent coke is left, which differs little in bulk from the original coal."

Nos. 5 and 6. From COAL CLIFF MINE, near Bulli.

Coke.—78·36 and 80·64 per cent. *Ash.*—Greyish white.

No. 7. GOULBURN.—Coal from near Goulburn.

No. 8. HEATHCOTE.—Coal from bore taken from distances of 1 foot in the 12-feet seam discovered 847 feet deep. Ignites quickly, appears to be of excellent quality for steam, smelting, and household purposes.

No. 9. HEATHCOTE.—Splint coal from top of seam struck in diamond-drill bore at a depth of 1513 feet, seam 4 feet 8 inches thick.

No. 10. HOLT SUTHERLAND, sixteen miles south from Sydney.—From second seam, 4 feet thick, struck in diamond-drill bore at a depth of 2228 feet. A short-flamed steam-coal.

Coke.—Swollen slightly, firm, but dark-coloured. *Ash.*—Dark grey.

No. 11. ILLAWARRA COAL COMPANY'S COLLIERY.—Coal from the Mount Pleasant Mine of the Illawarra Coal Company, Wollongong.

Coke.—76·84 per cent. *Ash.*—Grey.

MARULAN.—Two samples of weathered, dirty-looking coals, of high specific gravity, from near Hanging Rock, Marulan, were found to contain 28·09 per cent. and 39·76 per cent. of ash respectively, and were deemed unworthy of a more detailed examination.

No. 12. MARULAN.—This was a dull-coloured splint-like coal, of moderate firmness.

The coke obtained was dull-coloured and soft.

The ash was greyish white.

No. 13. MARULAN.—A moderately bright but firm coal. The coke and ash were similar to the last.

No. 14. MARULAN.—A rather dull-coloured coal, having somewhat the character of splint from Hanging Rock, near Marulan.

This coal scarcely forms a true coke, a very slight coherent black mass being left; but, as the specimen was evidently taken from an outcrop, where it would be more or less weathered, this character would probably be altered on opening out the seam.

The ash was greyish white.

No. 15. MITTAGONG.—From anthracite seam.

No. 16. MITTAGONG.—From borehole, Fitzroy Mine.

No. 17. MITTAGONG or NATTAI.—A hard, compact, lustrous, anthracitic coal, slightly stained in parts with iron oxide, which looks as if it had been derived from the decomposition of iron pyrites; but, contrary to what was expected, hardly a trace of sulphur was found to be present. Any pyrites which the coal may have originally contained must have practically undergone complete decomposition and removal.

Anthracitic coals generally occur in places where the coal-measures have been more or less disturbed or changed, i.e., in places where there is considerable contortion of the strata, and also where there are intrusive metamorphic or igneous rocks. Probably this particular specimen came from a portion of a seam which had been affected by one of the intrusions occurring in the district.

Coal containing pea-iron ore is abundant at Nattai. Another coal, from near to Nattai, is very brilliant in lustre, and breaks with a pitchy lustrous conchoidal fracture like albertite; it is also marked by the presence of thick layers of mother-of-coal or fibrous mineral charcoal.

No. 18. MOUNT KEIRA.—Possesses much the same characters as No. 16, only soils the fingers rather more readily.

Coke.—Hard, fairly lustrous, and much swollen up, with cauliflower-like excrescences. Ash.—Loose, brilliant white colour.

No. 19. MOUNT KEMBLA.—A coal of medium brightness, with laminated structure, breaking with a granular surface in places; splits readily along the planes of lamination. The bright layers are tender, and break into small pieces with conchoidal surfaces.

Coke.—Coal does not cake, therefore no true coke formed—a dull black fritted mass only is left. Ash.—Brilliant white colour.

Nos. 21, 22, and 23. SHOALHAVEN, Head of Clyde River.—Bituminous coal.

Nos. 24 and 25. ULLADULLA.

No. 26. WOLLONGONG.—Coal from Osborne-Wallsend Colliery. Coke.—77·12 per cent. Ash.—Grey.

No. 27. A rather dull-coloured coal, stained with ferric oxide, in some places iridescent. It was rather tender, and stained the fingers; fracture of the bright layers minutely conchoidal. From the southern district.

Coke.—74 per cent. ; bright and dense. *Ash.*—Greyish.

The coals in the foregoing tables are arranged in alphabetical order. With a few exceptions, such as the Teralba, Lake Macquarie, Anvil Creek, and the Cardiff Mine, it is rather interesting to note that the proportion of fixed carbon increases with the increase in the amount of ash ; the proportions of volatile hydrocarbons naturally undergo a corresponding diminution.

Speaking generally, the coals which yield a large percentage of volatile hydrocarbons may be said to be the best adapted for the manufacture of gas.

It will also be at once apparent that the specific gravity in most cases affords a very good indication of the quality of the coal. As a general rule, ordinary coals which possess a high specific gravity contain a large proportion of ash.

Although these tables show decided differences between the coals from the three districts, doubtless the examination of additional specimens will prove that the above means do not quite represent the average composition of the coals. Some of the analyses were necessarily made upon outcrop specimens, and such can hardly be regarded as truly representing the quality of the seams from which they were obtained.

It is noticeable that the quantity of ash yielded by the western and southern coals is much greater than is yielded by the northern ones, also that the specific gravity is higher as a rule.

The ash of western and southern coals is white and dense, whereas many of the northern coals yield ashes of a buff or red tint, which are often quite loose and flocculent.

It is a common opinion that the relative amounts of sulphur present in different coals can be approximately estimated by the redness of the ash, on the supposition that the whole of the sulphur exists in the coal in the form of iron pyrites, but such is not the case ; on referring to the analyses on the northern district coals, it will be seen that some of the coals which left pure white coloured ashes contained the largest amount of sulphur, and that others which left red ashes contained the smallest quantity of sulphur.

Sulphur may be present in coals in various forms, either in combination with iron as pyrites, which is the most common form of all ; as sulphuric acid in combination with the inorganic constituents of the coal, such as alumina, lime, magnesia, or potash ; or it may even exist in the form of organic compounds.

In order that an opinion may be formed with regard to the coals of New South Wales, it will perhaps not be amiss to compare them with some of those produced in various parts of Great Britain.

In the first place, the proportion of ash in a coal is a matter of the greatest importance; the value of coal as a fuel depends to a great extent upon the smallness of the quantity of non-combustible matter which it contains. A large proportion of ash, anything over 8 or 9 per cent., detracts very much from the value of a coal for most purposes; it is useless matter, which has to be handled and afterwards removed, which occupies valuable space (a subject of great consideration on board ship), and hinders the proper combustion of the coal. If the amount be very large the coal will be perfectly worthless; but for some purposes, as Dr. Percy states, "a certain amount of inorganic matter in coal is sometimes beneficial in preventing its too rapid combustion in the furnace. On this account a kind of coal called 'brasils,' which occurs in the middle of the Tenyard coal in South Staffordshire, is preferred for reverberatory furnaces by some smelters in Birmingham." * Neither must the quality or chemical composition of the ash be neglected, for if the ashes be easily fusible, as they usually are when a large quantity of iron is present, they tend to "clinker up" the grate, and thus cause great waste of heat, and the expenditure of much extra time and labour in stoking.

We have seen that the northern district coals yield on the average the smallest amount of ash, which is from 2·70 per cent to 11·51, with an average percentage of 6·80; the western district coals range from 6·24 to 12·91, and average 9·73 per cent.; and the southern district coals, omitting the samples which seem to be somewhat exceptional in character, yield from 4·41 to 13·52 per cent., and average 10·01 per cent. ash.

Percentage of Ash in some English Coals.†

COKING COAL.

Locality.	No. of Specimens.	Minimum.	Mean.	Maximum.	
Northumberland	4	0·79	1·68	2·49	...
Nottinghamshire	1	...	3·9	...	...
Blaina, South Wales	3	1·46	2·63	4·00	...

NON-COKING COAL.

Locality.	No. of Specimens.	Minimum.	Mean.	Maximum.	
South Staffordshire	10	1·03	2·58	6·44	} Rich in oxygen.
Scotland	3	1·43	3·38	6·75	
South Wales, Dowlais	4	1·20	3·42	7·18	Rich in carbon.

* Percy's "Metallurgy," 1875, vol. i. p. 281.
† *Ibid.*, vol. i. pp. 322 and 325.

From the above it will be seen that the New South Wales coals all contain more ash than the average of the English coals.

A matter to which it is necessary to pay careful attention is the proportion of sulphur present in a coal. The presence of a large amount of this element not only renders the use of the coal unpleasant and unhealthy for domestic purposes, but makes it useless for most manufacturing and metallurgical operations.

The quantity of sulphur existing in the New South Wales coals is by no means excessive, and they will in this respect compare not unfavourably with those of other countries.

Percentage of Sulphur.

	No. of Specimens.	Minimum.	Mean.	Maximum.
Northern coal-fields	20	·34	1·30	3·01
Western „	15	·02	·87	1·57
Southern „	15	trace	·72	1·88
Newcastle coal (England) *	4	·55	·97	1·51

The mean percentages of sulphur as given above for the New South Wales coals are probably too high, since, as has already been remarked, some of the samples were doubtless only outcrop specimens from seams not yet properly opened out.

The above was written in 1881; now that a greater number of specimens has been examined, the mean percentage of sulphur has been reduced in the northern and western coals.

	No. of Specimens.	Average of Sulphur, per Cent.
Northern coal-fields	42	·88
Western „	25	·73
Southern „	27	·76

Playfair and De la Beche found, during their investigation for the English Government, that the mean percentage of sulphur was as follows :—

	Per Cent. Sulphur.
Welsh coal	1·42
Derbyshire	1·01
Lancashire	1·42
Newcastle	0·94
Scotland	1·45

Most of the Secondary and Tertiary coals, on the other hand, contain a larger proportion of sulphur, usually 2·0 or 3·0, and sometimes as much as even 5·0 or 6·0 per cent.

Composition of the Ashes.—In the table showing the percentage composition of the ashes it will be noticeable there are great differences

* *Vide* Percy's "Metallurgy," 1875, vol. i. p. 322.

in the amounts of silica, alumina, and of iron sesquioxide. Some of the ashes, however, in the different groups seem to agree fairly well together, and although the samples came from different districts, yet it may be that they are from an extension of the same seam. The composition of the ashes as well as of the coals may help us to correlate the coal-seams of the different districts one with the other; i.e., assist in determining their positions in a geological section of the whole of the coal-measures as developed in different parts of the colony. Judging from the composition of the ashes, one would be inclined to say, that not only do certain of the coals in each district come from the same seam, but that the western coals from the Vale of Clwydd and Lithgow Valley belong to the same horizon as the southern coal from Berrima; but much importance cannot be attached to this matter. Certainly it would never do to allow the analysis of one specimen only from a given seam to have much influence, for although a sample of coal may appear to be free from foreign substances and to look perfectly uniform to the eye—in fact, appear to be homogeneous throughout— yet on analysis it is nearly always found that the different parts of one and the same piece yield different proportions of ash, carbon, hydrogen, &c. Hence, if different portions of the same lump vary, we may naturally expect that samples taken from different parts of the seam should also vary. But in spite of minor variations in different specimens of coal from any given seam, we find that on the average the coal will have a fairly uniform composition. To obtain uniform and truly representative samples, portions should be taken of the whole thickness of the seam from different parts of the working face. It would be well to take some tons weight of the coal, which should be broken up into pieces of moderate size and well mixed. From this heap portions should then be removed, in radial lines cutting down to the centre, and thrown into a smaller heap of a few hundredweights; after this smaller heap has been well mixed, portions should be again removed radially, and a third time well mixed; this last could then doubtless be regarded as a true sample and not a mere specimen, as a single lump of coal must necessarily be. Too much care cannot possibly be taken over the collection and preparation of samples.

Composition of Coal Ashes.
NORTHERN DISTRICT.

No.	Name of Colliery	Ash	Silica	Alumina	Iron Sesquioxide	Manganese (MnO)	Lime	Magnesia	Potash	Soda	Phosphoric Acid	Sulphuric Acid	Undetermined and Loss	Analyst
1 }	Anvil Creek	7·80	48·70	38·84	2·71	trace	5·20	·70	2·13	·43	trace	·85	·44	Liversidge
	" second specimen	...	50·16	40·50	2·00	trace	4·10	·32	2·02	·12	trace	·56	·22	"
6	Greta	2·72	48·14	39·99	4·40	...	5·95	trace	·82	·19	trace	·77	...	"
15	Cardiff Mine	4·94	38·36	35·57	9·27	2·60	8·05	1·08	·59	2·26	·24	2·25	...	"
26	Australian Agricultural	4·44	53·10	26·29	15·20	trace	1·98	trace	...	...	2·26	·97	...	Dixon.
27	Ferndale	5·35	50·05	34·90	13·81	...	0·56	...	·19	·02	1·25	1·06	...	Liversidge.
28	"	3·84	50·82	29·66	12·65	...	2·65	2·34	...	...	1·24	·74	...	Dixon.
29	Newcastle Coal Company	4·76	45·57	33·72	14·13	...	2·07	2·19	...	...	·12	·72	·36	"
30	Newcastle-Wallsend Company	4·64	39·30	25·24	26·02	1·03	4·35	·30	trace	trace	1·14	4·51	...	Liversidge.
31	"	4·28	50·21	28·73	14·51	...	3·37	1·72	...	...	1·28	·83	...	Dixon.
32	New Lambton (near Newcastle)	6·72	52·32	20·56	19·88	...	2·95	3·09	...	...	1·34	·72	·69	"
34	Co-operative-Plattsburg	4·20	49·32	29·24	12·50	...	3·71	2·48	...	...	1·72	·45	·18	"
35	Red Head Coal Company	6·84	69·65	18·62	6·71	...	1·96	·71	...	...	...	1·84	...	Liversidge.
37	Russell's Coal-mine	2·70	44·30	38·65	7·85	...	5·05	·49	1·37	·01	trace	·77	·12	Dixon.
41	Waratah	4·15	47·30	35·58	9·67	...	4·95	·30	1·92	·05	2·29	·71	...	"
42	"	4·64	56·17	26·90	10·42	...	2·41	·98	...	...	·96	·21	·05	Dixon.
	Clarence River	8·75	65·12	27·91	4·01	...	1·26	·48	...	...	...	...	...	"
	Average	5·04	50·50	31·23	10·92	·21	3·56	1·01	·53	·18	·81	1·05	...	...

Composition of Coal Ashes.

WESTERN DISTRICT.

No.	Name of Colliery.	Ash.	Silica.	Alumina.	Iron Sesquioxide.	Manganese (MnO).	Lime.	Magnesia.	Potash.	Soda.	Phosphoric Acid.	Sulphuric Acid.	Undetermined and Loss.	Analyst.
2	Bowenfels	11·40	69·15	29·65	·63	trace	·25	trace	·36	·32	·09	·22	...	Liversidge.
7	Eskbank	9·88	61·02	35·34	1·39	·78	·78	·61	...	...	·55	·16	...	Dixon.
8	"	12·91	62·15	29·43	1·20	trace	1·35	1·73	2·10	·19	·05	1·12	...	Liversidge.
11	Katoomba	10·84	59·58	38·49	·98	...	traces	·30	...	...	·56	·11	·09	Dixon.
16	Lithgow Valley	9·68	60·21	36·26	1·42	...	·74	·57	...	...	·64	·11	·05	"
17	"	10·12	59·10	38·95	·40	trace	·85	·30	trace	trace	·20	·43	...	Liversidge.
19	Vale of Clwydd	9·72	59·25	37·46	1·55	...	·81	1·05	...	trace	·59	·17	·12	Dixon.
20	" "	9·80	59·55	37·35	2·00	trace	·53	trace	...	trace	trace	·39	...	Liversidge.
	Weatherboard	...	83·80	14·43	·40	trace	·35	·95	·32	·12	trace	·46	...	"
	Average	9·37	63·64	33·04	1·11	·08	·63	·61	·31	·07	·29	·34	...	

SOUTHERN DISTRICT.

No.	Name of Colliery.	Ash.	Silica.	Alumina.	Iron Sesquioxide.	Manganese (MnO).	Lime.	Magnesia.	Potash.	Soda.	Phosphoric Acid.	Sulphuric Acid.	Undetermined and Loss.	Analyst.
1	Berrima	10·40	67·45	31·00	·40	·16	·15	...	·24	·18	traces	·06	·13	Liversidge.
2	Bulli	9·40	75·05	19·43	4·68	...	·58	...	...	...	traces	·13	·13	Dixon.
3	"	11·28	56·93	34·44	7·95	...	·67	traces	...	...	traces	·31	·14	"
6	Coal Cliff	10·80	57·41	35·17	5·33	...	·75	·60	...	...	·29	·31	...	"
11	Illawarra, Mount Pleasant	8·76	51·19	40·57	6·03	...	·82	traces	...	...	·32	·51	·56	"
18	Mount Keira	9·70	53·00	46·88	trace	...	trace	trace	...	·10	·17	...	...	Liversidge.
19	Mount Kembla	10·72	52·57	43·55	·95	trace	1·35	·60	·15	·27	traces	·79	...	"
26	Osborne Wallsend	10·20	54·76	35·55	8·68	...	1·18	traces	...	...	traces	·34	...	Dixon.
	Joadja Creek	6·71	48·86	33·38	11·55	...	·20	·63	4·05 (with chlorine) 0·85		·92	...	·41	"
	" "	22·28	68·63	27·83	1·84	...	·28	·36	...		·44	...	...	Dixon.
	Average	11·02	58·58	34·78	4·74	·01	·59	·21	...	...	·19	·07	...	

No.		Silica.	Alumina.	Iron Peroxide.	Lime.	Magnesia.	Phosphoric Acid.	Sulphuric Acid.	Undetermined and Loss.	Analyst.
1	Roof of Galley Way	75·56	18·99	2·77	·42	1·32	41	·21	·32	Dixon.
2	„ old No. 1 Way	84·73 (with alumina)	11·89 (sol.)		1·61	·93	·37	...	·47	„
3	Floor of Galley Way	75·12	21·84	2·21	·42	...	·65	...	...	„
4	„ old No. 1 Way	78·73	17·31	2·84	·53	·31	·16	...	·12	„

The above analyses show the differences between the composition of the ashes of the coal itself (Nos. 26 and 27), and of the upper and lower portions of the Australian Agricultural Company's seam, Newcastle.

No. 1 contained 9·97 per cent. of carbonaceous matter and water.
„ 2   „   7·70   „   „   „
„ 3   „   30·95   „   „   „
„ 4   „   4·30   „   „   „

It has already been mentioned that a large proportion of ash detracts very much from the value of a coal as a fuel, and this is also the case with the moisture and oxygen.

If a coal contains 10 per cent. of moisture, this means that it contains 10 lbs. of useless material in every 100 lbs., which has to be evaporated and driven off at considerable loss of heat; further, the oxygen present is useless matter, and if there be 15 to 20 per cent. of oxygen present, it means that another 15 or 20 lbs. of useless material is present in every 100 lbs. of coal. It is true that the oxygen present serves to assist in the combustion of the carbon and hydrogen of the fuel, but in most cases the atmospheric oxygen would serve equally well, and this oxygen in the air has not to be handled, and does not occupy storage space, a very important matter on board ship.

It is on account of the ultimate analysis showing the proportion of oxygen and hydrogen, and especially the amount of the latter over and above that necessary to combine with the oxygen in the coal, that the ultimate analyses afford much more valuable information than the proximate analyses. The proximate analyses are comparatively rough-and-ready tests approximately ascertaining the value of coals.

Hence, in a coal containing 5 per cent. moisture, 10 per cent. oxygen, and 10 per cent. of ash, none of which would be counted very excessive, we have 25 per cent. of useless incombustible matter.

In the following table the coals are arranged according to the order in which they have been described :—

Ultimate Analyses.

I. Northern District Coals.

No.	Locality.	Specific Gravity.	Carbon.	Hydrogen.	Oxygen.	Nitrogen.	Sulphur.	Ash.	Water per Cent.	Coke per Cent.	Calorific Intensity (Calculated).	Water Converted into Steam by 1 lb. Coal with Calorimeter.
						Composition per Cent. Exclusive of Water Only.						
1	Anvil Creek .	1·323	77·15	5·91	6·07	1·46	1·48	7·93	1·74	55·70	8009	12·65
6	Greta .	1·287	78·41	6·60	9·34	1·43	1·44	2·78	2·25	57·13	8208	13·21
15	Cardiff Mine .	1·286	82·25	4·38	6·95	1·03	0·35	5·04	1·853	54·43	7857	...
27	A. A. Co., Newcastle	1·297	78·76	6·34	7·28	0·79	1·36	5·47	2·20	62·87	8235	12·92
30	Wallsend .	1·333	79·96	6·26	7·08	0·68	1·25	4·77	2·75	61·86	8323	13·21
37	Russell's Mine .	1·274	77·37	6·48	10·46	1·51	1·43	2·75	1·85	52·65	8034	13·21
41	Waratah .	1·303	81·06	5·81	6·52	1·23	1·14	4·24	2·21	59·97	8271	14·30
	The mean .	1·300	79·28	5·97	7·67	1·16	1·21	4·71	2·122	57·80	8134	13·25

II. Western District Coals.

No.	Locality.	Specific Gravity.	Carbon.	Hydrogen.	Oxygen.	Nitrogen.	Sulphur.	Ash.	Water per Cent.	Coke per Cent.	Calorific Intensity (Calculated).	Water Converted into Steam by 1 lb. Coal with Calorimeter.
2	Bowenfels .	1·399	70·72	5·65	9·65	0·93	1·38	11·67	2·36	...	7245	12·65
8	Eskbank .	1·335	72·30	5·43	6·65	0·85	1·60	13·17	2·00	62·88	7426	12·65
17	Lithgow Valley .	1·329	69·41	6·10	11·70	1·03	1·44	10·32	1·95	62·46	7206	12·10
20	Vale of Clwydd .	1·323	69·86	5·82	11·89	1·02	1·40	10·01	2·10	63·18	7138	12·10
	The mean .	1·346	70·57	5·75	9·97	0·96	1·45	11·29	2·10	62·84	7254	12·37

III. Southern District Coals.

No.	Locality.	Specific Gravity.	Carbon.	Hydrogen.	Oxygen.	Nitrogen.	Sulphur.	Ash.	Water per Cent.	Coke per Cent.	Calorific Intensity (Calculated).	Water Converted into Steam by 1 lb. Coal with Calorimeter.
1	Berrima .	1·364	69·92	4·55	13·09	0·56	1·30	10·58	1·70	64·24	6653	11·82
4	Bulli (R. Smith) .	1·471	75·57	4·70	4·99 }		0·54	13·17	1·03	74·78	...	12·21
17	Nattai .	...	91·24	3·60	0·59	...	trace	4·56	3·28	92·37	8590	undetermd.
18	Mount Keira .	1·379	78·82	5·17	3·87	1·33	1·00	9·81	1·15	74·35	7983	12·92
19	Mount Kembla .	1·363	80·67	5·30	1·58	0·70	0·87	10·88	1·50	...	8276	13·21
	The mean .	1·394	79·401	4·675	4·833	0·52	0·74	9·829	1·733	76·436	7875	12·54

It is again apparent that the northern coals as a class are considerably superior to the southern coals, which in turn are better than those from the western districts; these differences are shown most plainly in the last two columns, viz., those showing the calculated calorific intensities and the proportions of water converted into steam by 1 lb. of each of the coals when burnt in Thomson's calorimeter.

As a class, the northern coals are brighter and more laminated than the southern and western; they yield a larger proportion of volatile hydrocarbons, and are therefore more suitable for making gas, and furnish bright, hard, sonorous cokes of extremely good quality.

The southern coals are not so bright, and, unlike the northern, they do not cake in an ordinary open fire, but yield a very good coke when treated in ovens.

The western coals are of a still drier character and duller appearance; they only coke when freshly raised from the mine.

Both of the latter burn with much less smoke than the rich bituminous northern coal.

RETURN showing the Quantity and Value of Coal produced in the Colony of New South Wales (from the *Annual Reports of the Department of Mines*, Sydney).

Year.	Quantity.	Value.	Year.	Quantity.	Value.
	Tons.	£		Tons.	£
Prior to 1829	50,000	25,000	1859	308,213	204,371
1829	780	394	1860	368,862	226,493
1830	4,000	1,800	1861	342,067	218,820
1831	5,000	2,000	1862	476,522	305,234
1832	7,143	2,500	1863	433,889	236,230
1833	6,812	2,575	1864	549,012	270,171
1834	8,490	3,750	1865	585,525	274,303
1835	12,392	5,483	1866	774,238	324,049
1836	12,646	5,747	1867	770,012	342,655
1837	16,083	7,828	1868	954,231	417,809
1838	17,220	8,399	1869	919,774	346,146
1839	21,283	10,441	1870	868,564	316,836
1840	30,256	16,498	1871	898,784	316,340
1841	34,841	20,905	1872	1,012,426	396,198
1842	39,900	23,940	1873	1,192,862	665,747
1843	23,862	16,222	1874	1,304,567	790,224
1844	23,118	12,363	1875	1,329,729	819,430
1845	22,324	8,769	1876	1,319,918	803,300
1846	38,965	13,714	1877	1,444,271	858,998
1847	40,732	13,750	1878	1,575,497	920,936
1848	45,447	14,275	1879	1,583,381	950,879
1849	48,516	14,647	1880	1,446,180	625,337
1850	71,216	23,375	1881	1,775,224	603,248
1851	67,610	25,546	1882	2,109,282	948,965
1852	67,404	36,885	1883	2,521,457	1,201,942
1853	96,809	78,059	1884	2,749,109	1,303,077
1854	116,642	190,380	1885	2,878,863	1,340,213
1855	137,076	89,082	1886	2,830,175	1,303,164
1856	189,960	117,906			
1857	210,434	148,158			
1858	216,397	162,162	Total . .		18,362,668

TORBANITE.—Cannel Coal *or* Kerosene Shale.

The name "kerosene shale," commonly applied to this mineral, is not at all appropriate. The substance does not sufficiently possess the properties of a shale, *i.e.*, it has not the characteristic lamellar or platy structure of a shale, but the reverse, being very compact, and breaking with large smooth conchoidal surfaces with equal readiness in every direction, and without any tendency to follow the planes of stratification. Ordinarily it is almost devoid of all traces of stratification, but occasionally indications can be seen where the mineral is in the form of sufficiently large blocks, or when it is *in situ*, but even then the planes of stratification are mainly rendered visible by the presence of layers or films of earthy matter. Near the top and bottom of the deposits the stratification layers are, however, usually better marked, *i.e.*, where the shale merges into the roof and floor. The planes of stratification are, however, shown by the weathered outcrops, and again when the shale is burnt, since it then splits up into more or less regular slabs along the lines of deposition.

This so-called "kerosene shale" does not differ very widely from cannel coal and torbanite. Like cannel coal, it usually appears to occur with ordinary coal in the form of lenticular deposits. Like cannel coal also, when of good quality it burns readily, without melting, and emits a luminous smoky flame. When heated in a tube it neither decrepitates nor fuses, but a mixture of gaseous and liquid hydrocarbons distils over. It is extremely valuable for manufacture of illuminating oils and gas. When of good quality it yields from 150 to 180 gallons of oil per ton.

In colour it varies from a brown-black, at times with a greenish shade, to full black. The lustre varies from resinous to dull. The fracture is usually broad conchoidal, but the concavities are sometimes very deep in proportion to their breadth, and at times long flexible concave-convex strips can be detached. When struck it emits a dull wooden sound. The powder is light brown to grey; the streak shining. It usually weathers to a light grey colour, and the surfaces of the joints also are often coated with a film of white clay; hence it has received the name of "White Coal."

It is easily cut into shavings. Thin sections under the microscope present a reticulated appearance. The network is black and opaque, enclosing brown and amber-coloured translucent particles.

The Hartley and Murrurundi shales are but slightly soluble, if at all, in alcohol, ether, carbon disulphide, petroleum, or caustic potash, even when boiled; but they gelatinise with boiling sulphuric acid, and evolve a sulphurous acid odour; with nitric acid they yield a yellow solution.

K

Professor Silliman has proposed the name of *Wollongongite* for the mineral, but this has not come into general use; neither is it an appropriate name, since the specimen sent to him was not from Wollongong, but from Hartley. All the Wollongong oil shales which I have seen are of different character; they are true shales, with well-marked lamination, black and fairly rich in carbon, and with a more or less greasy lustre, and often contain fossil ferns, especially the fronds of the glossopteris. No chemical examination has yet been made of any of them, although some of them yield a very large proportion of oil.

Unless it be decided to give the mineral a new name, it would be better to call it torbanite or cannel coal rather than kerosene shale, since the oil which it yields is probably not kerosene, and the substance itself is not strictly a shale, and, moreover, it is not very widely separated, either in physical properties or in chemical composition, from either torbanite or the cannel coals.

At Joadja Creek this mineral often contains impressions of the *glossopteris* and of the *vertebraria*. These fossil plant remains are usually best seen in the outcrops of the poorer portions of the shale deposits, and especially where somewhat weathered—the glossopteris fronds are generally found between the laminæ; the vertebraria run across them.

Large quantities of this mineral are now used for making illuminating, lubricating, and other oils, as well as for the manufacture of paraffin wax-candles. It is also shipped in considerable quantities to Europe and America for enriching ordinary coal-gas, as it yields a large volume of hydrocarbons which burn with high luminosity.

Discovery.—The occurrence of "kerosene shale" near Bathurst is mentioned in a book entitled "Two Years in New South Wales," written by P. Cunningham, surgeon, R.N., published in London in 1827. He says, p. 4:—"A singular species of coal has been found at Bathurst, resembling in some degree the Scotch cannel coal, . . . being nearly as light, and breaking with a similar fracture, while it burns almost with the steady brightness of a candle."

The following account of the rediscovery of the "kerosene shale" has been extracted from MSS. of the late Rev. W. B. Clarke, and placed at my disposal by Mr. C. S. Wilkinson:—

"It has been known for many years that in the vicinity of the Great Western thoroughfare, and very near to the now progressing line of railway from Sydney to Bathurst, at the base of Mount York, there exists a bed of coal, which is peculiar in its character and exceedingly inflammable. This property was discovered by the persons occupying the farm on Reedy Creek, who occasionally used it for fuel.

"In 1845 Count Strzelecki mentioned it in his 'Physical Description of New South Wales,' p. 129, as consisting of 'partial outcrops of coal observed in a small valley called the Reedy Valley (the Vale of Clwydd), north of Mount York and east of Mount Clarence, and which seemingly belong to the Newcastle basin—a probability, however, rather invalidated by the fact of the coals overlying masses of pure bitumen—a circumstance not discovered to exist elsewhere.'

"In 1847 the existence of coal in this position, as ascertained independently by myself in 1841, was introduced by me to the notice of the Legislative Council; * and in 1861, in a paper drawn up at the express desire of my brother commissioner of the International Exhibition, which was printed in the Sydney Catalogue, I further mentioned that the Reedy Creek coal, and a similar mineral at Colley Creek, on the north side of the Liverpool Range, would be likely to produce rock oil, and the specific gravity was alluded to as bringing it under close agreement with the Boghead coal of Scotland, viz., 1·204. It was stated that it was highly conchoidal in fracture, and lies in masses from six to twelve inches thick.† In reprinting the Catalogue in London the editor, without my authority, chose to strike out the paper in question, which was intended to afford information as to the extent of the coal-fields in New South Wales, and put the title at the head of a paper by another contributor, whose own appropriate description of his account was coal and collieries. Although this undeserved act had the effect of keeping out of view of the English reader the notice of the Reedy Creek cannel as oil-bearing, it did not prevent the turning of it to account by colonial manufacturers, and in 1865 a sample of the oil distilled from it was brought to me. Other persons had formed favourable opinions of its qualities, and amongst the rest Mr. Watt, an accomplished chemist, for a time attached to the University of Sydney, brought it still further under notice.

" As the circumstances connected with what may probably become a source of colonial wealth are of some historical interest in relation to the geology of New South Wales, the above-mentioned facts have been related.

" Oil-bearing products have already been found in the *third* and *fifth* divisions; of these, black cannel occurs in the latter at Stony Creek, near Maitland, on the Hunter River; brown cannel in the former at Reedy Creek; and shaly cannel on American Creek, at Illawarra; in various creeks running into the Wollondilly and Nattai Rivers, in the Grose River, in the Burralow Creek, a feeder of the latter, and in the Colo River. The Colley Creek cannel, which approxi-

---

* Report from the Select Committee on Coal, &c., ordered by the Council to be printed, 16th September 1847.

† " The Coal-fields of New South Wales," communicated by the Rev. W. B. Clarke; " Catalogue of Natural and Industrial Products of New South Wales, Sydney, 1861," p. 86.

mates to that of Reedy Creek, I believe will also be found to belong to the upper coal-measures."

Localities.—It is found in co. Brisbane, at Colley Creek; co. Camden, at Berrima, Broughton Creek, Cambewarra Ranges, Joadja Creek, Meryla, Mount Kembla, Saddleback, and Toonalli River, Burragorang; co. Clive, with graphite near Tenterfield; co. Cook, at Bathgate, Blackheath, Hartley, Katoomba, Milalong, on the Cox River, Mount Megalon, Mount York, and near Wallerawang; co. Gloucester, at Port Stephens; co. Hunter, at Capertee, and near Colo, on the Nattai River; co. Nandewar, near Narrabri; co. Northumberland, at Lake Macquarie and Greta; co. Phillip, at Gulgong; co. Roxburgh, at Rylstone; co. St. Vincent, at Mount Buddawang and Nerriga; co. Wynyard, at Stony Creek.

"KEROSENE SHALES" COMPARED WITH OTHER HYDROCARBONS.

Locality.	Moisture.	Volatile Hydrocarbons.	Fixed Carbon.	Ash.	Sulphur.	Specific Gravity.	Analyst.
1. Capertee . .	0·75	67·09	8·66	23·50	...	...	Gov. Analyst.
2. ,, . .	0·15	81·91	5·04	12·90	...	1·060	,,
3. ,, (near) .	1·04	72·56	18·40	8·00	...	...	,,
4. ,, ,, .	0·75	68·25	20·48	10·52	...	...	,,
5. ,, ,, .	0·20	70·48	12·92	16·40	...	1·120	,,
6. Greta Mine . .	0·48	61·18	25·13	13·21	...	...	Liversidge.
7. ,, . .	1·475	53·798	27·946	15·870	0·911	1·130	,,
8. Hartley Vale . · .	...	82·50	6·50	11·0	...	...	B. Silliman.
9. ,, . .	...	82·24	4·97	12·79	...	1·052	Liversidge.
10. ,, . .	0·55	64·62	8·71	26·12	...	1·238	,,
11. Joadja Creek . .	0·44	83·861	8·035	7·075	0·589	1·054	,,
12. ,, . .	0·04	82·123	7·160	10·340	0·337	1·229	,,
13. ,, . .	0·41	77·07	12·13	10·27	0·12	1·098	Dixon.
14. ,, . .	1·16	73·364	15·765	9·175	0·536	1·103	Liversidge.
15. Katoomba . .	...	85·35	9·80	4·85	...	1·017	Gov. Analyst.
16. Mudgee (near) . .	0·38	63·37	9·51	26·74	...	1·152	,,
17. ,, ,, .	2·96	53·70	25·04	18·30	...	1·270	,,
18. Murrurundi . .	1·165	71·882	6·467	19·936	0·549	...	Liversidge.
19. Shoalhaven . .	0·02	44·98	13·20	41·80	...	1·412	Gov. Analyst.
20. Wolgan River, Wallerawang . .	0·11	63·53	10·30	26·06	...	...	,,
21. Albertite, New Brunswick . .	...	57·490	42·086	0·424	...	1·105	Liversidge.
22. Cannel, Mold, Flints.	...	72·08	21·91	6·01	...	...	Percy.
23. ,, Scotland .	...	69·77	10·45	19·78	...	...	,,
24. ,, Wigan .	1·464	45·900	45·519	7·117	...	1·259	Liversidge.
25. Torbanite, Torbane Hill . · .	...	71·17	7·65	21·18	...	1·170	How.
26. ,, ,, .	0·720	69·695	9·045	20·540	...	1·316	Liversidge.

ULTIMATE ANALYSES OF "KEROSENE SHALES."

Locality.	Moisture at 100° C.	Carbon.	Hydrogen.	Oxygen.	Nitrogen.	Sulphur.	Ash.	Analyst.
7. Greta Mine . . .	...	65·610	7·507	9·851		·924	16·108	Liversidge.
9. Hartley Vale . . .	...	69·484	11·370	6·356			12·790	„
18. Murrurundi . . .	...	66·788	9·712	2·774		·555	20·171	„
Hydrocarbon, Waratah Mine	3·600	70·246	5·080	17·630	...	2·380	1·064	„

ANALYSES OF ASH OF "KEROSENE SHALES."

Locality.	Silica.	Alumina.	Iron Peroxide.	Lime.	Magnesia.	Potash.	Soda.	Phosphoric Acid.	Undetermined and Loss.	Total.	Analyst.
7. Greta Mine .	29·643	64·397	3·050	1·438	·250	·748	·355	·744	...	100·625	Liversidge.
13. Joadja Creek	77·12	20·14	·76	·30	·45	...	...	·65	·58	100·00	Dixon.

Nos. 1 and 2. CAPERTEE.

Nos. 3, 4, and 5. Near CAPERTEE.

Nos. 6 and 7. GRETA MINE, Newcastle District.—From the lower coal-measures.

The "kerosene shale" from this mine contains small specks of white clay.

No. 8. HARTLEY.—A specimen from Hartley was examined by Professor Silliman, and described as a new mineral, "Wollongongite" (*American Journal of Science and Art*, II., xlviii. p. 85), under the erroneous impression that the mineral came from Wollongong.

The Wollongong shale possesses none of the characters of a mineral, but only those belonging to carbonaceous shales.

No. 9. From the central part of a section taken from the Hartley seam, which is some five feet thick, where it is most free from mineral matter. Exhibited at the Agricultural Society's Show, 1873.

This block is in the Sydney University collection.

No. 10. This specimen of "kerosene shale" is said to have been found in New Caledonia; the physical and chemical properties are the same as that from Hartley; hence I have no doubt the specimen had been taken from New South Wales, and found its way back to Sydney as a New Caledonian product.

Ash.—White, with faint pink tinge. Does not yield a coke, only a black powder.

Nos. 11, 12, 13, and 14. JOADJA CREEK MINE.—Black, with a brownish shade; breaks with a large and well-marked conchoidal fracture. The ash is white or of a grey colour, with a slight reddish tinge. The coke of No. 13 was bright and lustrous.

On distillation this shale yielded 15·399 cubic feet of purified gas per ton of shale.

Illuminating power, 46·35 standard sperm candles.

Hydrocarbons condensible by bromine, 24·05 per cent.

Sulphur in coal, 0·49 per cent.

Tar per ton of shale, 40 gallons.

Liquor per ton of shale, 24 gallons.

The illuminating power of the gas ranges from 38·46 to 48·32 sperm candles.

Specific gravity, 1·060.

The above particulars are taken from the Catalogue of the Sydney Exhibition, 1879.

" Kerosene shale " of good quality yields about 150 gallons of oil per ton, which contains over 60 per cent. of refined oil, the remainder consisting of gasoline, benzine, lubricating oil, wood-preserving oil, paraffin wax, &c. If used for gas it should yield some 18,000 cubic feet of 38 or 40 candle-power gas; on this account it is very useful for mixing with other coals for gasmaking.

No. 15. KATOOMBA.

Nos. 16 and 17. Near MUDGEE.

No. 18. MURRURUNDI.—A specimen from this locality was of a dark grey, almost black colour, but spotted with small specks of a white clay-like substance.

No. 19. SHOALHAVEN.

No. 20. WOLGAN RIVER, Wallerawang.

The following analyses of torbanite, cannel coal, and albertite, were made to see how they compare in composition with the New South Wales kerosene shale :—

No. 21. ALBERTITE (New Brunswick).—Intensely black, highly lustrous, with well-marked conchoidal fracture. The ash is of a very pale brown colour. The coke is highly lustrous, much swollen, hollow like a bladder, with smooth outward surface.

No. 22. CANNEL (Mold, Flintshire).

No. 23. CANNEL (Scotland).

No. 24. CANNEL (Wigan).—Black, well-marked conchoidal fracture, shining streak and black powder. A bright lustrous coke is found, somewhat cauliflower-like in form.

No. 25. TORBANITE (Torbane Hill, Edinburgh).—Black-brown colour, light brown streak, flat conchoidal fracture. Scattered over with minute glistening particles. Does not form a coke; a black powder only is left.

Albertite from Victoria and Nova Scotia.

	I.	II.
Carbon	84·13	82·67
Hydrogen	9·23	9·14
Oxygen and nitrogen	6·64	8·19
	100·00	100·00
Coke per cent.	25·35	

I. From Coal Creek, Cape Paterson, Victoria. The mineral resembles the albertite of Nova Scotia, but is softer and less bright. It exhibits a conchoidal fracture. The analysis of this mineral was made by C. Tookey, in my laboratory, in 1865.

II. This is the beautiful mineral from Nova Scotia, of which magnificent specimens were shown in the International Exhibition in London in 1862. It is compact, homogeneous, black, very lustrous, conchoidal in fracture, and pitch-like. The analysis was made by Dr. Wetherill. I have a specimen from Rosshire, which exactly resembles in appearance the albertite of Nova Scotia.—*Percy's "Metallurgy,"* 1875, vol. i. p. 331.

The following table was prepared by Professor Chandler, of Columbia College, New York, to compare the Hartley mineral with Grahamite and Albertite, both of which are used for enriching gas :—

Mineral, &c.	Volatile Matter.	Fixed Carbon.	Ash.	Gas per Ton of 2240 lbs., in Cubic Feet.	Candle-power of Gas.	Coke, per Ton of 2240 lbs.	
						Lbs.	Bushels.
Grahamite, West Va.	53·50	41·50	2·00	15·000	28·70	1,056	44
Albertite, Nova Scotia	57·70	41·90	0·40	14·784	49·55	806	16
Hartley mineral, New South Wales . .	82·50	6·50	11·00	13·716	131·00	424	...

RETURN showing the Quantity and Value of Shale produced in the Colony of New South Wales.

Year.	Quantity. Tons.	Value. £
1865	570	2,350
1866	2,770	8,154
1867	4,079	15,249
1868	16,952	48,816
1869	7,500	18,750
1870	8,580	27,570
1871	14,700	34,050
1872	11,040	28,700
1873	17,850	50,475
1874	12,100	27,300
1875	6,197	15,500
1876	15,998	47,994
1877	18,963	46,524
1878	24,371	57,211
1879	32,519	66,930
1880	19,201	44,725
1881	27,894	40,748
1882	48,065	84,114
1883	49,250	90,861
1884	31,618	72,176
1885	27,462	67,239
1886	43,563	99,976
Total	441,242	995,412

HYDROCARBON.—Waratah Mine.

Amongst the specimens in the University collection is a piece of grey-coloured shale, containing a curious more or less rectangular pipe-like perforation filled with a carbonaceous mineral.

There is no history to this specimen, but it is labelled, "Over the Waratah Seam;" hence it doubtless came from the colliery of that name.

The mineral is jet black, highly lustrous, very brittle, breaking into long more or less regular four-sided prismatic pieces. These prisms run at right angles to two of the walls of the pipe. The cross fracture is conchoidal—the powder or streak is black.

The powdered mineral is insoluble in alcohol, carbon disulphide, benzol, ether, ammonia, caustic soda, and sodium hyposulphite, but it is partly soluble in boiling nitric acid, yielding a brown solution. Readily inflammable, does not fuse, burns with a smoky luminous flame and disagreeable smell. On platinum foil swells up but slightly. Hardness about 2.

Specific gravity, 1·30.

Proximate Analysis.

Loss at 100° C.	3·600
Volatile hydrocarbons, &c.	29·174
Fixed carbon	63·772 } Coke, 64·836 per cent.
Ash	1·064
Sulphur	2·380

99·990

Ultimate Analysis.

Moisture at 100° C.	3·600
Carbon	70·246
Hydrogen	5·080
Oxygen	17·630
Sulphur	2·380
Ash	1·064
	100·000

The ash is of a rich brown colour, light and spongy, No true coke is formed; the residue frits together and swells up slightly.

It does not quite agree with any described mineral, but on the whole it seems to resemble albertite more closely than any other. The composition does not yield a satisfactory formula. It is perhaps unnecessary to make a new mineral species of this substance.

JET.

A true jet, which takes a high polish and breaks with a conchoidal fracture, occurs as occasional layers in the "shale" at Hartley; Joadja Creek; Dubbo, and other places; but up to the present time no seams exceeding one-third of an inch in thickness have been found.

LIGNITE.—Brown Coal.

Chem. comp.: Carbon, hydrogen, oxygen, and ash. This substance may be looked upon as an imperfect coal, being intermediate in composition between wood and coal. In some cases it still retains the original fibrous woody structure; in other cases it is shaly or massive.

Found at Kiandra, where there is said to be a bed of lignite 30 feet in thickness. Brown, but black in parts, with a pitchy lustre; fracture subconchoidal; exhibits woody structure. On the Lachlan River, where it possesses a platy structure; Dubbo; found also on Mr. Berry's land, at the mouth of the Shoalhaven, at a depth of 12 feet; also at Turalla Creek, co. Argyle, where it retains original structure of the wood, and has much the same appearance as "bog oak."

At Chonta, between Tura and Boonda, about forty-two miles north of Cape Howe, there are beds of lignite, charged with iron pyrites, in association with kaolin; the clay containing the lignite is said to yield a fair proportion of lubricating oil. A so-called lignite occurs at Bowinda Cliff. Lignite is also found at Lake Macquarie.

In preparing the foundation for the bridge over the Parramatta River some wood was found at a depth of 44 feet, passing into the state of lignite. The colour was very dark, being almost equal to that of bog oak.

The air-dried specimen sank when immersed in water, being somewhat denser.

Proximate Analysis.

Moisture at 100° C.	20·82
Combustible matter	68·97
Ash	10·21
	100·00

The ash contains iron, alumina, lime, baryta, magnesia, potash, and soda, in combination with silica, sulphuric, hydrochloric, and phosphoric acids.

A specimen of lignite from Merimbula had the following composition :—

Hygroscopic moisture	19·40
Volatile hydrocarbons, &c.	30·49
Fixed carbon	32·84
Ash	10·33
Sulphur	6·94
	100·00

(Government Analyst.)

RESINITE.

Reported to occur on the Clarence River, and on the Macquarie Plains.

BOG BUTTER.

A soft, white, somewhat unctuous substance, like fat, only less greasy ; inclined to crumble to pieces when pressed. Probably a form of *adipocere.*

Found between Twofold Bay and Brogo.

Ultimate Analysis. Dried at 100° C.

Carbon	80·648
Hydrogen	5·618
Nitrogen	5·461
Oxygen	1·553
Ash	6·720
	100·000

The above results do not afford a satisfactory formula.

MINERAL WAX.—Ozokerite.

Chem. comp.: Carbon, hydrogen, and oxygen. Of a brown-grey colour. Breaks with a subconchoidal fracture. Coola.

BITUMEN.

Mr. Bolding, Commissioner for Crown Lands, informs me that bitumen oozes out of a sandstone rock at a place some fifteen or twenty miles from Coonanbarabran, on a creek which flows into the Castlereagh River.

ELATERITE.—Elastic Bitumen.

Chem. comp.: Carbon, hydrogen, and oxygen. At Reedy Creek or Petrolia there is said to be a band of thin and very elastic substance like elaterite.

CLASS II.

SULPHUR.

NATIVE SULPHUR.

Occurs in small quantities as a sublimate from the vents of Mount Wingen, the so-called "Burning Mountain," in association with iron sulphate and various other salts.

Also found in minute crystals in the cavities of auriferous quartz-veins on the Louisa Creek, co. Wellington, in association with hæmatite and zinc blende.

It is said to occur at Tarcutta, co. Wynyard.

CLASS III.

SALTS.

COMMON SALT.

Chem. comp.: Sodium chloride, NaCl. Common in most spring waters; occasionally found as an incrustation from the evaporation of lakes and water-holes. Found in rock crevices near Picton.

NATRON.

Chem. comp.: Hydrated sodium carbonate, $Na_2CO_3,10H_2O$. Said to occur as a deposit from the Mud Wells in the Namoi Scrub.

EPSOMITE.

Chem. comp.: Hydrated magnesium sulphate, $MgSO_4,7H_2O$. Occurs as an efflorescence in the caves and under overhanging rocks of the Hawkesbury sandstone; usually met with as masses of fibrous crystals, sometimes five or six inches in length, of a beautiful white, silky lustre. The crystals are usually curved at the free end; also in radiate groups of small crystals. Very fine specimens have been obtained from Dabee and Mudgee, co. Phillip; Wallerawang, co. Cook; also occurs at the Great Western Mines, Icely; Burragorang; and Pye's Creek.

With feather alum in caves in the coal-measures at Cullen Bullen; and the Turon District, co. Roxburgh; and Manero.

ALUNOGEN.—Halotrichite, Feather Alum.   Sulphate of Alumina.

Chem. comp.: Hydrated aluminium sulphate, $Al_2O_3,3SO_3$, $18H_2O$. Commonly called "alum," from its astringent taste, but potassium sulphate is usually present in but small quantity.

Commonly met with as an efflorescence in caves and under sheltered ledges of the coal-measure sandstone, usually with epsomite, as at Dabee, co. Phillip; Wallerawang and Mudgee Road, co. Cook; the mouth of the Shoalhaven River, and other places.   Also found in the crevices of a blue slate at Alum Creek, and at the Gibraltar Rock, co. Argyle.   Occurs as a deposit, with various other salts, from the vents at Mount Wingen, co. Brisbane, together with native sulphur in small quantities; and at Appin, Bulli, and Pitt Water, co. Cumberland.   At Cullen Bullen, in the Turon District, co. Roxburgh; at Tarcutta, co. Wynyard; Manero; Wingello Siding; and Capertee.

A specimen in the form of fibrous masses, made up of long, acicular crystals, white, silky lustre, like satin spar, found as an efflorescence in a sandstone cave near Wallerawang, was found to have the following composition :—

Analysis.

Water	47·585
Matter insoluble in water	1·079
Alumina	15·198
Sulphuric acid	34·635
Soda	·931
Potash	·337
Loss	·235
	100·000

The formula for the above is practically $Al_2O_3 3SO_3 + 18H_2O$.

Another specimen from the same place was found to contain a notable quantity of magnesium sulphate.

Analysis.

Water, by difference	47·388
Silica	1·908
Alumina	13·113
Sulphuric acid	33·067
Lime.	·798
Magnesia	3·726
	100·000

The formula for the above is also practically $Al_2O_3 3SO_2 + 18H_2O$.

Mr. W. A Dixon has also examined a specimen of this halotrichite as follows :—

" A yellowish-white porous mass, containing numerous tufts and masses of acicular crystals (hair salts), from Bungonia, gave on analysis:—

Analysis.

Sulphuric oxide	23·74	
Sulphurous ,, ...	traces	
Alumina	11·65	
Ferrous oxide	1·10	Soluble in water.
Magnesia	·99	
Potash	1·36	
Soda	traces	
Ferric oxide	1·91	Soluble in acid.
Magnesia.	traces	
Silica	32·25	
Water	27·12	

100·12

" It is somewhat difficult to state the proximate constitutents of this substance, as there is not enough sulphuric acid present to form normal salts, nor enough water to yield with the sulphate of alumina the usual crystalline salt. The probable contents are :—

Sulphate of alumina and potash (alum)	10·61
,, ,, magnesia (Epsom salts)	5·09
,, ,, iron (copperas)	3·58
,, ,, alumina ($Al_2O_3,3SO_3$)	23·06
Basic sulphate of alumina ($3Al_2O_3,SO_3$)	3·50 "

WEBSTERITE.

Chem. comp. : Aluminium sulphate, $Al_2SO_6 + 9H_2O$. Reported to occur on Brush Creek, Dumaresq River, co. Arrawatta.

ALUNITE.—Alumstone.

Large deposits of this mineral occur at Bulladelah, which are about to be worked. In composition it is a double sulphate of aluminium and potassium containing silica. It is used for making the best quality of alum, free from iron salts.

CLASS IV.

EARTHY MINERALS.

CALCITE.—Iceland Spar, Marble, and Limestone.

Chem. comp. : Calcium carbonate, $CaCO_3$. Hexagonal system. Sometimes well-developed crystals are met with. The usual forms are rhombohedra and their combinations, also combined with the terminal pinakoid or O.P. plane, and occasionally scalenohedra. I have not as yet observed the prism among the New South Wales forms.

The localities for calcite are extremely numerous, as it is not only met with wherever limestone occurs, but it is also a common substance in mineral veins.

Iceland spar occurs in small crystals near Dubbo.

Large well-developed flat rhombohedral crystals of calcite occur, associated with quartz, in the joints and cavities which exist in the basalt of the Pennant Hills, near Parramatta; at Gunnedah and Manilla. It is also met with in the quartz-veins in association with and as the matrix of gold, as at Gulgong and other places. It is sometimes present in the joints of the Hawkesbury sandstone, as at the Cataract River.

Calcite occurs in serpentine at Dungog and Orange; at North Point, Lord Howe Island.

It is found with metallic antimony in the North Redfern Gold-mine, Lucknow; in this district it is also found as the vein-stuff for gold (see Gold). Gold is also found in calcite near the falls on the Bowman River, and at Ti Tree Creek, Oaky, Barraba.

Opaque white calcite occurs at Capertee, co. Hunter; in serpentine at Jones' Creek, near Gundagai. Impure calcite in radiate groups of opaque white crystals occurs at Dunlop, Darling River. Good specimens have been obtained from Carwell. Crystals of black calcite have been found at Dayspring, Parkes, and Wollongong.

Marble.—Several beds of very fine marble or crystalline limestone occur in different parts of the colony, as at Wollondilly, whence one of the marbles used in paving the great hall of the University, the Post-Office, and other public buildings in Sydney has been obtained. Much of the Wollondilly so-called "white marble" is of a creamy tint, variegated with pale red and light blue streaks. A slate-coloured marble, used in the same buildings, is brought from Marulan, near Goulburn. There is a beautiful white saccharoid marble at Cow Flat, near Bathurst. A brecciated slate-coloured marble streaked with white calcite occurs at Wallerawang, co. Cook, under the following circumstances:—

"Between the iron ore deposits and the coal-seam outcrops there is seen an outcrop of limestone abutting against Devonian or Upper Silurian slates. Both the slates and the limestone are here standing at a high angle. The limestone does not show the dip so distinctly as the slates, for the lines of bedding have been almost completely obliterated, but the dip appears to be about 75° to the eastward, and the strike nearly N. and S. At the junction of the two the limestone has evidently undergone disturbance, and is much brecciated, and includes within it fragments of the slate. Some of the included slate contains small crystals of iron pyrites disseminated through it. In colour the limestone is of a bluish-grey or slate-colour, much veined with white calcite. The slate coloured portions break with a slight crystalline appearance, but the calcite veins show the rhombohedral cleavage of that mineral on a large scale. Its extension can be traced for a long distance to the north.—"Iron and Coal Deposits at Wallerawang," *Journal of the Royal Society of New South Wales,* 1874.

Beautiful marbles occur at Mudgee and Orange; also at Wellington, celebrated for its caves. At Bangalore, on the Goulburn Plains, there is found a white crystalline marble; at Yass and Queanbeyan, co. Murray; good grey and white crystalline marbles are found along the banks of the Murrumbidgee; the Belubula River and the Canomidine Creek, in the Orange District. Blue-grey limestone at Warialda, co. Burnett. The outcrops of small seams of grey crystalline limestone or marble are seen exposed in the Minumurra Creek, near Jamberoo, co. Camden, interbedded with the coal, shale, and sandstones of that district.

A specimen from a 2-inch band in the Minumurra Creek was slightly crystalline, of a grey colour, with a few thin streaks of a lighter colour. Small patches of a pale green mineral were detected in parts, something like glauconite in appearance.

It contained a considerable amount of impurity, and left a noticeable residue when decomposed with hydrochloric acid. Specific gravity, 2·679. See Analysis No. 4.

A jet black marble, traversed by veins of white calcite, occurs at Arnprior, Shoalhaven.

Variegated and white statuary marbles occur about four miles north of Parkes, co. Ashburnham.

The dark purple and red marbles from the Macleay and Tamworth Districts are very handsome when polished, and suitable for ornamental purposes.

In co. Roxburgh, at Mitchell's Creek; near Bathurst; in co. Argyle, at Marulan and Murrumbateman; at Bookham and Marsden, co. Harden; in co. Georgiana, at the Abercrombie Caves, and Rockley; at the Manning River, co. Macquarie; in co. Ashburnham, at Carrawabbity and near Forbes; at Port Stephens, co. Gloucester; at Tarrabandra, near Tumut, co. Wynyard, there is a richly variegated marble; Tarrago Creek, co. Argyle; Yass Plains, co. King; Havilah,near Mudgee, and Wellington, co. Wellington; at Wallabadah, co. Buckland.

A dark bluish-grey limestone, full of fossils, *Atrypa*, occurs in Windellama Creek, co. Argyle. See Analysis No. 10.

A white crystalline limestone is found at Wallerawang. See Analysis No. 8.

A grey crystalline limestone occurs at Wollongong. See Analysis No. 11.

A subcrystalline limestone occurs on the Wallerawang Reserve, containing fossils such as corals, encrinites, and other similar forms, which had weathered and become exposed on the surface. In colour almost white, mottled with pale grey, and further variegated by occasional brown streaks. Should polish well. See Analysis No. 9.

A subcrystalline limestone accurs at Tarrabandra, but rather more crystalline than that from Wallerawang Reserve. In colour almost white, possessing but a pale buff shade marked with bluish grey bands. It is probable that this marble would take a rather better polish than the former. See Analysis No. 6.

The limestone given under Analysis No. 5 is an argillaceous one, from the Myall River.

No. 7 is from Upper Muswell Creek, near Muswellbrook.

LIMESTONES.

Description.	No. 1. Bulli.	No. 2. Bulli.	No. 3. Bungonia.	No. 4. Minumurra Creek.	No. 5. Myall River.	No. 6. Tarrabandra.	No. 7. Upper Muswell Creek.	No. 8. Wallerawang.	No. 9. Wallerawang.	No. 10. Windellama Creek.	No. 11. Wollongong.
Hygroscopic moisture	}*3·95	*2·82	*1·30	0·73	}*2·33	...	1·10	...	...	0·071	}*3·71
Combined water				2·00		...	1·29	...	...	...	
Iron protoxide	...	...		3·52		...	...	...	...	...	
Iron peroxide	4·09	0·63	}+3·60	5·02	}10·21	1·750	7·65	0·75	}1·100	1·003	3·12
Alumina	5·06	1·02		0·46				traces			7·26
Silica	23·09	1·94	7·55	0·52 (sol.)	30·80	...	21·75	2·90	...	2·208	10·69
Magnesia	...	...	...	...	...	0·605	...	0·56	0·567	...	...
Carbonate of magnesia	0·36	1·32	0·85	...	2·04	...	1·66	...	...	...	0·82
Lime	0·84 (insol.)	...	...	‡38·27	...	54·600	...	53·42	54·096	54·602	...
Carbonic acid	...	...	...	35·70	...	42·898	...	42·23	42·704	42·369	...
Phosphoric oxide	...	...	traces	...	traces	...{	strong traces }	0·11	...	...	0·42
Sulphuric oxide	...	...	...	...	2·10	...	...	...	...	...	...
Silica and insoluble	...	...	...	13·08	...	0·160	...	...	...	0·720	...
Loss, undetermined, &c.	§0·17	§0·23	...	0·70	...	...	...	...	...	0·813	‖ ...
Carbonate of lime	62·44	92·04	86·70	...	52·52	...	66·55	...	...	...	74·28
Total	100·00	100·00	100·00	100·00	100·00	100·013	100·00	100·07	100·00	100·253	100·30
Specific gravity	...	...	...	2·679	...	...	...	...	...	...	...
Analyst	Dixon.	Dixon.	Gov. Analyst.	Liversidge.	Gov. Analyst.	Dixon.	Gov. Analyst.	Liversidge.	Dixon.	Liversidge.	Dixon.

In Nos. 1 and 11, 2·10 and 5·80 per cent. respectively of the alumina was insoluble.

* And organic matter.   † Traces of manganese protoxide.   ‡ Traces of strontia.
§ And alkalies.   ‖ Traces of potash, soda, and chlorine.

LIMESTONES.

Description.	No. 12. Capertee.	No. 13. Capertee.	No. 14. Single-ton.	No. 15. Single-ton.	No. 16. Singleton.	No. 17. Shoal-haven.
Hygroscopic moisture .	0·80 (with loss and alkalies)	0·15 (with loss and alkalies)	0·55	1·00	1·25	1·37
Combined water . .	...	...	...	...	...	...
Iron protoxide . .	...	...	...	...	...	...
Iron peroxide . .	0·50	3·40	1·72	3·00	2·33	2·28
Alumina . . .	...	trace	0·58	2·76	5·32	7·82
Silica and insoluble .	2·50	13·05	5·95	34·05	26·75	·37·74
Magnesia . . .	...	...	...	...	·93	...
Carbonate of magnesia .	trace	4·70	2·34	1·74		2·68
Lime	...	...	...	...	36·65	...
Carbonic acid . .	.:.	...	...	...	25·82	...
Phosphoric oxide . .	...	...	...	trace	...	...
Sulphuric oxide . .	...	...	...	...	...	...
Loss, undetermined, &c.	...	...	31·0	2·30	·95 (with combined water)	1·21
Carbonate of lime .	96·20	78·70	88·55	55·15	...	46·10
Total . .	100·00	100·00	100·00	100·00	100·00	99·20

(*Mines Report*, 1884.)

Nos. 15, 16, and 17 are argillaceous limestones.
No. 17 also contains ·80 of manganese protoxide.

Oolitic Limestones.—A limestone of this structure is said to occur on the Page River, co. Brisbane.

Concretions.—Calcareous concretions are common in the black and chocolate-coloured soils of igneous origin which occur in various ports of the colony, such as on the Liverpool Plains, New England, Gwydir District, Hunter River District, and at Scone, and in numerous other localities where there is a soil derived from the decomposition of a basaltic or other igneous rock.

Dana describes, in the "Geology of the United States Exploring Expedition Round the World," 1838–1842, some calcareous concretions of remarkable prismatic forms, occurring in clay at Glendon, probably pertaining to the sandstone rocks. Some of the crystals are twenty inches long, the average size being three or four inches. They have a rhombic form, and taper towards each extremity, the two ends curving slightly in opposite directions. Stars of four and six rays, and also globular masses, bristled on all sides with the ends of prisms, are common among them. They have a very rough brownish exterior,

L

like a fragment of sandstone; and within, instead of the regular cleavage structure of a proper crystal, the texture is crystalline granular. A surface of fracture glistens like a fine-grained statuary marble, though less bright. An attempt was made to burn them for lime, but they crumbled and so clogged the fire that it was abandoned.

"At one of the localities the specimens are coated with minute crystals of gypsum; they were probably formed through the decomposition of iron pyrites, this mineral giving rise to the sulphuric acid which united with the lime of the concretions. The rough surface of these rhombic concretions may have arisen from erosion by this process, or by the action of water percolating through the clay." Similar groups of crystals are found in the soil in various places along the course of the Darling River.

Thinolite.—Crystals composed of calcite, pseudomorphous after some unknown mineral, have been met with at Singleton.

ARAGONITE.

Chem. comp.: Calcium carbonate, $CaCO_3$.

Rhombic system. Good crystals of this form of carbonate of lime are perhaps more common than of the mineral calcite, especially upon stalactites in certain of the limestone caves and as enclosures within the amygdaloidal cavities of basalt.

Beautiful groups of crystals and bunches of *flos ferri* have been obtained from the limestone caves at Lob's Hole, the Coodradigbee, co. Cowley; the junction of Cotter's River and the Murrumbidgee, co. Murray; and from near Bungonia, co. Argyle. It also occurs at the Cataract River, and small specimens of stalactitic arragonite are to be seen at Port Hacking. The more or less spherical concretions termed "cave pearls" by Professor Boyd Dawkins, F.R.S., are also found in some of the above caves, notably those at the Coodradigbee, Bogenbung, Warrumbungle Mountains, and Uralla.

Aragonite occurs in vesicular basalt at Cherry-tree Hill, near Mudgee; groups of radiating crystals several inches in length are met with in a similar rock at Inverell; in serpentine on the Peel River, and on the Liverpool Plains; Jordan's Hill, Cudgegong, co. Wellington; at the Brick Kiln, Rock Flat, in radiate columnar crystals of variegated green and white colours; also in the basalt on Vegetable Creek.

Calcareous Tufa, Travertine, or Fresh-water Limestone.—At Burragorang, at Waibong, Picton, co. Camden; Quialago Creek, Goulburn Plains, and at Newstead Station, New England, co. Gough.

The fresh-water limestone at Newstead is of a greyish-white colour, and is, as shown by the following analysis, very impure.
Specific gravity = 2·69.

Analysis.

Moisture at 100° C.	·736	
Alumina	5·988	
Iron sesquioxide	1·760	
Manganese protoxide	·989	
Lime	10·571	Soluble in hydro-
Magnesia	·575	chloric acid.
Potash	·353	
Soda	·598	
Carbonic acid	8·450	
Silica	55·430	Insoluble in hydro-
Alumina	14·116	chloric acid.
Loss	·434	

100·000

A limestone is found in the neighbourhood of Mittagong which on the surface presents a peculiar pitted structure. The fractured surface shows an internal cup or cone-like arrangement, somewhat resembling certain coral remains; and in consequence of this, specimens have more than once been sent to me as fossils. The cups or cones fit closely into one another, and form vertical columns closely packed together. The limestone is very impure, and is probably stalagmitic in its origin.

FLUORSPAR.

Chem. comp.: Calcium fluoride, CaF_2. Crystallises in the cubical system.

Up to the present it has apparently only been found in the massive state, or in but very imperfect octahedral crystals. This mineral has been met with in several places in the New England District, near to Inverell, at Elsmore; at the Boundary, Sydney and Caledonian Tin-mines, on Cope's and Middle Creeks, co. Hardinge, where it is found in association with tinstone, a green steatitic clay, copper pyrites, galena, quartz, molybdenite, and other minerals, all of which may often be seen in one hand specimen.

It also occurs at South Wiseman's Creek, co. Westmoreland, in association with copper ores; on Mitchell's Creek, co. Roxburgh; in certain cases the fluor is much fissured, and the cracks are filled in with red oxide and blue carbonate of copper, which impart to the mineral a very pretty and ornamental appearance, and it would in consequence probably serve for inlaid work. At Woolgarloo Lead-mines and Silverdale it is found in the massive state as the matrix of galena; where it is usually opaque or but semi-translucent, white, with pale-bluish or purple veinings; the vein is 4 feet wide and also contains baryta.

Mr. Wilkinson reports its presence in the Devonian beds at Mount Lambie, co. Cook; also at Gow's Creek, near Wallerawang, where it occurs in small veins, traversing a felspathic rock. It has been found also on Vegetable Creek; at Deepwater and on the Grampians.

SELENITE.—Gypsum.

Chem. comp.: Hydrated calcium sulphate, $CaSO_4,2H_2O$. Rhombic system. Found crystallised in clay on the Darling River. Also on the Bogan River. Occurs near Singleton, and on Ash Island on the Hunter River, co. Northumberland; on the Cudgegong River, co. Phillip; Lake Cobham; the Grey Ranges, co. Evelyn; Bungonia, co. Argyle; at Cooma, co. Beresford; at Irrawang; and near Yass, co. King; near Tamworth; Mount Brown. Of commercial value for the manufacture of plaster-of-Paris and other cements.

Gypsum from Lake Tank, twenty miles N. of Bourke :—

Sulphate of lime	75·66
Water of crystallisation, hygroscopic moisture	20·20
Sulphate of magnesia	·61
Oxide of iron and alumina	·60
Silica	2·90
	99·97

Annual Report of the Department of Mines, 1886.

APATITE.

Chem. comp.: Chloro-phosphate of calcium, $3Ca_3P_2O_8,Ca(FCl)$. Crystallises in the hexagonal system, in the form of six-sided prisms. It is reported to occur in well-formed crystals with bitter spar on the Lachlan, between Boco Rock and Wog-Wog, and with graphite and quartz at the head of the Abercrombie River, co. Georgiana; also on the Clarence River.

This mineral is of considerable commercial value.

I have not yet met with any strontium minerals in New South Wales, nor do there appear to be any records of their discovery.

BARYTES.—Heavy Spar.

Chem. comp.: Barium sulphate, $BaSO_4$. Rhombic system. With fibrous and massive green carbonate of copper, copper pyrites, and galena, at Cambalong, Merinoo, co. Wellesley. Also with antimony ochre, near Kempsey, co. Dudley; with copper carbonates at Biben-luke, near Bombala, co. Wellesley; on the Euroka Creek with iron oxides; at Winterton Mine, Mitchell's Creek, co. Roxburgh; in more or less well-formed small tabular crystals associated with gold and other minerals; a vein of barytes twelve inches in width is said to exist at Croker's, on the Rocky Bridge Creek; also at Cobargo; near Carcoar, inclosing galena; Glanmire; Caloola Creek; Goulburn; near Germanton, and Albury, co. Goulburn; Barrier Range; with lead monoxide, Captain's Flat, Molonglo; and near Braidwood and Queanbeyan.

Dr. Hector, F.R.S., Director of the Geological Survey of New Zealand, reports having found a vein of barytes in 1877 on the Canobola Ranges, between the Lachlan and Belubula Rivers, near the junction of the Devonian limestone, with diorite schists intersected by porphyries and bands of serpentine. The barytes contains a little copper, and is associated with micaceous iron in lamellar crystals which are so thin as to be translucent.

BRUCITE.—Magnesium Hydrate.

Chem. comp.: MgO,H_2O or $MgH_2O_2 = MgO$ 69·0; water, 31·0 = 10·0

Crystallises in the hexagonal system in rhombohedral forms. Said to occur on Louisa Creek, co. Wellington.

HYDROTALCITE.

A hydrate of alumina and magnesia.

This is a soft white and pearly mineral, with a greasy feel. Said to occur in New South Wales.

MAGNESITE.

Chem. comp.: Magnesium carbonate, $MgCO_3$. Rhombohedral system. It is most commonly found massive, or in concretions, having a mammillated or botryoidal form.

$H = 4$ to 5. Specific gravity, 2·94.

It is found in New England in various places, and upon the diamond-fields at Bingera, co. Murchison; and near Mudgee. When impure it is of a grey or grey-brown colour, but when pure it is of a dazzling white; compact, tough, and breaks with a flat conchoidal fracture. It adheres to the tongue, and has a very cold feel like porcelain.

It effervesces with hydrochloric acid, but with difficulty.

At the diamond-diggings at Two-mile Flat, near Mudgee, pure white magnesite was observed to form by the spontaneous decomposition of the heaps of refuse from the miners' shafts; pebbles were quickly cemented together by it.

The late Dr. Thomson, of the Sydney University, found that the magnesite thus formed, and incrusting rubbish-heaps, timber, old tools, &c., had the following composition:—

Magnesia	46·99
Carbonic acid	49·78
Water	4·08
	100·85

Specific gravity = 2·94.

This magnesite sometimes contained calcite. It was also observed under the same circumstances on Cunningham's Diggings, on the east side of Cudgebegong Creek, and there with a peculiar vermicular or worm-like form.

Other localities are Kempsey; the Lachlan River, Mooby Gully; Scone, co. Brisbane; Louisa Creek and Lewis Ponds Creek, co. Wellington; Barraba, co. Darling; Tumut; Gulgong; and Warrell Creek, Nambuccra River.

Dolomite.—A double carbonate of lime and magnesia. Found at Carwell, Shoalhaven District, and Bowling Alley Point. On Jones' Creek, Gundagai, gold-bearing veins of dolomite were met with traversing the serpentine.

WAVELLITE.

Chem. comp.: A hydrated aluminium phosphate. A yellow mineral, reported to be wavellite, with a radiate structure, is found in the fissures of the felstone pebbles common in Rat's Castle Creek, Two-mile Flat, Mudgee.

SILICA.

QUARTZ.—Rock Crystal.

Chem. comp.: Silica. Hexagonal system. Found in nearly all parts of the colony, and in crystals more or less perfectly developed; the most common form is the prism combined with the pyramid. Occasionally the prisms are closed at both ends by planes of the pyramid; also as double pyramids; such crystals are, however, usually small, and generally occur in quartz porphyries, or are derived from the decomposition of such, found at Glenlyon, Home Rule, and Cooyal, co. Phillip; Solferino, co. Drake; and Peel River, co. Parry.

Occasionally some very large crystals are found, notably at New-stead Tin-mine, New England, where, in one of the shafts, crystals of nearly 1 cwt. were met with; within these, crystals of tinstone were often found disseminated.

Good rock crystal is said to occur at Cooyal and at Havilah.

Large crystals of smoky quartz are common almost throughout New England, as at Scrubby Gully; Bingera, co. Murchison; smoky-brown cairngorm, with limpid quartz crystals, are plentiful in Ranger's Valley, River Severn, and Inverell, co. Gough; Macintyre River, Middle Creek, and Byron's Plains, in the same county; and at Oban; Cope's Creek, co. Hardinge; Uralla, co. Sandon; Mudgee, co. Phillip. Some of the rock crystals found in the alluvial tin deposits present a very pretty appearance, from the presence of numerous minute fissures and internal films, streaks and patches of yellow, orange, and red colours. Most of the crystals from New England have one face of the pyramid much more largely developed, so much so in some cases as to almost obliterate the other faces.

Elongated pyramids containing disseminated crystal of cassiterite are common at the Albion Tin-mine; these crystals of quartz are dull and slightly rough on three of the faces, and bright on the opposite three.

White, colourless, and tinted quartz, pseudomorphous after calcite and other minerals, is abundant in some portions of the Yass District.

Quartz crystals, with rounded edges and dull surfaces, as if acted upon by hydrofluoric acid, occur in the coarse-grained granite on Mann's River. This is seen also in crystals from other places, as at Oban.

Quartz crystals are common near the junction of the Turon and Macquarie Rivers; at Bukkulla, co. Arrawatta, clear and brilliant crystals; the Diamond Mountain, Cudgegong, Macquarie River; in an amygdaloidal basalt, Deep Lead, Gulgong Rush, co. Phillip; at Carcoar, containing lamellar magnetite, also with a pale blue quartz. Well-developed and brilliant crystals from Bullamalite Creek, a tributary of the Mulwaree, near Goulburn, at Gurragangamore, and other places on the Goulburn Plains; the Lachlan River; at Cooma, co. Beresford; and Kiandra, co. Wallace; the Murrumbidgee; in the Naas Valley, co. Cowley, with tourmaline and schorl; between Pambula and Eden, with molybdenite.

Beautifully formed clear and transparent crystals of quartz occur on the Louisa Creek, co. Wellington. Also citrine, red, amethyst, and opaque white, remarkable in certain cases for the peculiar yellow and iridescent tarnish of many specimens. Peculiarly flattened forms are also found here, with four faces of the pyramid enormously developed; the remaining two being so much reduced, to a mere line almost, as to give the crystal the appearance of a rather acute rhombic pyramid.

Mr. D. A. Porter states that really good specimens or unusual forms of quartz can be obtained from the following places:—Oban, of all colours, Cope's Creek, Glen Elgin, Bowling Alley Point, and Hanging Rock, near Nundle; Garrawilla, Liverpool Plains, with stilbite and calcite; Puddledock, near Armidale, with stilbite; at Stannifer in rounded prismatic masses, sometimes five inches in diameter; at the Gulf, near Emmaville, where, in the Dutchman's Claim, very large groups are obtained.

Up to the present the number of substances which I have observed enclosed within quartz crystals found in this colony is not great, viz. :—

Endomorphs in Quartz Crystals.

1. Actinolite—Mowembah, Merrendee, on the Meroo, a tributary of the Cudgegong, co. Wellington.

2. Asbestos—Uralla, co. Sandon.

3. Cassiterite or tinstone—Albion and Newstead Mines, New England.

4. Epidote—Towamba and Manero, Morullan, on the Gwydir River.

5. Argentiferous Galena—New Summer's Hill, Bathurst.

6. Gold—Boro Creek, and other places, co. Murray; rough vein quartz is the commonest matrix of gold.

7. Graphite—Head of Abercrombie River, co. Georgiana.

8. Orthoclase felspar—Two-mile Flat, Mudgee, co. Phillip.

9. Molybdenite—Bullio Flat, near Goulburn, co. Argyle.

10. Rutile—Cope's Creek, co. Hardinge.

11. Schorl and tourmaline—Murrumbidgee.

Pseudomorphs, that is, quartz possessing the external form of other minerals. Quartz after calcite—Gulgong, Yass, and Bathurst; also after iron pyrites and mispickel.

Rose-Quartz.—Occurs with manganese on Hall's Creek, Moonbi Range; also at Stannifer, New England.

Amethyst.—A purple-coloured variety of quartz. It occurs as geodes in the basalt at Kiama. The crystals are usually small, not being more than three-eighths of an inch through. Found also at Dubbo. A quartz-vein containing amethystine quartz occurs near the top of Bullabalakit; also near Bathurst, near Twofold Bay, and Glen Elgin.

Agate.—Agates consist of mixtures of crystalline quartz and chalcedony, usually arranged in concentric layers and bands; their structure is caused by the peculiar mode of formation, viz., by the infiltration of silica into the amygdaloidal cavities of igneous rocks.

They are common in the basalt at Kiama, in a diorite, at Mittagong, co. Camden; near Scone, co. Brisbane; Inverell, co. Gough; and other places; and are very plentiful in the beds of many of the rivers and old drifts of New South Wales, as in the Macintyre, parts of the Gwydir, the Hunter, the Cookaboo, where they are derived from the basalt of the Western Range or Dewingbong Mountain, Gunningbland, Lake Cobham, Grove Creek, Trunkey, and Narrabri; also with opal on Breelong Creek; Cope's Creek, co. Hardinge; Nundle and Hanging Rock, co. Parry.

Agates are reported to occur at Mount Agate, near Mount Wingen, co. Brisbane, encrusted with native copper.

Agates and chalcedony are plentiful near Dubbo and Bald Hill, Wellington, Mount Wingen, Maitland, Cowriga, and other places.

Jasper.—Is very abundant and widely distributed throughout various parts of New South Wales. It is found of nearly all shades of colour—pure white, grey, slate, dull blue, olive and bright greens, brown, red, and black, both alone as simple colours and in varied combinations of stripes, streaks, and mottlings.

It is found mainly in the form of boulders and pebbles in river-beds, and it enters largely into the composition of nearly all conglomerates, gravelly alluvial deposits, and river-drifts. Much of it is evidently derived from the conglomerate of the coal-measures.

The peculiar variety known as Egyptian jasper does not appear to have yet been met with.

Amongst the principal localities are the Gwydir, the Macintyre, the Richmond, the Macquarie, Cudgegong, the Hunter, the Murrumbidgee, and many of their tributaries. There are large quantities of fine red jasper near Gobolion, Molong, co. Ashburnham; and at Scone, co. Brisbane. The drifts at Mudgee, co. Phillip; Bathurst; Bingera, co. Murchison; Lake George, co. Murray; Nundle, and with opal at Foley's Folly, co. Parry; Woolomon, and other places are rich in fine jasper specimens.

Ribbon Jasper.—At the junction of Pink's Creek with the Bell River a clay slate has been converted into ribbon jasper.

Eisenkiesel.—A variety of ferruginous quartz. Large masses of this mineral *in situ* occur near Bingera, co. Murchison; it also abounds between Guano Hill and the Bell River, at Carcoar, co. Bathurst; Mount Lindesay, Lowee, and at the junction of Cotter's River with the Murrumbidgee, co. Murray.

Lydian Stone.—A velvet black form of jasper, used by jewellers as a touchstone for gold alloys.   Mullion Range, Bathurst county.

Chert.—Common in seams and bands throughout the coal-measures. Its structure is often more or less lamellar, and the fracture conchoidal.

Abundant about Mount Victoria, Wallerawang, and Hartley, co. Cook; Jamberoo, co. Camden; Illawarra, and Lachlan River.

CHALCEDONY.

An amorphous or crypto-crystalline form of quartz.   There are several varieties of chalcedony.

Chalcedony proper.—Massive, translucent, pale-grey, blue, or brown; with waxy lustre, surface mammillated, and often of a stalactitic form.

Nodules of chalcedony are found near Carcoar, co. Bathurst; at Lowee, with resinite and chert; also at Gulgong, Home Rule, co. Phillip; Cowriga Creek, Wellington, Dubbo, Maitland, the Hunter River; and filling lines of small cavities in a green felstone on Rat's Castle Creek, six miles S.E. of Two-mile Flat, co. Phillip; Gunnedah, Newstead, Walcha; Monaltrie, on the Richmond River; at Nundle, bluish grey; also at Narrabri.

Found pseudomorphous after quartz at the Elsmore Mine, near Inverell.

Carnelian.—Is a bright red chalcedony, but the ornamental white varieties of chalcedony are also usually included under the same name by jewellers.

Red and white carnelians are rather common in the Hunter River, at Maitland, and other places; also near Wellington; in Pond Creek, near Inverell, co. Gough; also near Narrabri.   Mr. D. A. Porter states that veins of carnelian occur at Mount Misery, near Nundle.

Carnelian in quartz prophyry, on Nymboi River, Clarence River. Beautifully coloured carnelians also occur in the basaltic country about the Tweed River.

Onyx is said to occur in the neighbourhood of Narrabri.

Cat's-eye.—A variety of chalcedony, which, from the presence of capillary crystals of asbestos, shows a peculiar chatoyancy or glare when cut and polished *en cabochon.*

A polished specimen in the University collection from the western districts of New South Wales weighs 1·2636 grammes, and has a specific gravity of 2·6703 at 18·5° C.

The Oriental cat's-eye of Ceylon is a variety of chrysoberyl, and is distinguished by its much higher specific gravity, which is about 3·7 to 3·9.

OPAL.

This mineral consists of silica, with usually from 5 to 12 per cent. in water.

Precious or Noble Opal.—The precious opal of New South Wales has the milky body colour usually possessed by this mineral, and the same brilliant play of colours; the dominant colours of the scintillations are metallic green, pink, and red. Some of the best specimens form, when polished, very fine gem-stones; but here, as elsewhere, the valuable specimens obtained bear but a small proportion to the whole. The best have been obtained from Rocky Bridge Creek, Abercrombie River, co. Georgiana; the matrix is a fine-grained, bluish-grey, amygdaloidal trachyte, some thirty feet thick, which is so much altered that it can be abraded by the thumb-nail; the opal has filled by infiltration certain of the vesicular cavities and crevices in this rock; it is associated with much common opal free from any play of colour and hyalite.

Some cut and polished specimens of opal from Trunkey were found to have the following specific gravities:—

No. 1, weighing ·3610 gramme, had a specific gravity of 2·1488 at 17° C.
No. 2, „ ·1146 „ „ „ „ 2·1300 at 18° C.
No. 3, „ ·1860 „ „ „ „ 2·1703 at 17° C.

The appearance and mode of occurrence of the opal found at Bulla Creek, in Queensland, is very different; the body colour of the Queensland opal is usually deep ultramarine blue or green, and the reflections are usually metallic green and red; the matrix is in this case a brown mottled clay ironstone, in which the opal occurs as small veins and strings. This variety of opal occurs for the most part in films too thin to cut *en cabochon*, but it yields beautiful specimens when cut as cameos.

Opal is also found in a similar clay ironstone in the Wellington District; but up to the present I have only seen small particles of the precious opal diffused through much valueless opal; it also occurs at Louisa Creek, at Bland, near Forbes; at Coroo, with chalcedony, agates, &c.; and at Bloomfield, near Orange.

Fire Opal or *Girasol, i.e.,* an opal with a red or orange tint; occurs at Wellington. Of no value hitherto.

Specific gravity of one specimen, 2·106

Common Opal, Semi-opal, and *Wood Opal* are common in all the basaltic districts, usually of pale shades of pink, brown, green, and varying from translucent to opaque; Louisa Creek, Tambaroora, Lowee, Carwell; Uralla, co. Sandon; Inverell, co. Gough; Richmond, Hunter, Lachlan, and Castlereagh Rivers; Trunkey and Cowra, co. Bathurst; Kiama, co Camden; Hookanvil Creek, below Hanging Rock, and

Nundle, co. Parry; Home Rule and Gulgong, co. Phillip; Wellington, co. Wellington; at O'Connell, co. Westmoreland, there is a vein running through Silurian slates; Carcoar, Cowra, Cobar, Braidwood; on Lawson's Creek, a tributary of the Cudgegong River; also from Breelong Creek, Warrumbungle Mountains; Orange; Gundagai.

Cacholong.—The Tumut River, co. Selwyn. Of an opaque porcelain-white colour passing into white opal, with conchoidal fracture. Adheres strongly to the tongue. Hardness, 5–6. It was found to have the following composition:—

Specific gravity, 1·884.

Analysis.

Water lost at 195° C.	2·553
,, combined	5·185
Silica	88·811
Alumina and traces of iron sesquioxide	1·206
Lime	1·134
Carbonic acid	traces
Magnesia	·485
Loss	·626
	100·00

Hyalite.—Muller's glass.

Found coating the joints in basalt, Jordan's Hill, Cudgegong, co. Phillip; of a blue colour at Ororal. Also near Nundle and Bowling Alley Point, and Scrubby Gully, Vegetable Creek.

Siliceous Sinter.*—Most of the specimens of this material which I have had the opportunity to examine exhibit many of the appearances which are usually presented by the deposit thrown down from hot springs or geysers.

Although no such hot springs or geysers are known to exist at the present day in the colony, yet I understand from Mr. W. Wilson of Monaltrie, to whom I am indebted for my specimens, that the district in which they occur presents many features which lead him to consider that it had been the scene of comparatively recent (*i.e.*, in a geological sense) active volcanic phenomena.

The district has not, I believe, been examined in detail by any trained and experienced geologist, but, judging from Mr. Wilson's account, it must be one of remarkable interest.

Basaltic and trachytic rocks are the principal surface rocks occurring in the neighbourhood. The basalt is remarkable for containing very large and well-developed amygdaloids of chalcedony, agate, arágonite, and certain of the commoner zeolites. Of the amygdaloidal and other minerals, together with specimens of the matrices, Mr. Wilson sent a large series to the Commissioners for the Philadelphia Centennial

* "Fossiliferous Siliceous Deposit from the Richmond River, New South Wales." By A. Liversidge. *Journal of the Royal Society of New South Wales*, 1876, p. 237.

Exhibition, the collection of which must have entailed the expenditure of much time and labour.

In the interior of the mass the siliceous deposit is usually of a more or less pale wax colour, and in certain respects closely resembles *wood opal*. Wood opal is actually present, and in parts streaks of common opal occur. Occasionally, on breaking open a specimen, jet black patches are met with; the colouring matter apparently contains carbon, as it is slowly burnt off before the blowpipe flame, which shows that the whole of the original woody matter has not been removed.

On the surface the mineral weathers white, and the decomposition passes in to a depth of from ¼ to ½ inch.

It adheres strongly to the tongue.

WEATHERED PORTION.

Analysis.

Moisture, given off at 100°	4·16
Combined water (loss on ignition)	1·78
Insoluble silica	89·74
Soluble silica	·47
Alumina and iron sesquioxide	1·13
Lime	·48
Magnesia	1·98
Loss	·26
	100·00

Specific gravity, 2·046 at 66° Fahr.

UNWEATHERED PORTION.

Analysis.

Water, given off at 100°	4·08
Combined water (loss on ignition)	·58
Insoluble silica	91·67
Soluble silica	·30
Alumina and iron sesquioxide	1·56
Lime	·36
Magnesia	·55
Loss	·90
	100·00

Specific gravity, 2·330 at 66° Fahr.

The composition shows that it answers to the common siliceous sinters or geyser deposits.

It will be seen that the weathered specimen has a lower specific gravity and contains rather more water, also more lime and magnesia.

In places the structure is more or less distinctly lamellar, evidently due to the manner of its deposition in successive layers. The fracture is more or less distinctly conchoidal across the planes of deposition, but where the lamellar structure is less strongly marked, or altogether obliterated, the fracture is conchoidal in all directions.

The weathered surface is usually marked with the remains of ferns, which stand well out in relief; with the ferns and stems are the fruit and seeds of other forms of vegetable life.

Within the substance of the mass occasional layers of a brilliant white colour are met with, and along these layers it splits into flakes and slabs with ease; these white layers are much softer than the other portions, and they are found to be composed almost exclusively of the casts of vegetable tissue; the fern fronds and stems are especially well preserved; also scattered irregularly through these layers and the solid substance of the mineral the remains of certain fruits and seeds are met with, belonging to a new genus (*Liversidgea*, F. von Mueller).

Silicified Wood.—Is very abundant over nearly all the basaltic districts. Much of it has doubtless been derived from trees overwhelmed by old lava-flows. The remains of these trees have become silicified, and have since, by the disintegration and removal of the enveloping rock, been set free as "fossil wood."

The following note * upon a specimen of partially fossilised wood may help to show how this has been brought about :—

The specimen forming the subject of this note was found by Mr. C. S. Wilkinson, F.G.S., at Inverell, where the Macintyre River has cut through the basalt and formed a river-cliff; by the formation of this section the included fragments of wood and trunks of trees are exposed to view.

In the *Mines and Mineral Statistics,* published by the Mining Department in 1875, Mr. Wilkinson gives the following description of the manner in which the fossilised wood occurs, and on the same page (p. 76) he gives a diagram showing the position occupied by the particular tree-trunk from which this specimen was taken :—

"An interesting cliff-section of basalt may be seen on Mr. Colin Ross's property on the bank of the river at Inverell. The following is a sketch of it :—

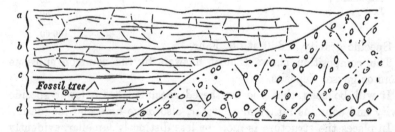

" *a, b,* amygdaloidal basalt, much decomposed; *c,* friable cellular basalt, enclosing fragments of wood and pieces of earth; *d,* dense

* " On the Composition of some Wood Enclosed in Basalt." By A. Liversidge. *Journal of the Royal Society of New South Wales,* 1880.

columnar basalt; *e*, volcanic breccia, composed of fragments of basalt of various sizes embedded in an indurated volcanic mud, much stained with peroxide of iron, which imparts to the rock varying shades of deep red and yellow. This breccia is older than the *a*, *b*, *c*, *d*, and evidently formed the side of a hill on which plants were growing at the time of the basalt eruption; for at the junction of the basalt and breccia lies a thin bed of red clay, the former surface soil, in which I discovered numerous stems of plants. Some of these stems are in an upright position, and even penetrate a few inches into the basalt rock above, and several I found with the woody matter but little altered. These facts are very singular, as proving the viscid state of the overflowing basaltic lava, to have thus surrounded the small plants without destroying them, and how rapidly it must have cooled. Another interesting relic of the Newer Pliocene period that this section reveals is the trunk of a tree about two feet in diameter, embedded in the layer of basalt marked *c* in the above sketch.

" The wood, though much changed, yet retains its fibrous structure most completely. It somewhat resembles the stringy-bark, and may possibly be a species of eucalyptus; but this is difficult to decide without the aid of the microscope.

" Surrounding the tree is a soft substance two inches thick, which was probably the bark."

As pointed out by Mr. Wilkinson, the woody structure has not been destroyed, and it is still visible to the unassisted eye, but with the aid of a microscope the structure of the cellular tissue is much more clearly seen; patches of white carbonate of lime and of yellow oxide of iron are also observed deposited within its substance.

The specimen seems to have been considerably crushed and broken; in general appearance it looks as if a number of angular fragments of charcoal had been pressed together. This brecciated structure was probably set up after the trunk was enveloped by the fluid lava, and was doubtless caused by the contraction of the rock round the wood, as it solidified and cooled.

When heated in a closed tube much water is given off; when ignited on platinum foil it does not inflame or glow like a carbonaceous substance, but quickly burns to a pale brownish-grey ash; the carbon, which has apparently been converted into graphite, is present in very small quantity, and barely sufficient to impart a black colour to the substance.

It effervesces with acids, is fragile, and sufficiently soft to be scratched with the thumb-nail.

Analysis.

Water lost at 100° C.	12·54
Combined water—by direct weighing	·46
Silica	35·57
,, ...	·31
Carbon	5·14
Iron sesquioxide	1·76
,, protoxide	3·67
Manganese	traces
Alumina	15·79
Lime = (16·42.CaCO₃)	9·20
Magnesia = (7·24.MgCO₃)	3·45
Potash	·22
Soda	·27
Sulphur	traces
Sulphuric acid	traces
Carbonic acid	11·29
	99·67

The lime and magnesia evidently exist as carbonates; a small quantity of the protoxide of iron may also exist in combination with carbonic acid, as there is ·28 per cent. of carbonic acid left after converting all the lime and magnesia into carbonates. The alumina and iron probably exist in the form of silicate, as the amount of silica is nearly sufficient to form a silicate of the formula $R_2O_3,3SiO_2$, or if the water also be taken into account, $R_2O_3 3SiO_2 + 4H_2O$.

As it contains traces of sulphur and of sulphuric acid, small quantities of iron pyrites are probably present.

The combined water was determined by heating the powdered substance in a combustion tube and collecting the water in a weighed chloride of calcium tube, and the carbon by combustion with lead chromate in a current of oxygen, the silica by fusion with the mixed alkaline carbonates, and the alkalies by Dr. J. Lawrence Smith's process with calcium carbonate and ammonium chloride.

Masses of silicified wood are very common in nearly all basaltic areas over all parts of the world, and they are very noticeable in many parts of this colony. This particular specimen is different from the above, inasmuch as, instead of being composed almost exclusively of silica or of hydrated silica, as is the case with ordinary silicified wood, it has been mineralised by a mixture of various substances.

On account of the mineralised wood having such a complex constitution, it may be thought that it may have been merely replaced mechanically—*i.e.*, it might be supposed that the wood has been burnt or rotted away and the mould left by it filled in with earth and charcoal, but such is not the case. There is no doubt that the mineral matter has been deposited from solution; the woody tissue, which was doubtless much charred, has been almost completely replaced, particle by particle, by the deposition of mineral matters from infiltered water holding them in solution. This process must have been a very slow one; the cavities of the cells were probably filled first, the cell walls were next gradually removed, except those portions represented by the small remaining quantity of graphite-like carbon, and replaced by mineral matter as the decay went on, but so slowly and quietly that no violence was done to the microscopic structure of the woody tissue.

Silicified wood is very abundant, also, throughout the coal-measures in New South Wales. Large boulders of such fossilised wood are met with in most of the drifts and river deposits.

Fossil Forests.—I am inclined to think that some of the so-called fossil forests found in different parts of the world, as at Lake Macquarie, Antigua, in California and Egypt, have been formed in much the same way as the specimen just described, *i.e.*, they are derived from forest trees which have been overwhelmed by a flow of lava or basalt, and silicification has taken place as in the instance at Inverell, the basalt or lava has undergone decomposition and removal, and the much less perishable silicified trees or portions of trees have been uncovered and exposed as we now see them. This, of course, is not the only way in which trees have been fossilised, and I do not wish to apply this explanation to the occurrence of fossil trees in the coal-measures and other sedimentary rocks.

Tripoli or Infusorial Earth.—Abundant in several places in the colony, notably at Barraba, where it is made up almost entirely of the remains of diatoms resembling *melosira*.

A specimen of tripoli, supposed to be meerschaum, obtained about forty miles from Tamworth, gave Mr. Dixon the following results:—

Analysis.

Water	12·84
Silica	80·56
Alumina	4·15
Oxide of iron	1·77
Carbonate of calcium	·31
„ magnesium	·21
Alkaline salts, and loss	·16
	100·00

CLASS V.

ANHYDROUS SILICATES.

WOLLASTONITE.—Tabular Spar.

Chem. comp.: A silicate of lime. $CaSiO_3$. Oblique system. Found at Duckmaloi with garnets and epidote.

M

CHRYSOLITE.—Peridot, Olivine.

Chem. comp.: Magnesium silicate. Rhombic system. Transparent bright-green-coloured specimens of chrysolite are common in many of the gold-drifts. Found in the Shoalhaven and Hunter Rivers; Louisa Creek; Old Trigomon. Associated with the various gems in Great Mullen Creek, which falls into the Cudgegong, co. Phillip; also at Two-mile Flat; Bingera, co. Murchison; the Barrier Range; Nundle; in a trap dyke on the Upper Murray, about seventy miles above Albury; and other places. The exterior often has a white opaque enamel-like crust. At Inverell in an amygdaloidal basalt with zeolites.

AUGITE.—Pyroxene.

Chem. comp.: A silicate of magnesia, iron, lime, &c. Oblique system. There are several varieties of augite, which range from white, or almost white, to dark green, black, and opaque minerals.

Well-formed short columnar crystals of augite are not uncommon. They are abundant at Cameron's Creek, co. Hardinge; and Newstead and Middle Creeks, co. Gough; near Guntawang, co. Phillip; Pretty Plains, near Molong; and near to the Pigeon House. At Bruno Waterfall, Callalia Creek, with mesotype and aragonite in a vesicular and amygdaloidal basalt, which rests upon columnar basalt. Found at Barraba, co. Darling; Murrurundi, co. Brisbane; and Lucknow, co. Bathurst.

A specimen from Oberon, of a green colour, more or less decomposed, only traces of the previous crystallisation are left. Soft and fragile. Collected by Mr. C. S. Wilkinson, F.G.S. Was found to have the following composition:—

Analysis.

Water lost at 100° C.	·210
Silica	35·319
Alumina	5·922
Iron sesquioxide	28·557
„ protoxide	1·809
Manganese protoxide	4·056
Lime	22·751
Magnesia	absent
Potash	·378
Soda	·221
Loss and undetermined	·777
	100·000

Specific gravity, 3·48.

Chondrodite.—Said to occur at Gulgong, co. Phillip.

Specific gravity, 3·349.

Smaragdite, containing native copper, occurs in a hard elvan porphyry at Molong Creek, co. Ashburnham; and near to Dowagarang (the Old Man Canobolas), co. Wellington.

DIALLAGE.

Chem. comp.: Calcium and magnesium silicate. Occurs in small bronze-green-coloured crystals in the serpentine of Bingera, co. Murchison; Warialda, co. Burnett; and Kelly's Creek, Gwydir River, with chrome iron. Also, at Bowling Alley Point, near Tamworth, co. Parry, with bronzite; and with copper ores near Locksley; the crystals are thin, translucent, and more or less brittle.

HYPERSTHENE.

Chem. comp.: Calcium, magnesium, and iron silicate. An outcrop of a hypersthene rock is said to occur near the Lagoons, west of Gulgong, and Cooma, co. Beresford.

Diaclasite, from Bingera, co. Murchison.

Bronzite, associated with small colourless garnets, crystallised in rhombic-dodecahedra, and saussurite, occurs near Nundle and Tamworth.

HORNBLENDE.—Amphibole.

Chem. comp.: Magnesium, calcium, iron, and manganese silicate. Oblique system. Different varieties of hornblende vary extremely in colour, form, and composition.

1. *Tremolite.*—A white, or nearly white, variety occurs at Cooma, in long slender crystals.

2. *Actinolite.*—A dark green fibrous actinolite occurs at Mowembah; in quartz at Giant's Den, Bendemeer, and near Trunkey; also in quartz at Merrendee, on the Meroo Creek, a tributary of the Cudgegong. Large radiate groups occur at Cow Flat, near Bathurst; also at Manar and Jejedzeric Hill.

3. *Sahlite.*—Crystals of this form are said to occur in a compact augite paste on the Cowridga Rivulet and near Scone.

A light grey sub-translucent hornblende mineral was collected by Mr. C. S. Wilkinson, F.G.S., at Mount Walker, on the Mudgee road, which breaks in places something like a very fine-grained quartzite or jade, with somewhat conchoidal surface; in other places there is a fibrous structure due to the presence of bright acicular crystals. The weathered portions are stained brown with oxide of iron, and show the cavities left by fossils. It seems to have been highly charged with the shells of spirifera.

Partly soluble in acid.

Extremely tough. Hardness, 6–7. Specific gravity, 3·003.

Analysis.

Loss on ignition .	·60
Silica .	50·44
Alumina .	6·19
Iron sesquioxide .	1·25
Lime .	28·70
Magnesia .	11·14
Soda .	1·16
Loss .	·52
	100·00

The sedimentary rock in which the fossils were originally embedded must have been highly metamorphosed to account for the present character of the matrix. This has been brought about by the intrusion of a vein of igneous rock at the spot. The formula is practically $2(\frac{2}{3}CaO\frac{1}{3}MgO)3SiO_2$. Physically this mineral resembles pectolite; but in chemical composition it is more closely related to some of the hornblende group.

Large crystals of common hornblende occur at Uralla, co. Sandon; Tenterfield, co. Clive; in the New England District, and in other places. In quartz with lamellar magnetite, at Merrendee, on the Meroo, a tributary of the Cudgegong, and on the road from Jungemonia to Uranbeen, co. Phillip; also at Cooma, co. Beresford; Bendemeer; and Cope's Creek. Also near Tamworth; Vegetable Creek; Tintin Hull; Tallebung Mountains, near Forbes. At the Floreston Gold and Asbestos Mine, $2\frac{1}{2}$ miles north-west of Gundagai, hornblende occurs with talc and calcite, with paint-like films of gold on the surface of the crystals.

Good crystals are reported by Mr. Ranft near the junction of granite with metamorphosed rocks at a copper-mine, Locksley.

4. *Asbestos* (Amianthus).—Chem. comp.: Essentially a magnesium silicate. A fibrous variety of hornblende.

Localities.—Said to occur in veins at Bukkulla, co. Arrawatta; Guyong, co. Bathurst; and Burraba Creek, co. Wellington; in the basalt at Pennant Hills, co. Cumberland; with auriferous quartz in diorite at Gulgong, King's Plains, co. Phillip; also at Wentworth, co. Wentworth; Lucknow, with gold; Icely, Trunkey, Caloola, and Mount Lawson, co. Bathurst; Lewis Ponds Creek, co. Wellington; the Lachlan River; Briar Park, Sewell's Creek, and Tintin Hull, near Rockley, with marmolite and schiller spar in serpentine, and Abercrombie Range, co. Georgiana; Barnard River; Carangara, near Orange; Wiseman's Creek; and Jones' Creek, near Gundagai, co. Clarendon. Abundant at Cow Flat Copper-mines, but not of the best quality; with serpentine, Briar Creek, Campbell River; also at Mount Macquarie, near Carcoar. The asbestos from near Gundagai appears to be found in long silky white fibres, and is apparently of very good quality.

A dark, olive-green coloured, and imperfect asbestiform mineral from near Cow Flat was found to have the following composition :—

Analysis.

Hygroscopic water	1·084
Combined „ by difference	1·941
Silica	49·447
Alumina	9·688
Iron sesquioxide	16·330
Iron protoxide	5·151
Manganese protoxide	4·389
Magnesia	traces
Lime	11·970
	100·000

Specific gravity, 3·02.

The value of the asbestos raised in the colony of New South Wales during 1885 and previous years is given at £3216 (*Annual Reports of the Department of Mines*, Sydney).

5. *Glaucophane*, a bluish-coloured mineral placed with the hornblende group, is said to occur at Bingera.

PYROPHYLLITE.

A hydrated aluminium silicate.

Occurs at the Asbestos Mine, Jones' Creek, Gundagai, whence fine specimens have been obtained.

KYANITE.—Disthene.

Chem. comp.: Aluminium silicate. Anorthic system. Occurs near to Kangaloolah, an arm of Tuena Creek, some ten miles south of Tuena, and at Bingera. In colour it is nearly white, the lustre pearly, in slender flattened brittle crystals.

Rhœtizite, a white variety, occurs at Belwood, Abercrombie River.

STAUROLITE.

Chem. comp.: Aluminium silicate. Rhombic system. Occurs in a talcose schist near Bathurst, in the form of small brown prismatic crystals.

ANDALUSITE.

Chem. comp.: Aluminium silicate. Rhombic system. A vein of this mineral, crystallised in rhombic prisms of a pinkish-grey colour, is said to occur in the slate rock to the east of Bungonia.

Chiastolite (a variety of Andalusite).—Chem. comp.: Aluminium silicate. Rhombic system. Occurs in granite rock at Arnprior, Boro, near Goulburn, and in small imperfect crystals in the slate near Modbury, Shoalhaven; and near Tumut, in a dark-coloured micaceous slate or schist.

Zoisite.—Found at Avisford, co. Wellington.

Fibrolite.—Mr. D. A. Porter met with fibrolite on the Barrier Ranges. It is also said to occur at Rocksbury, co. Westmoreland.

EPIDOTE.

Chem. comp.: Silica, alumina, lime, iron, &c.   Oblique system.

Occasionally well-developed columnar crystals have been met with, but I have seen none of large size; also massive. Usually various shades of green.

Epidote is found on Diamond Hill, Sidmouth Valley, in altered Silurian schist, near to its junction with diorite and granite, in association with wollastonite, garnets, specular iron ore, brown hæmatite, and black oxide of manganese.

With garnets at Duckmaloi in wollastonite; also near Parkeston, Jones' Creek, Gundagai.

Found in the Murrumbidgee District, near Mount Tennant; at Goree, near Mudgee; at Bundian, with glassy felspar and quartz; at Manilla, co. Darling; at Oberon, co. Westmoreland; the Windindingerie Cataract; Jejedzerick; between Jingery, Bobbera, and Pambula, co. Auckland; "The Gap," Lewis Ponds, co. Wellington; the Shoalhaven River, co. St. Vincent; to the east of Bungonia, co. Argyle; Gulgong, co. Phillip; Bathurst; and in the bed of the Gwydir River and of the Ora Ora.

At Camberra Mr. Ranft reports that it occurs massive in association with limestone, and that some of the Silurian fossils have been converted into epidote.

TOURMALINE.—Schorl.

Chem. comp.: Very complex, but mainly composed of silicate of alumina, iron, lime, and soda, with usually some 3 or 4 per cent. of boracic acid; other substances, such as lithia, are often present.

Crystallises in the hexagonal system, usually in the form of prisms having a more or less triangular section, and strongly striated parallel to the principal axis. Large prisms are met with in the New England District, and also in the Murrumbidgee. When the crystals are small and more or less aggregated together into bundles the mineral is termed schorl; this form of it is common in the granite of the New

England tin district; Cope's Creek; Scrubby Gully; at Bendemeer, Never Never, Nundle, Bulanamang, and in veins and nests in granite, with large mica crystals, at Wombat, near Young. Tourmaline is also found at Capertee and to the north-west of Barmedman.

Large crystals are found in the south with pegmatite between Morowat and Burramungee; with tremolite at Jejedzeric in granite; at Tarcutta, co. Wynyard; and Jingellic Tin-mines.

It is also commonly found associated with gold, diamonds, and other gems in drifts and river deposits, more or less rolled; at times all trace of the original crystallised form is removed.

Large crystals of tourmaline at Oban, co. Clarke. Some of the large crystals of tourmaline from the Uralla District are large and well formed, and closely resemble the black rhombohedral crystals from Bovey Tracy, in Devonshire. Masses weighing as much as 20 lbs. have been met with. Mr. D. A. Porter of Tamworth records finding part of a crystal at Oban 2 inches long by $6\frac{3}{4}$ inches in circumference, and weighing $9\frac{1}{4}$ oz.; also a fragment from Mount Gulligal, near Bendemeer, 2 inches by $8\frac{1}{2}$ inches circumference, weighing $14\frac{1}{4}$ oz.; and another from the same place weighing $14\frac{3}{4}$ oz., 4 inches by 6 inches in circumference. Balola; Cooma, co. Beresford; Orara; at Albury and Mount Tennant; in laminated granite at Oura, in the Wagga Wagga District.

AXINITE.

In composition this is a complex silicate of calcium, aluminium, and other bases, containing boron. It crystallises in the anorthic system, and is remarkable for the sharp axe-like form of its crystals. Mr. D. A. Porter of Tamworth records its occurrence at a place five miles south-east of Moonbi railway station. It possesses the usual clove-brown colour of the mineral. Specific gravity, 3·11. The crystals are fairly large and well developed.

FELSPAR GROUP.

ORTHOCLASE.—Common Felspar.

Chem. comp.: Aluminium and potassium silicate. Oblique system. There are several varieties of this mineral: *Common* or *Orthoclase Felspar* includes all the common non-transparent varieties; *Adularia*, the sub-transparent forms; opalescent adularia is termed *Moonstone;* and *Glassy Felspar*, or *Ice Spar*, comprises the clear and transparent forms.

Fine well-formed crystals of felspar have not yet been obtained here, although fairly large and moderately well-developed crystals are not uncommon in the coarse-grained granites of the New England, Bathurst, and Southern Districts. Simple and compound crystals of an inch or so in length, exposed by weathering, are common in the granite of New England. Dark grey felspar at Mount Walker. At Lawson's Creek in fairly well-formed, large isolated crystals, and at Oban, co. Clarke; Balola, co. Hardinge; in large crystals on the Cudgegong River, and at Home Rule, co. Phillip. Medium-sized crystals of *glassy* felspar are reported at Benada Creek; also near Naas, co. Cowley; and with quartz at Lanyon, to the west of Mount Tennant. Again near "The Pass," Bundian. With mica chlorite and quartz at Windindingerie Cataract. Acicular crystals of glassy felspar occur in compact felspar at Mount Wingen, near the burning coal seam, co. Brisbane.

A porphyry occurs near Tumut, in which red and white felspar crystals are diffused through a dark green felspathic paste; this rock would form a very attractive ornamental stone.

Crystallised adularia felspar is plentiful on Mount Lindsay. It is also reported from Montague Island in granite.

ALBITE.

Chem. comp. : Aluminium, sodium, and potassium silicate. Doubly oblique system.

Occurs massive and in the form of white crystals in New England, as at Bingera. co. Murchison; also in one or two places near Gulgong, co. Phillip, at one of which it is said to be found in association with calcite, opal, asbestos, epidote, sphærosiderite, mispickel, blende, galena, pyrites, and copper pyrites in an auriferous vein traversing a diorite; at Rylstone, co. Roxburgh. It occurs crystallised with translucent quartz at Mount Dixon, Dewelamble, Murrumbidgee, and with quartz, chlorite, and green mica on the Coolalamine Plain and at the head of the Yarralumla.

OLIGOCLASE.

Chem. comp. : Aluminium, sodium, and calcium silicate. Doubly oblique system.

Reported by Mr. Wilkinson in basalt with olivine and augite at Collingwood, and in the Lachlan and Fish Rivers.

NEPHELINE.

Chem. comp. : Aluminium, sodium, and potassium silicate.

Hexagonal system. Occurs in amygdaloidal porphyry at "The Pinnacle," co. Forbes; Dowagarang, and the Old Man Canobolas, near Wellington, co. Wellington.

LEUCITE.—Amphigene.

Chem. comp. : A potassium aluminium silicate. Silica $= 55\cdot0$; alumina, $23\cdot5$; potash, $21\cdot5 = 100$. Cubical system.

Mr. T. W. Edgeworth David, F.G.S., of the Geological Survey, has found this mineral in basalt near Byrock. An account of it was given to the Mineralogical Society by Professor Judd, F.R.S., in October 1887.

SPODUMENE.

Chem. comp. : Aluminium and lithium silicate. Oblique. Mr. Wilkinson reports its probable occurrence at Oura Station, near Wagga Wagga, co. Wynyard.

HAUYNE.

Chem. comp. : Silica, alumina, soda, lime, and sulphuric acid. Cubical system.

The Rev. W. B. Clarke discovered some small specimens of a blue-coloured mineral, which he believed to be haüyne, below the Windindingerie Cataract, in association with flesh-coloured felspar, adularia, quartz, and epidote.

MICA.

MUSCOVITE.—Potash Mica.

Chem. comp. : Aluminium and potassium silicate. Oblique system.

Large tabular crystals of mica are met with in the coarse-grained granite of the Bathurst District, as at Broadwater and other places on the Macquarie River, and at Cooma and Wheeo, co. Beresford; crystals of a golden-coloured mica are also obtained from the same place, and at Orange with crystals of felspar in a pink-coloured granite.

Green mica is common in the granite of New England; the mica entering into the composition of the greisen at Elsmore and Newstead, co. Gough, and other places is greenish. Green mica also occurs in the granite of Yarrangun and Ororal.

In the Naas Valley, co. Cowley, mica is found in large crystals, associated with quartz, felspar, hornblende, tourmaline, and chlorite.

Large plates of mica are found near Albury and at the Gulf.

A mammillated, bright, golden-coloured mica is found in white quartz at Kiandra, co. Wallace : this has very much the appearance of rolled gold, for which, in fact, it has been mistaken; yellow mica also occurs in Frazer's Creek, co. Arrawatta.

A bright-coloured mica with silvery lustre is met with in a manganiferous cement at Buckley's Lead, Two-mile Flat, co. Phillip.

Large groups of beautiful plumose crystals of mica occur at Oura Station, Wagga Wagga, co. Wynyard.

LEPIDOLITE.—Lithium mica.

Is reported by Mr. Ranft to occur at Black Swamp, Mole Table-land, New England, in well-developed, six-sided plates.

CLASS VI.

HYDROUS SILICATES.

PREHNITE.

Chem. comp.: Hydrous silicate of alumina and lime.   Rhombic system.   Occurs at Emu Creek, New England, of a green colour; and, in association with orthoclase felspar and copper ores, at Reedy Creek, co. Murchison; Molong, co. Ashburnham; also at Prospect Hill, co. Cumberland.

Gismondine.—A hydrated calcium-aluminium silicate, crystallising in rhombic forms resembling the tetragonal pyramid, present with zeolites in the Murrurundi Tunnel.

ALLOPHANE.

Chem. comp.: Hydrous silicate of alumina, Al_2O_3, SiO_2, $6H_2O$. Occurs as amorphous masses and incrustations of a bluish and opaque white colour at the Great Blayney Copper-mine, near Blayney, associated with native copper.   The surfaces are mammillated in part.

ZEOLITE GROUP.

This group of minerals is distinguished by the property which most of them possess of fusing with intumescence before the blowpipe, *i.e.*, they boil up, the name being derived from ζέω, to boil, and λίθος, a stone.   They are usually found filling the amygdaloidal cavities and crevices in igneous rocks, and never as crystals disseminated through the mass of the rock, like pyrites, garnet, or mica.   In chemical composition they consist essentially of compound hydrated silicates of alumina, the alkaline earths and alkalies; and when treated with acids gelatinous silica is separated.

Zeolites are reported as having been found at Muswellbrook, co. Durham; on the Conical Hills, Bando Plains; and with green earth at Wallabadah, co. Buckland.   Also near Vegetable Creek; Narrabri;

near Tamworth; Murrurundi; Prospect Hills, Parramatta River; in fact, wherever there are more or less decomposed amygdaloidal rocks. Amongst those which have been examined are:—

THOMSONITE.

A hydrous calcium-aluminium silicate, crystallising in the rhombic system.

Comptonite. — A variety of Thomsonite, found at Dabee, co. Phillip.

STILBITE.

Chem. comp.: Hydrous silicate of alumina and lime. Rhombic system. Reported to occur in metamorphic Silurian shales at Adelong, co. Wynyard; at Gunnedah, co. Pottinger; in the neighbourhood of Tenterfield; Tamworth; Gulgong, with analcime; Werris Creek; and Carroll.

HEULANDITE.

Chem. comp.: Hydrous silicate of alumina and lime. Oblique system.

Found at Hartley, co. Cook, in small red crystals, seated on a bluish-grey schistose rock or slate.

LAUMONITE.

Chem. comp.: Hydrous silicate of alumina and lime. Oblique system. This mineral occurs in the form of white crumbly prismatic crystals in association with black and white parti-coloured calcite crystals in the cavities of an amygdaloidal rock on the road between Geringong and Kiama, co. Camden.

This mineral was also observed by Mr. C. S. Wilkinson, the Government Geologist, and obtained by him from a cutting on the Bathurst Road, near the Cox River.

It occurs as small irregular veins, of a pleasing salmon-colour, running through a soft bluish-grey shale; the veins together with the included plates of shale are sometimes six inches thick, but usually smaller; the actual veins of the mineral itself being only about one-eighth of an inch thick. Some difficulty was on this account experienced in obtaining sufficient of the sample pure enough for analysis.

The mineral appears to be partially crystallised; nothing definite could be made out, but some of the confused crystals had somewhat the appearance of rhombic prisms. It apparently cleaves parallel to the long axis, and less perfectly at right angles to it. Translucent; lustre pearly.

Specific gravity, 2·5. Hardness, about 2·5; can be crushed by the thumb-nail, being very tender. Streak pink, but paler than the mineral itself.

Heated in the closed tube it gives off water, and at a red heat becomes grey, but reacquires a pink colour on cooling, which is rather paler than the original colour. On platinum foil, when strongly heated, it fuses to a whitish mass. Does not impart any distinctive tint to outer flame. With nitrate of cobalt gives a blue colour. Soluble in HCl, with separation of much gelatinous silica.

Analysis.

Combined water	12·646
Silica	53·266
Alumina and traces of iron	22·833
Lime	11·000
Magnesia	·479
	100·224

It also occurs as a white powdery mineral in a soft grey-coloured amygdaloidal trachytic rock at Myralla; also on calcite near Tamworth. This mineral may at once be recognised by its tendency to decompose.

APOPHYLLITE.

Chem. comp.: A hydrated calcium silicate, containing potassium fluoride. Crystallises in the tetragonal system.

Found on the Talbragar River, co. Bligh; and in the Murrurundi Tunnel, co. Brisbane. Also at Strathbogie, near Vegetable Creek.

NATROLITE.—Mesotype.

Chem. comp.: Hydrous silicate of alumina and soda. Rhombic system.

In radiate groups of long acicular crystals; found in amygdaloidal basalt in the Murrurundi Tunnel, co. Brisbane; and near Inverell, co. Gough. Also at North Point, Lord Howe Island, with calcite, in decomposed igneous rock.

SCOLEZITE.

Same chem. comp. as the above. Rhombic system. This mineral is found with cylindrical masses of bitter spar in a basalt, Emu Creek, New England. It is distinguished by curling up like a worm before the blowpipe—hence the name, from σκώληξ, a worm.

ANALCIME.—Cubical Zeolite.

Chem. comp.: Hydrous silicate of alumina and soda. Cubical system. Occurs at Inverell, co. of Gough.

Analcime in grey amygdaloidal rock, with laumonite and apophyllite, on the Talbragar River, co. Bligh. Also with stilbite in amygdaloidal basalt at Gulgong, and at Inverell in basalt with Herschellite, arragonite, and olivine.

CHABASITE.

Chem. comp.: Hydrous silicate of alumina, lime, and potash. Hexagonal system; commonly assumes rhombohedral forms. This is perhaps the most abundant of the New South Wales zeolites, and the crystals are often very well developed. It occurs in basalt with delessite at Muswellbrook, co. Durham; and in well-formed rhombohedra in trachyte on the Lachlan River; also in an amygdaloidal basalt at Reedy Creek, Sutton Forest, co. Camden; with calcite in a similar rock at Coroo. In the Murrurundi Tunnel, with other zeolites, in a decomposing amygdaloidal rock, also halloysite associated with a nepheline basalt; also near Tamworth, in amygdaloidal cavities with other zeolites. It also occurs in the cavities of a puce-coloured rock at Fountain Head in simple rhombohedral crystals of a wax-yellow colour, and is associated with a bright orange-coloured powdery mineral and a grey-green steatitic substance; the matrix can be readily cut with a knife, and leaves a shiny streak.

It is also reported from the Talbragar, co. Bligh, and Abercrombie Rivers, co. Georgiana; and is present in the basalt of the Illawarra District, and at Red Point Bay, Lord Howe Island.

Gmelinite.—This is a variety of chabasite, and occurs, crystallised in double hexagonal pyramids, with calcite and analcime, at Inverell, co. Gough.

The name of Herschellite has been given to this mineral both in Victoria and New South Wales.

A specimen from Inverell, crystallised in double hexagonal pyramids, of a cone-like appearance, from the faces merging one into the other. Transparent and colourless to opaque white. Dr. Helms, of the University of Sydney, has analysed the best-formed and most transparent crystals of this specimen with the following results:—

Specific gravity, 2·100.

Analyses.

	No. 1.	No. 2.	Mean.
Water at red heat	20·67	...	20·67
Silica	47·59	47·81	47·70
Alumina	19·51	19·06	19·31
Lime	10·83	10·87	10·85
Magnesia	·36	·50	·43
Potash	1·15	1·21	1·18
Soda	·29	·49	·39

100·53

Corresponding to the formula $CaOSiO_2$, $Al_2O_33SiO_2$, $6H_2O$.

The composition is thus proved to be really that of chabasite; hence it was quite unnecessary to make the new species for some time known as Herschellite.

In amygdaloidal rock, in the Murrurundi Tunnel, with laumonite, &c.; also near Tamworth and near Inverell, with analcime, aragonite, and olivine.

An account of some zeolites and other minerals from New Holland is given by F. Alger in *Silliman's American Journal of Science* for 1840, but no information is given as to the localities; hence the paper is not so valuable as it otherwise would have been.

SERPENTINE GROUP.

There are several varieties of the mineral serpentine met with in New South Wales. The rock of the same name is also found very largely developed, both in the northern, western, and southern districts.

SERPENTINE.

Chem. comp.: Hydrous silicate of magnesia.   Amorphous.

Of an oil-green colour, semi-transparent, on the Murrumbidgee; at Bingera, co. Murchison; Warialda, co. Burnett; Barraba, Manilla, co. Darling; and Stony Batta, co. Hardinge.   Different varieties of red-veined serpentine, steatite, and other similar minerals are reported in the Upper Peel River, at Mount Misery, near Nundle, Bowling Alley Point, and other places.

It also occurs at Coolac and Jones' Creek, near Gundagai, co. Clarendon, and on the Clarence River.   Serpentine occurs near Young, Orange, Dungog, and Lucknow, where it is associated with gold.

As in other parts of the world, serpentine is often associated with chrome iron, nickel, asbestos, copper, gold, and other valuable minerals.

Williamsite.—Apple-green, translucent, somewhat greasy to the touch, takes a very fair polish, and forms very pleasing ornamental stone.   H = 3.

From Tuena, co. Georgiana; also Middle Creek, Jocelyn, co. Westmoreland.

Marmolite.—A foliated variety of serpentine; occurs on the Murrumbidgee, of a yellowish colour, associated with dull-red and green serpentine rock; and at Cowarbee, forty miles from Wagga Wagga, with leaf gold.

The late Mr. Stutchbury mentions the occurrence of an orbicular serpentine on the Apsley, Manning, and Hastings Rivers or Creeks.

Marmolite, schiller spar, and asbestos occur in serpentine on the Peel, co. Parry.

Picrolite.—Chem. comp.: Hydrous magnesium silicate. A fibrous variety of serpentine. Found at Kelly's Creek, Gwydir River, and in the serpentine at Bingera, co. Murchison, with meerschaum. It occurs also as a green striated mineral at Lucknow, co. Wellington, and Wentworth, near Orange, co. Bathurst.

TALC.

Chem. comp.: Hydrous magnesium silicate. Rhombic system. Occurs in the form of hexagonal crystals between Gudgeby River and Naas Valley, co. Cowley; also about Bathurst; and between Jungemonia and Uranbeen with steatite and large hornblende crystals. Also near Bathurst, Gundagai, near Newbridge, and on the Barrier Range.

Steatite.—A massive indurated form of talc or hydrous magnesium silicate, near Cow Flat, co. Bathurst. A green steatite occurs near Tenterfield.

Occurs in Ranger's Valley, Severn River, co. Gough, at Elsmore, and the Bolitho Tin-mine, associated with tinstone. At Jungemonia and Uranbeen, Icely, and Trunkey, co. Bathurst; and Sewell's Creek, co. Georgiana.

Soapstone, Saponite.—Williams River, Icely, and Lowee.

Agalmatolite, or Chinese Figure Stone.—In chlorite schist, Nurembla, Callalia Creek.

Meerschaum.—Chem. comp.: Hydrous magnesium silicate. Said to occur near Bingera and on the Richmond River. All the specimens of so-called meerschaum which I have yet seen from the latter district have proved to be cimolite; hence the statement requires confirmation.

CHLORITE.—Green Earth.

Chem. comp.: Hydrous silicate of alumina and magnesia, with more or less oxide of iron.

In a confused mass of various crystallised substances, Gulgong, Lachlan River; on Pine Ridge, Copperhannia Creek, in an auriferous quartz-reef; Queanbeyan, Yass. With a white crystalline marble near Wagga Wagga. At Paradise Creek, Mole Tableland, Mr. Ranft reports a dyke of chlorite thirty feet thick in granite, containing tinstone and other minerals.

It is met with at Bolivia, Emmaville, Oberon, Glen Creek, Nymagee, Solferino, and Orange, and is often present in auriferous veins.

De Lessite.—A ferruginous chlorite. Its occurrence is mentioned by the Rev. W. B. Clarke. It is found with chabasite in basalt near Muswellbrook.

A pink schistose mineral was found embedded in the slates and other rocks at the south-east corner of Rocky Ridge by the late Dr. Thomson, Professor of Geology in the University of Sydney, and Mr. Norman Taylor.[*]

The mineral is somewhat friable, earthy and meagre to the touch; emits an argillaceous odour when breathed upon; adheres to the tongue; is decomposed by hydrochloric acid with separation of granular silica; yields a very pleasing bright pink-coloured powder; before the blowpipe does not fuse, but darkens slightly; heated in a tube it evolves moisture, darkens, but reacquires its original colour on cooling. As the mineral is evidently only a non-crystallised decomposition product it is unnecessary to give it a name; it is therefore provisionally placed with the chlorite group.

Analysis.

Water lost at 105° C.	1·335
Silica	61·951
Alumina	24·120
Iron protoxide	1·222
„ sesquioxide	3·400
Lime	7·850
Magnesia	trace
Loss	·122
	100·000

PINITE.

The following account of a mineral occurring in serpentine at Hanging Rock is by Mr. W. A. Dixon, F.C.S. (*Report of the Department of Mines*, Sydney, 1879):—

"It is massive, translucent, with a sea-green colour, waxy lustre, and unctuous feel; gives a white streak and powder. In a sealed tube it gives off water and becomes white; before the blowpipe it is infusible, but becomes opaque and reddish-white, and is not acted on by hydrochloric acid.

"Hardness, 2; specific gravity, 2·68.

Analysis.

	No. 1.	No. 2.
Silica	35·72	36·10
Alumina	38·60	38·41
Oxide of iron (FeO)	8·64	...
Magnesia	5·40	5·64
Lime	·61	...
Water	10·96	...
	99·93	...

---

[*] The "Mudgee Diamond-fields," by Thomson and Taylor, *Jour. Roy. Soc. N. S. W.*, 1869.

"The mineral seems to be new, and the ratio of the oxygen in $\dot{R}\ \ddot{R}_2\ \ddot{S}i\ \dot{H}$ is 1 : 4·2 : 4·5 : 2·3, which would give a formula approximating to $= 4$ ($\dot{F}e\ \dot{M}g\ \dot{C}a$) $6\ddot{A}l$, $9\ddot{S}i$, $9\dot{H}$."

Although the mineral does not agree with any of the pinites, yet it may be provisionally classed with them, until further examined.

CLAYS.

Kaolin, or China Clay.—Is derived from the decomposition of granite, and is not uncommon in many parts of the colony. A deposit of kaolin suitable for the manufacture of the best porcelain is reported to occur at Lambing Flat, King's Plains, co. Bathurst; and another of a dazzling white colour on a hill near to Rocky Ridge, which is in association with a bright and pretty-coloured lavender clay derived from decomposed basalt; also found near Barraba, co. Darling; at Uralla, Gulgong, and other places where there is decomposing granite.

In the *Philosophical Transactions of the Royal Society of London* for 1798 there is an account of an earthy substance by Mr. Charles Hatchett, brought from Sydney by Sir Joseph Banks, and variously named *Sydneia, Australa, Terra Australis,* and *Austral Sand.* The substance is of no importance, but there is a certain amount of interest attached to the paper, since it contains probably the first analyses of any mineral from this colony.

It had previously been examined in 1790 by the celebrated Mr. Wedgwood,[*] also by Professor Blumenbach of Göttingen, by Dr. Klaproth, and by Professor Haidinger of Vienna.

One of the specimens consisted of "a white transparent quartzose sand, a soft opaque white earth, some particles of white mica, and a quantity of dark lead-grey particles, which have a metallic lustre."

SYDNEIA.

Analyses.

No. 1.		No. 2.	
Silica . . .	·30	Silica and mica .	77·75
Silica combined .	75·25	...	...
Alumina . . .	7·20	Alumina . . .	6·50
Oxide of iron . .	3·20	Oxide of iron . .	3·0
Graphite or plumbago	10·25	Plumbago . .	10·0
Water . . .	2·20	...	...
	98·40		97·25

As the result of his examination Mr. Hatchett came to the conclusion that the substance had been derived from a decomposed granite, and recommended the removal of Sydneia from the list of minerals,

* *Phil. Trans.,* vol. lxxx., 1790, Part ii. p. 306.

since it did not contain any new primitive earth, nor did it possess the characteristic properties previously ascribed to it.

Cimolite.—There is a deposit of very white and porous hydrous silicate of alumina * on the Richmond River, which has often been sent down to Sydney as meerschaum. Probably this is partly due to its low specific gravity, for when first immersed it floats upon the water. It sometimes contains leaf-impressions ; colour, dead white; breaks with more or less well-marked conchoidal fracture; shows traces of stratification ; very porous, and adheres strongly to the tongue ; hardness, 2–2·5 ; can be scratched by the thumb-nail, and leaves a mark on cloth, but not readily.

The specific gravity after immersion in water for some time is 1·168.

Before the blowpipe it blackens slightly at first, and becomes harder after ignition; it is infusible, and yields a blue mass when ignited after moistening with cobalt nitrate ; this at once distinguishes it from meerschaum, which would under those circumstances afford a pale pink-coloured mass.

Analysis.

Water, given off at 100°.	3·28
Combined water (loss on ignition) .	4·34
Insoluble silica	51·35
Soluble silica .	·11
Alumina	37·72
Iron sesquioxide	·46
Lime	·34
Magnesia	1·25
Alkalies .	traces
Carbonic acid	1·54
	100·39

The low specific gravity is very characteristic of this mineral, but in other respects it answers to the mineral *cimolite.*

Fire-Clays of good quality are common throughout the coal-measures ; and in the shales, claystone nodules which would probably yield high-class cement are plentiful.

Brick-Clays.—Large deposits of clay, which burn to red, white, and intermediate colours, are common in the county of Cumberland, derived from the disintegration of the Waianamatta shale.

Doubtless many of the deposits of clay met with in many parts of the colony would be useful for pottery and other purposes, but up to the present but little use has been made of them. Some of the clays contain small quantities of vanadium compounds, which show as yellow, green, brown, and red stains on the bricks made from them.

---

* The so-called meerschaum from the Richmond River. A. Liversidge, *Journal of the Royal Society of New South Wales*, 1876, p. 240.

Clay from Capertee :—

Silica	62·05
Alumina	28·00
Iron peroxide	trace
Lime	0·30
Magnesia	0·47
Loss on ignition	7·50
Alkalies, iron, &c.	1·68
	100·00

(*Government Analyst, Mines Report*, 1883.)

HALLOYSITE.

Chem. comp.: Hydrous silicate of alumina.

This is an amorphous earthy mineral, resembling steatite, derived from the decomposition of igneous rocks. Adheres to the tongue, can be scratched and polished by the nail; of various colours—black, brown, grey, green, and red; the black often contains small brilliant white veins. When placed in water the mineral usually falls to pieces, and the edges become translucent.

Specimens of black halloysite are from time to time brought from various parts of the colony as samples of graphite.

A specimen collected by Mr. C. S. Wilkinson, F.G.S., from near Berrima had the following properties and composition :—Black, black streak on paper; somewhat greasy feel; does not adhere to the tongue; soft, readily scratched by nail, leaving shiny streak; brittle; conchoidal fracture.

Analysis.

Water lost at 105° C.	3·047
„ combined	12·840
Silica	45·289
Alumina	38·547
Lime	trace
Loss	·277
	100·000

Pale green and white halloysite occur in decomposing amygdaloidal rocks, with zeolites, in the Murrurundi Railway Tunnel, co. Brisbane.

Occurs in a railway cutting through decomposed basalt containing chabasite at Reedy Creek, co. Murchison; and Stony Creek, co. Wynyard; Sutton Forest, co. Camden; at Two-mile Flat, co. Phillip, of a pretty green colour; Carcoar, co. Bathurst; and on the Lachlan River.

LITHOMARGE.

A variety of lithomarge, a hydrated silicate of alumina, of a pale bluish colour, more or less translucent, occurs as the matrix of native copper at the Great Blayney Mine, near Blayney. The metallic copper is scattered through it in small granular crystalline masses.

Breaks in places with a somewhat conchoidal fracture, but earthy in others. Soft and greasy feel.

CLASS VII.

GEM STONES.

CORUNDUM.

There are several forms of this substance—alumina. The blue is known as the sapphire, the green as the oriental emerald, the red as the ruby, the yellow as oriental topaz, the hair-brown as adamantine spar, the magenta-coloured as barklyite, and the common dark-coloured ones as corundum and emery. Corundum is said to occur in basalt at Bald Hill, Hill End, co. Wellington, with olivine.

The rolled pebbles of corundum from the diamond drift on the Cudgegong River were found by Dr. A. M. Thomson to have a specific gravity of only 3·21 to 3·44, but with a hardness of 9. as usual.

SAPPHIRE.

Chem. comp.: Alumina or aluminium sesquioxide, Al_2O_3. Hexagonal system. The usual forms met with in New South Wales are double pyramids, sometimes combined with the basal pinakoid; the prism is less common. Perfect crystals are, however, rare, the majority of the specimens being either fractured or waterworn. There appears to be no record of their having been found *in situ*. In certain cases it would appear from their sharp and unworn edge that they had not travelled very far.

H. = 9. Specific gravity = 3·49 to 3·59.

The New South Wales sapphires, in common with those from other parts of Australia, are usually rather dark in colour; they, however, are found varying from perfectly colourless and transparent, through various shades of blue and green, to a dark and almost opaque blue. One or two green-coloured sapphires or oriental emeralds are almost always met with in every parcel of a hundred or so specimens, also blue and white parti-coloured.

Asteria or sapphires which show a six-rayed star of reflected light are by no means uncommon.

Sapphires are almost invariably met with by the miners as an accompaniment of alluvial gold.

They are widely distributed over the New England District, as at

Bingera, co. Murchison; and near Inverell, Rose Valley, Swanbrook, Vegetable Creek, Scrubby Gully, Mole Tableland, Glen Elgin, Dundee, Ben Lomond, Mann's River, and Newstead, co. Gough, with tin, adamantine spar, zircons, topaz, and bismuthite; in Cope's Creek and Tingha, co. Hardinge; Oban, co. Clarke; Nundle Creek and Peel River, co. Parry; Gwydir River; in co. Sandon, on the Wollombi River and at Uralla; on the Namoi River; on the Abercrombie River; blue and green sapphires, with pleonaste, zircons, gold, &c., near Mount Werong, co. Georgiana; on the Cudgegong River, co. Phillip; at Two-mile Flat, Bell's River, and Pink's Creek, co. Roxburgh, with white topaz, almandine garnets, epidote, spinelle, chrysoberyl, chrysolite, hyacinth, &c.; at Tumberumba, co. Wynyard, with tinstone, gold, diamonds, emeralds, and other minerals; in the Shoalhaven River, co. St. Vincent; and the Snowy River, co. Wallace. Also, with other gems, in the Wingecarribee River, at Berrima, Mittagong, Puddledock, &c.

Blue and green sapphires are found with gold, zircons, and other gems on Native Dog Creek, an eastern branch of Sewell's Creek, Oberon District.

Small sapphires in the form of prisms have been found in the Severn River, Furrucabad Creek, and Glen Elgin.

Some specimens of cut and polished sapphires were found to have specific gravities as follows:—

		Weight.	Sp. Gr.	Temperature.
No. 1.	Royal blue colour . .	·1400 gramme	4·1170	at 18° C.
„ 2.	Dark „ „ . .	·2332 „	4·2326	„ 18° „
„ 3.	„ „ „ . .	·4776 „	3·9115	„ 16° „
„ 4.	„ „ „ . .	·6488 „	3·9404	„ 19° „
„ 5.	Four small dark sapphires	·5050 „	4·1124	„ 18·5° „
„ 6.	Five „ „ „	·6255 „	4·0225	„ 17·5° „
„ 7.	One large „ „	·9738 „	4·0206	„ 17·5° „
„ 8.	Oriental emerald . .	·9674 „	4·0041	„ 19° „
„ 9.	„ „ . .	·5996 „	4·0733	„ 18·5° „

The late Dr. A. M. Thomson, Professor in the Sydney University, detected a variety peculiar to the Mudgee District, which occurs in uniformly small slightly barrel-shaped hexagonal crystals of about $\frac{1}{4}$ inch long and $\frac{1}{20}$ inch diameter—opaque, and of a peculiar lavender colour, with a few dark blue spots. He made out the composition as follows:—

Analysis.

Alumina	98·57
Iron sesquioxide	2·25
Lime	·45
	101·27

H. = 9. Specific gravity = 3·59.

RUBY OR RED SAPPHIRE.

This is much more rare than the blue gem. The late Mr. Stutchbury reports its occurrence with sapphire, chrysolite, hyacinth, amethyst, and other gems in the Cudgegong, between Eumbi and Bimbijong, and in Mullen's and Lawson's Creeks, co. Phillip, which fall into the Cudgegong; and the Rev. W. B. Clarke found it at Tumberumba, co. Wynyard, with similar gems. It is found, too, at Mudgee, but is not common, and usually of small size; also from a small creek about two miles from the head of the Hunter River, as well as in the Peel River; also at Bald Hill, Tumberumba, with diamonds and other gem stones. Dr. A. M. Thomson determined the composition, hardness, and specific gravity of a specimen from Two-mile Flat to be as follows:—

Analysis.

Alumina	97·90
Iron sesquioxide	1·39
Magnesia	·63
Lime	·52
	100·44

H. = 9.   Specific gravity = 3·59.

Small and imperfect fragments of ruby have been occasionally met with in most of the localities yielding sapphires and other gems.

Barklyite.—This name has been given in Victoria to the more or less opaque magenta-coloured variety. A specimen from Two-mile Flat, uncut, weighing ·5884 gramme, had a specific gravity of 3·7382 at 18·5° C.

ORIENTAL TOPAZ.

Some small specimens have been found near Mudgee, and colourless sapphires have occasionally been found in different places with other gem stones.

ADAMANTINE SPAR.

The brown variety of alumina. Found at Two-mile Flat, co. Hardinge; Uralla, co. Sandon; Bingera, co. Murchison; and Inverell, co. Gough.

Some cut and polished specimens of adamantine spar were found to have a specific gravity of 4·0306 at 17° C.

When cut and polished *en cabochon* this forms a very handsome ring stone.

EMERALD.—Beryl.

Chem. comp. : Silicate of aluminium and glucinium. Hexagonal system.

The name emerald is usually reserved for the deep-green-coloured stones fit for jewellery, while the less beautiful and pale varieties are termed beryls.

It has been found at Bald Hill, Tumberumba, with other gems. The emerald is said to occur mixed with granite detritus in Paradise Creek, near Dundee, co. Gough, and on the Mole Tableland. Also in gneissiform dykes on the summit of Mount Tennant, and at Lanyon, to the west of that mountain; in the granite at Cooma; and in Mann's River and Kiandra with other gems. In some cases the beryl is probably meant.

The beryl is much more common. It is found at Elsmore in veins associated with quartz and crystals of tinstone. The beryl crystals, which are often very thin and fragile, are seen interlaced with, and seated upon, the crystals of tinstone. It also occurs *in situ* in a tin-lode on the Gulf Stream Company's ground, Mole Tableland, associated with chalcedonic quartz.

At Ophir, co. Wellington, the beryl occurs in white felspar with quartz and white mica; one crystal from Ophir, $\frac{5}{8}$ inch through, of a pale transparent yellow-green colour and vitreous lustre, had a specific gravity of 2·708.

The beryl is found on Kangaroo Flat, Emmaville; at Tingha, Cope's Creek, and Scrubby Gully in alluvial tin deposits.

A greenish-coloured opaque beryl in small hexagonal prisms has been found in the Shoalhaven River, east of Bungonia; the crystals are associated with mispickel, and in some cases they penetrate it.

A specimen of beryl from Australia (probably New South Wales) was examined by Schneider (" Ramm. Min. Ch.," p. 555, and quoted in Dana's " Descriptive Mineralogy," p. 247), and found to have the following composition :—

Analysis.

Silica	67·6
Alumina	18·8
Beryllia, or Glucina BeO	12·3
Iron sesquioxide	·9
	99·6

CHRYSOBERYL.—Cymophane.

Chem. comp.: Glucinium aluminate, BeO, Al_2O_3. Rhombic system.

The late Mr. Stutchbury mentions that he found a fragment of this gem in the Macquarie River.

ZIRCON.—Hyacinth, Jacinth, or Jargoon.

Chem. comp.; Zirconium silicate, $ZrSiO_4$ Pyramidal system.

The transparent red varieties are known as hyacinths, the smoky as jargoons, while the grey, brown, &c., are known as zircons.

This mineral is found in granite on the Mitta Mitta, and on the Moama River, some four miles west of Jillamalong Hill, co. Cadell.

Zircons are very common in the auriferous river sands and drifts, as at Uralla and near Wollombi, co. Sandon; Bingera, co. Murchison; the Cudgegong River, co. Phillip; the Macquarie River; the Abercrombie River, co. Georgiana; the Rocky River and Two-mile Flat, co. Hardinge; the Shoalhaven River, co. St. Vincent; they are common, with iron pyrites, in the granite on which Kiandra is built; on the Talbragar River, co. Bligh.

Also at Swanbrook near Inverell, in Scrubby Gully, Vegetable Creek, with topaz; and with tinstone, titaniferous iron, spinel, &c., in the Y Waterholes. At Mount Misery small zircons are abundant; with gold and native bismuth, Mount James, Rocky River near Armidale. Lately they have been found with diamond, sapphires, &c., and a little gold near Berrima and Mittagong, along the course of the Wingecarribee River.

They are, of course, usually more or less rolled, but occasionally the crystalline form is well preserved: they vary much in colour, from more or less colourless and transparent through pale-red to crimson, brown, and opaque; they are also found of a clear transparent green, but these are rarer than the others.

Dr. Helms kindly examined for me some specimens in the form of small rolled pebbles, of good colour, fairly transparent, fit to cut, and obtained the following results:—

Specific gravity, 4·675.

Analysis.

Silica	32·99
Zirconia, ZrO_2	66·62
Iron sesquioxide	·43
Lime	·14
	100·18

The above corresponds to the formula $ZrSiO_4$ or ZrO_2SiO_2.

When cut and polished some of the New South Wales zircons form very beautiful gem stones of a hyacinth-red colour. The following determinations of the specific gravities were made upon such specimens :—

Cut and polished . .	·3118	gramme in weight	. .	Sp. gr.=4·7822 at 18° C.
,, ,, . .	·4023	,, ,,	. .	,, =4·697 ,, 17° ,,
,, ,, . .	1·8145	,, ,,	. .	,, =4·7191 ,, 18·5° C.
Uncut	2·4580	,, ,,	. .	,, =4·6838 ,, 17·5° ,,

TOPAZ.

Chem. comp.: Alumina, silica, and fluorine. Rhombic system. Occasionally met with in well-formed columnar crystals capped with planes of numerous pyramids. Some of the crystals are perfectly clear, colourless, and transparent. Some very large crystals have been met with; a portion of a large bluish-green-coloured crystal found at Mudgee, and now in the Melbourne Technological Museum, weighs several pounds; and others weighing several ounces are by no means rare; they are sometimes two to three inches long, and broad in proportion, especially those from Uralla.

A specimen of clear, transparent, pale purple topaz from New England, weighing 4 oz., was found to have a specific gravity of 3·5.

One found at Gundagai of a pale blue-green tint, measured 3 by 1½ inches, with a weight of 11 oz. 5 dwts. Another of a similar colour from Gulgong weighed 18 oz. avoirdupois; unfortunately it had been broken into two pieces.

The pale bluish-green tint is the most common colour; sometimes they are slightly yellow.

The specific gravities of two cut and polished specimens of colourless topaz from the New England District were determined as follows :—

No. 1.	Weight = 1·523	gramme	. . .	Sp. gr. = 3·5666 at 17° C.;
No. 2.	,, = 11·6010	,,	. . .	,, = 3·5640 ,, 19° ,,

It is comparatively abundant all over the granite region of New England; it occurs associated with tinstone in veins traversing the eurite, greisen, and granite near Elsmore, The Gulf, and other parts; some of the small crystals found with the tin ore are beautifully developed.

Found also on Glen Creek, Scrubby Gully, Mole Tableland, where they are plentiful, and have been found of good quality and large size; Vegetable Creek and near Inverell, co. Gough; Dundee; Oban, co. Clarke; in the Rocky River, Uralla; Balala; Bingera, co. Murchison; Two-mile Flat, co. Hardinge; Bathurst, co. Bathurst; Bell River, co. Roxburgh; Macquarie and Lachlan Rivers; the Shoalhaven and Abercrombie Rivers. Found at Cooyal. It is met with on Boonoo Boonoo Creek with tourmaline and variously coloured chalcedony.

SPINELLE.—Spinel Ruby.

Chem. comp.: Magnesium aluminate, $MgAl_2O_4$. Cubical system. Small well-formed octahedra are by no means rare; the colour varies from pale brown, red, deep crimson, green, to black, when it is known as pleonaste.

It is found in most river deposits containing gold, as in the sands of the Severn and its tributaries at Uralla, and with stream tin at Tingha, co. Hardinge; Y Waterholes, co. Gough; and Oban, co. Clarke; Bingera, co. Murchison; at Werong with gold, zircons, blue and green sapphires, and other gems; Two-mile Flat, co. Hardinge; Bathurst, Macquarie, Peel, and Cudgegong Rivers.

Spinel is said to occur in the sandstone on the road near the Fitzroy Iron-mines, Nattai; and at Inverell.

W. B. Clarke also mentions occurrence of minute spinel rubies in carboniferous sandstone at Kayon, Richmond River, but states that they are probably derived from the igneous rocks of which most of the beds in the Richmond River District are the decomposed materials.

Pleonaste.—Fairly well-formed large crystals of pleonaste with well-marked conchoidal fracture are found in the Lachlan River; also in Scrubby Gully, Mole Tableland, New England. One very well-formed octahedron from the Muntabilli River, Monaro District, was remarkable for its channelled faces.

The amorphous black vesicular pleonaste occurring on the Mudgee diamond-fields was examined by the late Dr. A. M. Thomson, who found it to have the following composition:—

Analysis.

Silica and undecomposed	2·75
Alumina	64·29
Sesquioxide of chromium	4·62
Magnesia	21·95
Protoxide of iron	4·49
	98·10

Specific gravity = 3·77. Hardness, 8.

The colour is dull black, the surface vesicular; no cleavage, but a highly lustrous well-marked conchoidal fracture; streak grey.

GARNET.

Chem. comp.: There are several kinds of garnet, and they vary in composition, but the most common are silicates of alumina, lime, iron, manganese, and other bases.

Cubical system. The rhombic dodecahedron and the icositetrahedron are the most common forms here, as in other parts of the world.

It is the alumina-lime or common garnet which is most generally met with, especially in the granite ranges, as at Hartley, co. Cook; it is found also at Bingera, co. Murchison; Pond's Creek and other places near Inverell, co. Gough; at Uralla, co. Sandon; in a talc schist at Bathurst, Trunkey Creek, and Coombing Creek Copper-mine, co. Bathurst; with mica schist in Washpool Creek, co. Drake; on the Abercrombie River, co. Georgiana; in co. Cadell, on the Old Trigomon, Moama River, four miles west of Jillamalong Hill, with hyacinth and gold; at Hardwicke, near Yass, co. King; red translucent garnets are found at Gulgong, co. Phillip, and in Sidmouth Valley, co. Bathurst.

The garnets from Duckmaloi are dull brown and crystallised in combinations of the rhombic dodecahedron and icositetrahedron, with large irregular crystals of epidote, in association with wollastonite in schist.

A dark greenish-brown garnet occurs in large quantities, with magnetic iron ore, at Wallerawang, well crystallised in rhombic dodecahedra.

Small colourless crystals and massive garnet with a variety of diallage or bronzite occur near Tamworth.

Small brown garnets crystallised in rhombic dodecahedra occur in a mica schist near Sofala.

Occurs at Sunny Corner Creek, near Locksley, associated with epidote, and near Tarana with copper pyrites and magnetic pyrites; also at Tingha; near Forbes with malachite; also at Silverton with malachite in a micaceous schist.

Found too at Mount Gipps; Tintin Hull; and at Pye's Creek, Bolivia, in an altered shale, on the Abercrombie River.

Andradite, Common Garnet.—Lime-iron garnet. Found associated with magnetite at Wallerawang; of a brown colour, rather dull. Crystallised in rhombic dodecahedra. The composition of the massive garnet is given under the head of Magnetite,* the mineral with which it is associated. The following shows the composition of the crystals:—

* See p. 89.

Analysis.

Hygroscopic moisture	·322
Carbonic acid	1·982
Silica	34·164
Alumina	3·251
Iron sesquioxide	29·435
„ protoxide	·931
Manganese protoxide	·553
Lime	28·303
Magnesia	absent
Potash	·341
Soda	·186
Loss	·532
	100·000

Grossularite.—Lime-alumina garnet.

From near Mudgee; of a rich dark brown colour; translucent. Imperfectly crystallised in groups of large rhombic dodecahedra.

Analysis.

Silica	40·517
Alumina	19·906
Iron sesquioxide	·285
„ protoxide	3·165
Manganese protoxide	3·700
Lime	32·245
Magnesia	traces
Carbonic acid	·254
	100·072

Also at Bowling Alley Point, Nundle. Mostly in small but well-defined rhombic dodecahedra; some are white and colourless.

IDOCRASE, OR VESUVIANITE.

Chem. comp.: A calcium aluminium silicate. Tetragonal system.

Said to occur in the Snowy Mountains with epidote, diopside, and garnets.

Is found at Bowling Alley Point, with common garnet, both crystallised and massive in veins traversing serpentine. The crystals are small but well formed, transparent, of a dark green colour with vitreous lustre.

MINERAL WATERS

Occur in several places in the colony, but at present little is known about them; doubtless in course of time they will be chemically investigated, and some of them utilised for medicinal and other purposes.

Chalybeate springs are not at all uncommon in the coal-measure districts, and some of them have a local reputation, notably the Lady Mary Fitzroy Spring at Mittagong, others at Joadja Creek, Kiama, &c. The first of these yielded—

Total fixed residue	11·20 grains per gallon.
Chlorine	2·67 ,, ,,
Iron oxide	1·50 ,, ,,

(*Mines Report*, 1885.)

And it is doubtless worthy of a more detailed examination.

Many of the springs and wells in the Waianamatta shale about Parramatta, Sydney, Camden, Picton, &c., yield water containing considerable quantities of magnesium sulphate (Epsom salts); and this also occurs in wells in other districts.

Others are charged with carbonic acid, alkalies, &c., as in the Cooma District.

Water from Sodawater Spring, Rock Flat, ten miles east of Cooma:—

Total solids	134·02 grains per gallon.
Soluble solids	59·96 ,, ,,
Insoluble solids	74·06 ,, ,,
Chlorine	29·51 ,, ,,
Equal to common salt	48·60 ,, ,,

Slightly turbid. Contains carbonic acid, and when heated had a slightly sweetish odour. The insoluble solids consist of carbonate and sulphate of lime, silica, and some iron carbonate. The soluble solids consist chiefly of chloride and carbonate of soda. Although a very hard water, much of the hardness may be removed by boiling.

Water from bank of Sodawater Creek, Rock Flat, ten miles east of Cooma:—

Total solids	133·98 grains per gallon.
Soluble solids	62·77 ,, ,,
Insoluble solids	71·21 ,, ,,
Chlorine	14·28 ,, ,,
Equal to common salt	24·35 ,, ,,

Turbid suspended matter. Contains a good deal of free carbonic acid, and in many respects is similar to the previous water. It has only half the amount of sodium chloride, and is still very hard. There is a comparative absence of organic matter.

Water from a depth of 540 feet at Ballimore. Artesian mineral water is now flowing at the rate of 1000 gallons per hour, and will flow 30 feet above the surface, and higher if required:—

Carbonate of lime	14·00	grains per gallon.
Magnesium chloride	12·05	„ „
Iron oxide	1·02	„ „
Alumina	trace	
Silica	·21	„ „
Alkaline carbonates	199·38	„ „
Total fixed residue	226·66	

It has all the properties of a mineral water, and is highly charged with carbonic acid.

Water from Spring Ridge, Liverpool Plains, from a well 22 feet deep, in centre of black soil plain, about 1100 feet above sea-level :—

		In 1000 parts.
Sodium chloride	864·83	12·146
Magnesium chloride	170·25	2·389
Sodium sulphate	54·80	0·769
Carbonate of soda	60·57	0·850
„ lime	18·15	0·255
„ magnesia	33·25	0·467
Silica, iron, and alumina	2·10	0·029
Organic matter	12·90	0·182
	1216·85	17·087

Total chlorine	669·65	grains per gallon.
„ sulphuric acid	30·88	„ „

Colour yellow, with a strong odour of sulphuretted hydrogen. The above are by the Government Analyst (*Mines Reports*).

APPENDIX.

THE DENILIQUIN OR BARRATTA METEORITE.*

History.

WE are mainly indebted to the energy and perseverance displayed by Mr. H. C. Russell, the Government Astronomer, for the discovery and preservation of this most interesting meteorite. His account of it is as follows:—

"While at Deniliquin on the 9th April 1871, on my way overland from Melbourne to Sydney, I learned from Mr. Thomas Robertson that a meteorite was to be found at a station thirty-five miles north-west of Deniliquin. He very kindly took me the next day to the station, which is called Barratta, and is the property of Mr. Henry Ricketson, who treated us with great hospitality, and gave me the meteorite. We found it lying in the yard close to the gate, where it had evidently been for years as a curiosity. Many small pieces, fragments of larger ones which had been broken off as specimens, were collected near the stone, and weighed altogether 2 lbs. The weight of the specimens so taken it is difficult to estimate, but probably in the aggregate it was less than 10 lbs.; the weight of the solid part is 145 lbs., so that originally its weight must have been from 150 to 157 lbs. There had also been two other pieces found with the larger piece. The smaller of these, weighing about 4 lbs., had been for years at the homestead, but was lost. The other piece, estimated to weigh between 60 and 70 lbs., had been taken to the editor of the *Pastoral Times* newspaper at Deniliquin, and there lost. The total weight, therefore, of the three pieces originally found together must have been nearly 2 cwts.

"Barratta station is situated on a vast plain, on which no sign of rocks can be seen; under the surface there are a few small stones, but the largest I could find only weighs 2 ounces. The homestead is surrounded by a few stunted trees, from 30 to 35 feet high. The

* Read before the Royal Society of New South Wales, 1872. Second notice, December 1880.

only man at the station at the time of my visit who professed to have any personal knowledge of the fall of the meteorite was a stockman named Jones, who had been there for many years. His account was as follows :—' About dusk one evening in the month of May, ten or twelve years since, I was standing in the yard, and heard a great noise, as if a storm were coming over the plain ; looking up, I saw a large body like a bush on fire, and making a loud hissing or roaring noise, coming from south-east. It passed very obliquely just above the trees on the north side of the homestead, and I then lost sight of it behind the trees.'

" The next day some fencers who were camped about four miles north-west of the homestead came in and said they had seen a thunder-and-lightning stone fall into the ground near their camp. They said it frightened them considerably, because they saw it coming directly towards their camp, but after that it went into the ground; they walked down to see it, and found it was about a quarter of a mile from their tent. A few days after this Jones went to the place, and the other men pointed out to him where the stone fell; he found the meteorite about half buried in the ground, which it had ploughed up for a considerable distance. The stone was cracked in several places, but he could not remember particularly, as it was so long since. Subsequently, in correspondence about the meteorite, I learn from Mr. F. Gwynne, of Murgah, the next station to Barratta, that he found the stone when riding over Barratta Plain, about the year 1845. As far as he can recollect, the meteorite was about 30 inches in diameter, and about 12 inches thick, and must have weighed between 2 and 3 cwts. It was lying flat on the ground; nor was there any indentation which might lead to the supposition that it had been dug up by the blacks. So far as Mr. Gwynne could judge, it might have been where he saw it for many years. The blacks could not give any account of it. It thus appears that the history of the Barratta meteorite is not satisfactory, and all my efforts to clear the matter up have so far proved useless. Mr. Robertson, who from the first has taken very great interest in the matter, has most ably assisted me in my endeavours to find the fencers who are said to have seen the stone fall into the ground, but we have not been able to find them, and the history must, for the present at least, be left as here given."

The meteorite being Mr. Russell's private property, it is still in his possession.

By way of introduction, I will now, with your permission, and in the briefest possible manner, pass in review some of the best-known facts relating to meteorites generally, so that whatever I may have to say respecting the Deniliquin meteorite may be made as clear as possible.

ORIGIN OF METEORITES.

With regard to the origin of meteorites very little indeed is known with any degree of certainty, but numerous hypotheses have been put forth from time to time.

One is, that they have been ejected from lunar volcanoes; another, that they are ejections from the sun; a third regards them as fragments of a former satellite or moon of the earth's which has undergone destruction; and a fourth would account for them by supposing them to be fragments of a destroyed planet now represented by the asteroids.

But against these hypotheses there are more or less weighty objections. One great objection is, that some meteorites revolve in a direction opposite to that taken by other heavenly bodies round the sun; and, moreover, the chemical composition of many renders it utterly impossible that they could have emanated from the sun and passed through the fiery ordeal of its chromosphere.

COMPOSITION OF METEORITES.

For convenience we classify meteorites under three heads, as determined by their chemical composition :—

(*a*.) *Metallic Meteorites,* composed mainly of metallic iron and nickel. These have been termed *siderites.*

(*b*). *Non-metallic Meteorites or Meteoric Stones.* These are mixtures of several minerals, chiefly silicates, such as olivine, felspar, augite, &c. These are also known as *aerolites.*

(*c*.) *Mixed Meteorites.* This class includes all those which are mixtures of metallic iron with various silicates. This mixed kind is far more abundant than that containing only earthy matter. They are known also as *siderolites.*

Meteoric Iron is always mixed with other elements; nickel is the one most commonly present, and in amounts varying from 1 to 15 or 20 per cent. Some cobalt is usually present also, and occasionally manganese, copper, chromium, tin, magnesium, &c.

Amongst the non-metallic elements we find silicon, carbon, sulphur, and phosphorus.

The presence of carbon is remarkable, and more especially as it has been proved to exist in two states; as free carbon or graphite, and also in chemical combination with the iron, just as in cast iron. Meteoric iron emits the same fœtid-smelling hydrogen gas when treated with hydrochloric acid as does common pig iron, due in both cases to the hydrogen being contaminated with traces of some evil-smelling liquid hydro-carbon.

Most specimens of metallic meteorites, when polished and then etched with an acid, show remarkable more or less crystalline markings known as "Widmanstädt's" figures. Berzelius attributed the formations of these figures to the presence of an alloy of nickel and iron disseminated through the mass, and this alloy being less soluble in the acid, resisted its action more than the iron itself, and thus stood out in relief on the removal of the iron by solution.

But since certain masses of meteoric iron do not exhibit these figures, although rich in nickel, it has been thought that they owe their origin to a phosphide of iron and nickel (schreibersite) which forms the major part of the residue which is left on dissolving the iron in acid.

Masses of iron found on or a little below the surface of the earth are usually regarded as meteoric when containing nickel, although the actual fall of the specimen may not have been observed.

The phosphorus present usually exists combined with iron and nickel, forming the mineral *Schreibersite*, a magnetic steel-grey mineral of H. 6·5, and with a specific gravity of 7·2; its chemical composition is represented by the formula $(Fe_4Ni)P$.

The sulphur also is usually in combination with the iron, either as magnetic pyrites (Fe_7S_8) or the monosulphide of iron (FeS), known as *Troilite;* both may be present in the same meteorite.

Troilite is of a brownish colour, while the magnetic pyrites is more of a bronze yellow, brittle, and attracted by the magnet.

In some cases the iron sulphides are visibly disseminated throughout the mass in the form of grains; in others they exist in such a fine state of division that they can only be detected by the evolution of sulphuretted hydrogen on treating with acids.

The *non-metallic* constituents of meteorites are usually crystalline. This can at once be proved, in such as are transparent, by the colours which they in thin sections display under the polariscope.

Amongst the minerals found in them are—

> Olivine.
> Bronzite or Enstatite.
> Augite.
> Anorthite.
> Labradorite.
> Calcium sulphide.
> Asmanite, a form of silica crystallising in the rhombic system.

The first two are essentially silicates of magnesium; augite is a silicate of aluminium and calcium; anorthite and labradorite are felspars, and in composition are silicates of aluminium, calcium, and sodium.

But perhaps the most surprising constituents of some few

meteorites are peculiar compounds of carbon and hydrogen.   Thus, the Bokkeveld meteorite from Cape Colony yielded a bituminous and waxy substance to alcohol; and when heated it furnished a sublimate containing the readily volatile salt sulphate of ammonium.   This meteorite contains, in round numbers, some 80 per cent. olivine, about 7 per cent. nickeliferous iron, 5·5 per cent. other silicates, and 2 per cent. carbon and bituminous matter.

About one-third of the known elements have been found in meteorites, as follows:—

Aluminium, calcium, carbon, chromium, cobalt, copper, iron, magnesium, manganese, nickel, oxygen, phosphorus, potassium, silicon, sulphur, titanium, tin, and probably in some instances antimony, arsenic, chlorine, hydrogen, and lead.   All of them occur in minerals constituting the crust of the earth.

If we arrange the elements according to their importance or abundance in meteorites, they will take up the following order, thus:—Magnesium, then iron, silicon, oxygen, and sulphur; next will come calcium, aluminium, nickel, phosphorus, carbon, &c. &c.

I need hardly remind you that but a few years ago Professor Graham, of the English Mint, exhibited specimens of hydrogen gas which he had pumped out of a meteorite, and which had been occluded within its pores.

The nature and constituents of meteorites prove beyond a doubt that they have been formed under conditions which either did not furnish sufficient oxygen to combine with the iron, a most readily oxidisable metal, or that the circumstances were unfavourable to such combination.

We are likewise justified in inferring that water was absent also.

Usually meteorites are hot externally when picked up soon after their fall, but at times they are not.   In one well-authenticated case the mass was hot externally, but excessively cold internally, the interior still retaining the intense cold of space; while the surface had become heated from the friction in its passage through the earth's atmosphere.

Meteorites are always coated with a more or less enamel-like crust. When there is an absence of iron silicates this crust is sometimes glassy and colourless; at times it is coloured, but usually it is black.

There is no doubt that the crust has been formed during the passage of the meteorite through the air, since the fragments which have become detached by an explosion are also thus coated.   And when we recollect that the meteorite enters our atmosphere with a velocity of from seven hundred to two thousand five hundred miles per minute, or from twelve to forty miles per second, we need not be at all surprised at its surface being thus fused, for on entering our atmosphere its velocity is at once retarded; in fact, its motion is virtually arrested by the resistance of the atmosphere, and, as a natural consequence of this, an intense degree of heat is immediately generated quite sufficient to fuse the surface, from which fluid matter flies off. The brilliant light emitted by a falling meteor is due to this, and the luminous streak left behind in its path is caused by the streaming of incandescent matter from it.

Explosions are frequently heard during the fall of a meteorite. These are doubtless due to the high state of tension in which the different portions must necessarily be; for the intense heat generated must greatly expand the outer portion, while the inner still remains as contracted as in the cold of space; when, therefore, a portion of the shell has become so expanded that it can overcome the resistance of cohesion, it will violently detach itself from the cold inner kernel, and fly off with explosive violence.

The pitted surface so common on meteorites is probably due to the same action on a smaller scale.

I will now pass on to the more immediate subject of this note.

THE DENILIQUIN METEORITE.

It is one belonging to the third class, *i.e.*, it is a mixture of earthy silicates with metallic substances.

Externally it is coated with a blackish fused skin; this has changed to a rusty brown in parts, from the formation of oxide of iron.

From the fractured surfaces it is at once noticed that the outer layers have a strongly marked laminated structure to the depth of about $\frac{3}{4}$ of an inch to 1-inch. Deeper in than this the structure is much more compact, and shows no traces of lamination at all, but is granular, and contains numerous spheroidal bodies, *i.e.*, the structure is chondritic. Most of the grains consist of bronzite or enstatite, grey or brownish in colour, and earthy-looking. The majority vary in size from $\frac{1}{16}$ to $\frac{1}{8}$ of an inch in diameter. Under the microscope these are seen to have an imperfect crystalline structure. Intermingled with them are a few particles of a green mineral like olivine, and of others which are not readily recognisable.

The freshly broken surface of the outer and laminated portion is somewhat granular between the laminæ, and presents numerous bright yellowish-white metallic-looking particles of irregular outline, some as distinct grains, and the rest forming a fine network enclosing non-metallic matter within the meshes. Brittle. Fragments readily scratch glass.

Under a 1-inch objective this part of the meteorite presents the appearance of an iron-black material of sub-metallic lustre, with disseminated grains, and enclosing fibres of the previously mentioned metallic-looking substance.   Small grains of green mineral resembling olivine are to be seen, also microscopic crystalline particles of a yellow mineral, which also passes into brown.

The metallic-looking portions are attracted by the magnet.

This outer part of the meteorite is magnetic, but not polar.

Its powder is grey.

The specific gravity of a portion of this crust was 3·382.

As before remarked, the inner portion is not so fine-grained as the crust; it has a distinctly chondritic structure; some of the grains are

Fig. 1.—THE BARRATTA METEORITE, one-sixth of natural size.

comparatively large, being ordinarily about from $\frac{1}{16}$ to $\frac{1}{8}$ of an inch across; a few are a little larger.   Many are grey and earthy-looking; others are brownish, and present a somewhat crystalline appearance.

The enclosing matrix has a slate-blue colour.   The white metallic-looking particles are present here also.

The specific gravity of a fragment of this inner part of the meteorite was 3·503, hence its specific gravity was ·221 higher than that of the outer portion.   This difference is not surprising when we consider that the exterior has been subject to rather different conditions, and has probably, amongst other changes, undergone a greater degree of oxidation.   Mr. Russell took the specific gravity of the main mass, weighing 145 lbs., and found it to be 3·387.

Grey Granules.—A portion was freshly broken off from this central part, as far as possible from the fused crust; was very tough and difficult to break; struck fire with the hammer, and emitted the usual empyreumatic odour of rocks when struck.   Out of this some of the small grey granules were then carefully and laboriously picked and cut out by means of a hard steel penknife; a small quantity of a light grey powder was thus obtained, which, on analysis, proved to be a silicate of magnesia, containing some iron, and is probably a bronzite. The amount obtained, after much trouble, was very small indeed.

FIG. 2.—THE BARRATTA METEORITE, reduced to one-sixth.

Nickeliferous Iron.—To ascertain roughly the proportion of this, a mass was broken off and powdered in a steel diamond mortar, and the powder passed through a fine muslin sieve, to retain the flattened scales of malleable metal.

In the first experiment 12·0618 grammes of powder were obtained, and only ·0768 gramme of metal, or about ·063 per cent.

In the second experiment 16·3170 grammes of powder were furnished, and ·1408 gramme of metal, or about ·86 per cent.

But in a third experiment 86·716 grammes were powdered in a hard steel diamond mortar, and 3·3983 grammes of metal were obtained, equivalent to 3·93 per cent. Hence the metal is very irregularly scattered through the meteorite; some portions are almost devoid of it altogether.

Magnetic Portion.—In order to ascertain the proportion of the non-magnetic to the magnetic portions of the meteorite, 25 grammes were powdered; this gave 2 grammes of magnetic powder, or about 8 per cent. This included the total magnetic mineral. It was found that the magnet did not afford a completely satisfactory means of separation.

In the main this part was found to consist of sulphide of iron, with the nickeliferous iron.

Non-magnetic Portion.—Gave off a copious evolution of sulphuretted hydrogen on the addition of hydrochloric acid. Mainly consisted of silicates of magnesium, aluminium, and iron and sulphide of iron. Calcium absent. Not entirely soluble in hydrochloric acid. The residue contained silicates of iron, a little aluminium, and a large quantity of silicate of magnesium.

From the foregoing it will be seen that this particular meteorite consists of about 92 per cent. of silicates of magnesium, iron, and aluminium, and about 8 per cent. of magnetic minerals. The proportion of nickeliferous iron is small in the extreme.

After prolonged digestion in strong hydrochloric acid, only 47·47 per cent. of the meteorite was found to be soluble.

On an analysis of the whole, it was found to have the following

Chemical Composition.

Silica	40·280
Copper	·182
Tin	absent
Iron	14·966
„ sesquioxide	3·930
Alumina	1·843
Chromium	traces
Nickel	4·219
Cobalt	traces
Manganese monoxide	·734
Lime	1·400
Magnesia	23·733
Potash	1·024
Soda	·997
Sulphur	2·288
Phosphorus	·617
Carbon	traces
Oxygen, by difference	3·787
	100·000

The alkalies were determined by Professor Lawrence Smith's process. All the iron is stated as metallic iron, except that existing as sesquioxide, on account of the difficulty of accurately estimating the amount in the free state, in the presence of protoxide, sulphide, and phosphide of iron; the sulphur was determined by the potassium chlorate and nitric acid process, the phosphorus was also oxidised by means of potassium chlorate and nitric acid, and the phosphoric acid precipitated by ammonium molybdate in the usual way. The other constituents were determined after fusion with the mixed carbonates of potassium and sodium.

Composition of the Metallic Portion.—Another portion was crushed, and the flattened metallic particles separated by means of a fine sieve of nearly 8000 holes per square inch, and analysed separately.

This metallic portion was first fused with pure caustic potash, to remove as far as possible any adherent earthy matter, and carefully washed with distilled water until the washings no longer gave an alkaline reaction.

Analysis.

Silica, &c., insoluble in HCl .	6·617
Iron .	79·851
Nickel .	7·340
Cobalt .	·431
Phosphorus .	·240
Sulphur.	traces
Oxygen, &c. .	5·521
	100·000

The iron in the above became somewhat oxidised during the washing and drying.

A second specimen of the metallic portion from a different part of the meteorite yielded the following results:—

Iron .	91·25
Nickel and cobalt .	7·20
Undetermined, silica, &c. .	1·55
	100·00

In the first case the cobalt and nickel were separated by the potassium nitrite process.

Attempts were made to estimate the amounts of schreibersite, troilite, &c., but none were sufficiently satisfactory to warrant their publication.

Much still remains to be done to fully work out the constitution of this meteorite.

When at Paris, in 1878, I submitted sections of the meteorite to Professor Des Cloizeaux, Director of the Mineralogical Museum, Jardin des Plantes; and in a letter dated December 9, 1878, he says, after having examined them under the microscope:—"Malheureusement elles n'offrent pas de forme nettement définie, et tout ce qu'on peut dire, par analogie avec des autres pierres, c'est qu'il existe de l'enstatite à structure fibreuse avec de l'olivine."

Professor Daubrée, Director of the School of Mines, Paris, has provisionally classified this meteorite, of which there are some sections

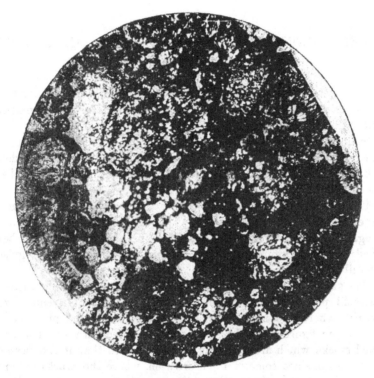

FIG. 3.—Section enlarged 18 diameters.

and a fac-simile in the collection of meteorites in the Paris Natural History Museum, with those from Tadjera, Orvinio, and Koursk.

The meteorite is essentially a mixture of the silicates of magnesia and iron (enstatite) with smaller quantities of other silicates, together with some nickeliferous iron, sulphide and phosphide of iron.

PRELIMINARY NOTE ON THE BINGERA METEORITE, NEW SOUTH WALES.*

This meteorite was found by some gold-miners in the course of their work at Bingera, and was placed at my disposal for examination by the Mining Department in Sydney.

In form this meteorite roughly resembles a pear; it is about 2 inches in length and $1\frac{1}{2}$ inch across at the thick end, tapering down to about half an inch at the other; it does not possess any sharp angles, the corners being well rounded off.

The surface is covered with a black closely adherent skin of magnetic oxide of iron, formed by fusion during its rapid course through the earth's atmosphere, except in one place where it had been rubbed off by the miners. This burnished spot is readily recognised in the illustration as a white patch. (See fig. 2.)

The skin is hard, brittle, of about the thickness of stout writing-paper, and possesses a laminated structure. At the thin end of the meteorite, where the scale is thicker, it is wrinkled, and puckered, as shown in the figure; in other places it is marked with small sharply defined cracks, which are evidently connected with the internal crystalline structure, since some of them are seen, where the crust or skin is scaled off, to pass through the crust into the mass of metal itself. These cracks are clearly the outlines of Widmanstädt's figures, probably developed in the crust by the contraction of the fused scale of oxide after cooling.

There are also a few pits on its surface, the most remarkable being a double depression, rather obscurely shown in fig. 2.

It has fairly well-marked polarity, the thin end repelling the south pole of the needle.

The total weight of the meteorite was 240·735 grammes, at 22·8° C., and its specific gravity as a whole 7·834; the specific gravity of some fragments cut off with a cold chisel was found to be slightly higher, viz., 7·849.

It does not appear to "deliquesce" at all like the Greenland and some other masses of meteoric iron; this is probably due to the entire absence of chlorine.

* Read before the Royal Society of New South Wales, December 8, 1880.

Fig. 2

Fig. 4.

THE BINGERA METEORITE, N.S.W.

Fig. 1.

Fig. 3.

Minerals of N.S.W., 1887.—A. Liversidge

The flat side of the meteorite was carefully chipped away with a cold chisel, to obtain material for analysis and to prepare a smooth surface for the development of the Widmanstädt figures. It proved to be very tough and difficult to cut. On analysis the following results were obtained :—

Chemical Composition.

	I.	II.
Carbon	·137	·668
Matters insoluble in hydrochloric acid	·553	
Tin	traces	
Copper	,,	
Iron	93·762	
Nickel	4·391	
Cobalt	·668	·484
Phosphorus	·195	·270
Sulphur	absent	
Sodium	trace	
	99·706	

The matter insoluble in hydrochloric acid consisted mainly of iron, carbon and with a trace of silica. The second determinations were made on a separate set of chippings.

The amount of tin present is apparently slightly larger than that of the copper; but both exist in such small quantities that it would have been necessary to have cut up much more of the meteorite in order to estimate the amounts, and this was considered undesirable.

The carbon given in the analysis represents the portion left on prolonged treatment with hydrochloric acid; some may also enter into solution, but this I have not yet had time or opportunity to ascertain. Under the microscope this insoluble carbon presents the appearance of opaque black particles bearing a rough resemblance to crystal forms, but with very jagged and irregular outlines; its amount was estimated by direct weighing. On incineration it was found to still contain a trace of iron, in spite of the continued treatment with acid.*

It has been urged that if carbon and combustible gases be present in a meteorite it cannot have been in a state of fusion; but the argument does not carry much weight, since whether the combustible matter were burnt or not during the fusion would depend entirely upon the conditions; it might just as well be said that cast iron has not been fused because it contains carbon.

The cobalt and nickel were precipitated together as sulphides, and

* Mr. Fletcher, Keeper of the Mineral Department, British Museum, has recently found a new variety of graphite in a meteorite from Youndegin, Western Australia, crystallising in the cubical system, which he has named *Cliftonite*; and it is just possible that the Bingera and Thunda meteorites also contain this, but at the present time I have no opportunity of examining them, but will do so on my return to Sydney.

the cobalt estimated by the potassium nitrite process; the phosphorus, by ammonium molybdate; and the other constituents in the usual ways, every determination being made in duplicate.

From the illustration of the polished and etched surface, fig. 4, it is at once noticeable that the structure is not uniform throughout, Widmanstädt's figures being fairly well marked, although by no means so well developed as in many of the larger meteoric metallic. masses which have been met with elsewhere. Some of the crystals are much elongated ("beam iron"), whilst others are but small, and are arranged in a tesselated form.

None of the crystals are large; hence it would appear that the crystalline structure had been set up in the mass whilst quite small. Of course it may have been a fragment of a large meteorite with a small-grained internal structure, but most of the specimens which have come under my notice with largely developed crystals have been cut from comparatively large masses; others again do not present the structure at all.

In places the acid has eaten freely into the metal, and more or less deep cavities have been formed; in some cases these are much more numerous on one side of the "beam iron" crystals than on the other.

As soon as an opportunity presents itself I wish to continue the examination of this meteorite, especially with reference to the kinds and amounts of nickeliferous irons which are present, as well as to the question of occluded gases.

METALLIC METEORITE, QUEENSLAND.*

This meteorite was found at Thunda, Windorah, in the Diamantina district, Queensland, and was kindly lent to me for examination by Mr. C. S. Wilkinson, F.G.S., Government Geological Surveyor of New South Wales.

Mr. Wilkinson was informed that this specimen was broken off a larger mass weighing a hundredweight or more, and it certainly has every appearance of having been recently detached. The large piece is said to be buried about 4 inches in the ground, and the natives had covered it with stones, so that they evidently regarded it as something of importance or value. The weight of the fragment was 258·7 grammes, and its specific gravity was  found to be 7·77 at 16° C., being the mean of two determinations made on separate pieces, viz., 7·75 and 7·79. In form it is very irregular; the internal crystallised structure is well shown by the fractured surface, the plates standing out in bold relief and meeting one another at fairly regular angles, which are apparently those of an octahedron. In the hollow on one side a distinct pitted structure is seen, showing that this apparently formed one of the external surfaces of the meteorite, although the usual well-marked skin of fused magnetic oxide is not present.

The woodcut represents the natural size of the fragment.

Up to the present I have not had time to make more than a preliminary qualitative examination, but this shows clearly that this specimen has the usual composition of the metallic group of meteorites. It consists mainly of iron, with nickel, and a trace of cobalt; both sulphur and phosphorus are present, together with some carbon, and I think it will be found not to differ materially from the New South Wales meteorite found at Bingera. (See *Journal of the Royal Society of New South Wales*, 1882, p. 35.)

* Read before the Royal Society of New South Wales, December 1, 1886.

ON THE CHEMICAL COMPOSITION OF CERTAIN ROCKS, NEW SOUTH WALES, &c.*

Freshwater Limestone.—Newstead, New England District, New South Wales. Of a grey colour, breaks with an earthy granular fracture, and shows particles of included clay—evidently very impure. On analysis the following results were obtained :—

Chemical Composition.

Hygroscopic moisture at 100° C.	·736
† Silica	55·430
† Alumina	14·116
Alumina	5·998
Iron sesquioxide	1·760
Manganese protoxide	·989
‡ Lime	10·571
Magnesia	·575
Potash	·353
Soda	·598
Carbon dioxide (CO_2)	8·450
	99·576

Limestone.—Windellama Creek, co. Argyle. Dark bluish-grey in colour, and containing *atrypa* and other fossils.

Analysis.

Hygroscopic moisture at 100° C.	·071
Silica	2·203
Alumina with trace of iron	1·003
Lime	54·602
Manganese	absent
Magnesia	trace
Potash	,,
Soda	,,
Chlorine	,,
Carbon dioxide (CO_2)	42·369
	100·253

Slate.—Cox River. Bluish-grey colour, somewhat weathered, slightly fissile. Of Devonian age.

Specific gravity, 2·706 at 21° C.

---

* Read before the Royal Society of New South Wales, December 8, 1880. The plates of microscopic sections accompanying these analyses are not reproduced, and the remarks. referring to them are also omitted.

† Insoluble in HCl.　　　　　　　　　　‡ Soluble in HCl.

Analysis.

Hygroscopic moisture at 100° C.	·861
Combined ,, by difference	5·106
Silica	61·012
Alumina	21·343
Iron sesquioxide	3·704
,, protoxide	2·109
Manganese protoxide	·729
Lime	1·176
Magnesia	·887
Potash	1·223
Soda	1·850
	100·000

The combined water is estimated by difference, on account of the difficulty of completely driving off and collecting the whole of the water in the hydrated silicates of alumina.

Shale.—Wallerawang, of a slate-grey colour, full of white impressions of *glossopteris* roots, fronds, &c.; fairly hard, and somewhat slaty. Contains a little carbonaceous matter.
Specific gravity, 2·304 at 20·6° C.

Analysis.

Hygroscopic moisture at 100° C.	1·115
Combined water and organic matter	6·391
Silica	71·854
Alumina and traces of iron	17·736
Lime	1·777
Potash	·466
Soda	·383
	99·722

Slate.—Wollondilly River, from above Goulburn, of a dark bluish-grey colour, imperfectly fissile.
Specific gravity, 2·58 at 18° C.

Analysis.

Hygroscopic moisture at 100° C	·301
Combined ,, with carbonaceous matter	3·990
Silica	75·566
Alumina and traces of iron	16·466
Manganese protoxide	traces
Magnesia	·106
Lime	·708
Potash	2·274
Soda	·820
	100·231

A second specimen yielded silica 77·54, and only 10·76 per cent of alumina.

Granite.—Gunning, co. King. Rather fine-grained. Composed of white felspar, quartz, hornblende, and black mica.
Specific gravity = 2·779 at 20° C.

Analysis.

Hygroscopic moisture at 100° C.	·269
Silica	69·793
Alumina	14·693
Iron sesquioxide	3·148
„ protoxide	3·371
Manganese protoxide	trace
Lime	4·861
Magnesia	trace
Potash	2·610
Soda	1·970
	100·715

Granite.—Hartley. Composed of white and dull pink orthoclase felspar, with a little plagioclase felspar, quartz, black mica, some small crystals of hornblende, and a few minute crystals of staurolite, with a small quantity of auriferous iron pyrites.

Another specimen of granite from the same district is, however, much coarser in structure, and the crystals of felspar are fairly large and pink in colour; but in other places the felspar is granular or powdery and quite white, and bears galena in place of pyrites.

The galena is for the most part present in small cavities, which are usually lined with quartz crystals. Both gold and silver are present in small quantities in the galena, but the granite itself yielded no indications of either metal. These specimens were collected by Mr. C. S. Wilkinson, F.G.S., the Government Geologist.
Specific gravity, 2·712 at 20° C.

Analysis.

Hygroscopic moisture at 100° C.	·257
Silica	70·302
Alumina	18·845
Iron sesquioxide	·730
„ protoxide	1·855
Manganese protoxide	trace
Lime	1·336
Magnesia	trace
Potash	3·361
Soda	3·174
	99·860

Granite.—Moruya. Coarse-grained, composed of white felspar, quartz, and black mica, in rather large crystal groups, with some hornblende and other minerals.
Specific gravity, 2·678 at 21° C.

Analysis.

Hygroscopic moisture at 100° C.	·168
Silica	67·557
Alumina	16·391
Iron sesquioxide	1·246
„ protoxide	1·858
Manganese protoxide	·794
Lime	5·075
Magnesia	1·484
Potash	1·770
Soda	3·540
	99·883

The polished granite pillars at the General Post-Office, Sydney, are of this rock.

Granite.—Pomeroy, co. Argyle. Red in colour. Collected by the late Professor A. M. Thomson, D.Sc.

Specific gravity = 2·60.

Analysis.

Silica	72·200
Alumina	11·399
Iron sesquioxide	6·172
„ protoxide	absent
Manganese protoxide	traces
Lime	2·000
Magnesia	trace
Potash	4·490
Soda	3·910
	100·171

Graphic Granite.—Co. Bligh, on the road from Talbragar to Two-mile Flat. Collected by the late Professor A. M. Thomson, D.Sc.

The felspar is a mixture of orthoclase and plagioclase varieties. Not yet analysed.

Syenite.—Boro Creek, co. Argyle. Composed of grey orthoclase felspar, with some plagioclase felspar, hornblende, and dark-coloured mica.

Specific gravity, 2·74.

Analysis.

Silica	64·27
Alumina	16·40
Iron sesquioxide	7·86
„ protoxide	trace
Manganese protoxide	·81
Lime	3·88
Magnesia	trace
Potash	3·16
Soda	4·19
	100·57

P

Syenite.—Reevesdale, Bungonia. A dark green compact rock.
Collected by the late Professor A. M. Thomson, D.Sc.
Specific gravity = 2·64.

Analysis.

Silica	66·876
Alumina	19·640
Iron sesquioxide	4·060
„ protoxide	trace
Manganese protoxide	·188
Lime	1·471
Magnesia	trace
Potash	2·677
Soda	4·887
	99·799

Quartz-porphyry.—Lumley Creek, co. Argyle. Dark grey, with
scattered crystals of hornblende. Collected by the late Professor A.
M. Thomson, D.Sc., who states that it underlies a fossiliferous limestone.
Specific gravity = 2·67.

Analysis.

Silica	67·714
Alumina	18·530
Iron sesquioxide	4·488
„ protoxide	traces
Manganese protoxide	„
Lime	2·857
Magnesia	traces
Potash	2·920
Soda	3·230
	99·719

Quartz-porphyry.—Gurragangamore. Somewhat friable in places
from decomposition; of a pale grey colour, almost white, with disseminated quartz grains.
Specific gravity = 2·58.

Analysis.

Silica	75·195
Alumina, with trace of iron	17·603
Lime	1·313
Magnesia	traces
Potash	2·343
Soda	4·016
	100·470

Quartz-porphyry.—Mount Lambie, Rydal. Consisting of a green
base, containing small opaque white felspar crystals, of about one-eighth

of an inch in length. Contains some carbonic acid and combined water. From a dyke cutting through Devonian rocks. This rock varies much in different parts.
Specific gravity = 2·727 at 15° C.

Analysis.

Hygroscopic moisture at 100° C. .	·355
Silica .	61·504
Alumina .	16·792
Iron sesquioxide .	3·483
,, protoxide .	2·225
Manganese protoxide .	1·222
Magnesia .	1·958
Lime .	5·436
Potash .	2·380
Soda .	4·780
	100·135

Felsite.—Two-mile Flat, Cudgegong River. A fine-grained greenish-grey rock, collected by the late Dr. A. M. Thomson, D.Sc.
Specific gravity = 2·706 at 20·4° C.

Analysis.

Hygroscopic moisture at 100° C. .	·104
Silica .	72·120
Alumina .	9·750
Iron sesquioxide .	4·105
,, protoxide .	3·224
Manganese protoxide .	1·833
Lime .	2·989
Magnesia .	trace
Potash .	2·756
Soda .	3·420
	100·301

Dolerite.—Waimalee, Prospect Hills, Parramatta River. This is the rock spoken of as a magnetic *Diorite* by the late Rev. W. B. Clarke, M.A., F.R.S., in his " Southern Gold-fields."

It occurs in association with intrusions of basalt, largely quarried for road metal, which have forced their way through the Waianamatta shale.

It possesses a coarsely crystalline structure, but so decomposed as to be easily broken up, and is in parts more or less friable. Some portions contain so much magnetite as to readily deflect the needle.
Specific gravity, 2·780 at 18° C.

Analysis.

Hygroscopic moisture at 100° C.	·991
Combined water, by direct weighing	7·009
Silica	46·498
Alumina	17·620
Iron sesquioxide	8·251
„ protoxide	5·238
Manganese protoxide	traces
Lime	9·303
Potash	1·612
Soda	3·476
	99·988

Basalt.—Pennant Hills, Parramatta. Of a specimen of this rock, eruptive through the Waianamatta shales, the late Professor A M. Thomson made the following analysis:—

Silica	60·42
Alumina	10·29
Iron oxides	14·10
Lime	2·66
Magnesia	·96
Potash	1·81
Soda	2·39
Water, by difference	7·37
	100·00

From the large quantity of silica and the small proportion of alumina, lime, &c., in the above, it looks as if the basalt had not been wholly brought into solution prior to analysis.

" *Greenstone.*"—Gympie, Queensland. Of a green colour, compact and hard, breaking with a somewhat conchoidal fracture.

This is really an altered ash or breccia, as is shown by the section, made up of particles of various rocks, with fragmentary crystals of felspar, augite, magnetite, &c., mingled with chlorite.

For a description of the mode of occurrence and analyses of other specimens of this rock, see Daintree on the geology of Queensland, *Quarterly Journal of the Geological Society*, 1872, p. 272, &c.

Specific gravity = 2·86.

Analysis.

Silica	54·952
Alumina	16·643
Iron sesquioxide	2·410
„ protoxide	7·849
Manganese protoxide	traces
Lime	8·645
Magnesia	trace
Potash	1·540
Soda	6·647
Water, carbonic acid, &c., not estimated	1·314
	100·000

Trachyte.—Gladstone, Port Curtis, Queensland. From a dyke cutting through Devonian rocks. It consists of a grey crystalline felspathic base containing embedded sanidin crystals—some scales of red hæmatite and other minerals in smaller quantities. Very vesicular in parts, also compact and close-grained in others. Specific gravity = 2·23.

Analysis.

Silica	66·932
Alumina	19·902
Iron sesquioxide	2·410
,, protoxide	trace
Manganese protoxide	,,
Lime	·797
Magnesia	trace
Potash	5·290
Soda	4·820
	100·151

The following analyses of rocks have since been added to this paper :—

Decomposed Basalt from Goulburn District.

Analysis.

Loss on ignition	11·46
Silica	41·05
Iron peroxide	14·35
Alumina	13·20
Titanic acid (TiO₂)	2·20
Lime	8·97
Magnesia	7·00
Phosphoric oxide	1·28
Alkalies, loss, &c	·49
	100·00

This mineral, if mixed with a proper proportion of lime, and properly ground, burnt, and otherwise manipulated, might probably make a good cement. (*Mines Report*, 1885.)

Pitchstone from Port Stephens.

Analysis.

Loss on ignition	8·18
Silica	73·45
Alumina	13·25
Ferric oxide	·95
Lime	2·85
Magnesia	·52
Alkalies, &c.	·80
	100·00

(*Mines Report*, 1885.)

Leucite-basalt has been found by Mr. T. Edgeworth David at Byrock, near Bourke, in a lava-stream about 500 feet above the sea.

The following analysis of a quartz-porphyry is from Professor J. M. Thomson's paper on the geology of the district round Goulburn, *Journal of the Royal Society of New South Wales* for the year 1869 :—

" *Quartz-porphyry.*—It consists of quartz fragments of irregular form, seldom larger than a pea, set in a compact homogeneous matrix, generally of a dull green colour. The following analysis shows the chemical composition of an average sample collected at the Marulan railway station.

" Specific gravity, 2·75.

Silica	66·50
Alumina	15·58
Iron peroxide	5·13
Lime	5·29
Magnesia	2·14
Potash	2·46
Soda	2·64
Loss by ignition	1·31
	101·05

" The dull green colour is due to iron; as the rock weathers, the iron separates and gives it a ferruginous surface, while the quartz-grains fall out and strew the surface of the ground. Its disintegration affords a cold muddy soil. Instances are met with where the matrix is black, and in others white."

Indurated shale from Ballimore, near Dubbo :—

Loss on ignition	1·50
Silica	65·52
Alumina and traces of oxide of iron	28·05
Lime	1·20
Magnesia	0·41
Alkalies, loss, &c.	3·32
	100·00

(*Mines Report*, 1883.)

NORFOLK ISLAND.

The following analyses are from the *Annual Report of the Department of Mines*, Sydney, 1885.

Compact coral rock from water-level, Emily Bay, Norfolk Island :—

Hygroscopic moisture	0·22
Carbonate of lime	86·85
Carbonate of magnesia	8·72
Iron oxide and alumina	1·10
Silica	·90
Phosphoric acid	trace
Alkalies and undetermined	2·41
	100·00

At the old Government lime-kiln at Emily Bay the rock is divided into layers of about half an inch or more in thickness. At a depth of a few feet these are found cemented together into a compact stone, quarried for building. The union of the particles is effected by calcareous matter, and in the compact varieties each particle can be seen to be enveloped in a husk of pellucid carbonate of lime. The cementation takes place rapidly, owing to the permeating water being charged with carbonate of lime in solution, derived from the waste of the coral reef.

Brown volcanic tuff, from the landing-place, Phillip Island, three miles south of Norfolk Island :—

Moisture	10·73
Combined water	4·08
Lime	1·06
Iron oxide	27·95
Organic matter	·32
Phosphoric acid	·05
Insoluble siliceous matter	54·81
Loss and undetermined	1·00
	100·00

Volcanic ash from Anson's Bay, Norfolk Island :—

Moisture	21·20
Gangue (insoluble in acids)	47·45
Iron peroxide	11·75
Alumina	15·30
Lime	1·80
Magnesia	1·89
Phosphoric oxide	trace
Undetermined, &c.	·61
	100·00

ON A REMARKABLE EXAMPLE OF CONTORTED SLATE.*

This specimen of more or less imperfect slate which I now have the pleasure to lay before the Society is, I think, a most remarkable example of true contortion, accompanied by slaty cleavage, but contortion on such an extremely small scale that it in certain aspects appears to resemble the well-known cone-in-cone structure seen in coal and many rocks.

The specimen was obtained by Mr. Fielder from the Peelwood Copper-mine, near Tuena. Mr. Fielder succeeded in detaching this

FIG. 1.—One-half natural size.

most interesting and beautiful example from a projecting point of weathered rock, but only after the expenditure of much time and trouble, for the slaty rock was far too tough, and also too fissile, to admit of its being broken off in large blocks by blows from a hammer or pick, so he had to saw it off—a very tedious and laborious operation.

It will be equally observable in the specimen and the photographs which I lay before you, that some of the plications are not more than, even if so much as, an eighth of an inch across, whilst the widest of

* Read before the Royal Society of New South Wales, December 6, 1876.

them do not exceed 2 inches, and the depth of the largest cleavage plane in the specimen barely reaches 3 inches; its extent in the direction of from before backward I have no means of telling, as the specimen sawn off had a thickness of but about 2 inches. The dark lines *l, m, n,* and *u, v,* in fig. 2 represent fractures in the specimen, and their plications beautifully indicate the cleavage planes to which they are parallel. Whether the cleavage planes extend over any length of country I do not know, as I have not visited the locality whence the block was brought, neither have I been able to obtain particulars on this point. The contortion is probably of quite a local character, as it does not appear to have been noticed elsewhere in the district.

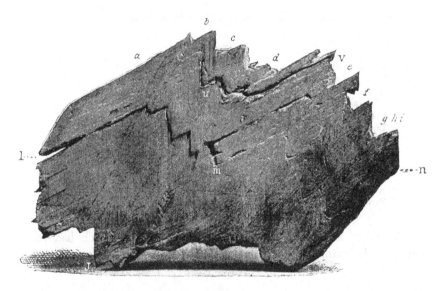

FIG. 2.—One-half natural size.

The rock or slate has the appearance of the grey killas of Devon and Cornwall; it is in all probability of Devonian age.

As I have before mentioned, the specimen has somewhat the appearance of the familiar cone-in-cone structure (see figures, one-half of natural size—No. I. shows the weathered surface, and No. II. a smooth and imperfectly polished one). The surface, which has been carefully rubbed down and smoothed, presents a series of alternating light and dark bands, similar to the banded or ribboned appearance exhibited by a well-kept English lawn cut by a mowing-machine which has been worked in lines alternately up and down the length of the lawn.

This banded effect is due to the manner in which the light is reflected from the cut edge of the cleavage planes. When held in one position the smooth surface presents a fairly uniform grey tint, but at a certain angle to the light it appears to be made up of alternate light and dark bands, and when reversed in position the light bands become the dark ones, and *vice versâ*.

Thus, in one position the bands *a, b, c, d, e, f, g, h, i, j,* appear light grey, but when the specimen is turned upside down they exhibit a dark grey shade.

Even if subsequent examination made on the spot should prove this to be a case of cone-in-cone structure, the specimen will still, I think, be of equal value and interest.

The chemical composition of the slate is as follows :—

Analysis.

Hygroscopic moisture . . . 00·48 ⎫ . . . . . . 3·85		
Combined water (by difference) . 3·37 ⎭		
Silica . . . . . ·22 ⎫		
Alumina . . . . 3·63		
Ferric oxide . . 4·47		
Lime . . . . . ·19 ⎬ Soluble in acid . . 10·51		
Magnesia . . . . traces		
Soda . . . . . 1·16		
Potash . . . . ·84 ⎭		
Silica . . . . . 67·64 ⎫		
Alumina . . . . 16·77 ⎬ Insoluble in acid . 85·64		
Ferric oxide . . . 1·23 ⎭ ——————		
		100·00

Specific gravity = 2·75, given by small fragments which had been immersed in water for some time at 75° Fahr.

I reprint the following paper in response to requests made to me from time to time by those interested in these matters:—

THE BINGERA DIAMOND-FIELD.*

In the following note I purpose giving a few facts concerning the recently opened diamond workings in the neighbourhood of the town of Bingera. Bingera is situated some 400 miles north of Sydney, on the Horton, or, as it is more popularly termed, the Big River; this river runs into the Gwydir River, the Gwydir in turn losing itself in the Barwon or Darling River.

Being on my way last winter (June 1873) to visit the tin districts of New England, I turned aside and availed myself of the opportunity to pay a hurried visit to the above diamond workings. The trip was not a satisfactory one, for, owing to the persistent rains and floods, travelling was at times quite impracticable, and at all times done under difficulties; hence, in the limited time at my disposal, all hopes of anything like a thorough geological examination of the spot had, to my great regret, to be relinquished. However, I was enabled to acquire a certain amount of information, which I venture to lay before you this evening, in the hope that it may not prove to be altogether devoid of interest and value.

But, in the first place, I may, perhaps, be permitted, *en passant*, to preface my remarks upon Bingera by briefly mentioning a few of the facts relating to the other and longer-known diamond-bearing localities of Australia, but only so far as they throw light upon the Bingera deposits. For fuller information I must refer you to the Rev. W. B. Clarke's addresses to the Royal Society of New South Wales in the years 1870 and 1872, and to the very complete account of the Mudgee diamond district by Mr. Norman Taylor and the late Dr. Thomson, read before this Society in 1870.

DIAMONDS IN AUSTRALIA.

As early as 1860 the Rev. W. B. Clarke mentions the discovery of diamonds in the Macquarie River, but no information is furnished as to the conditions under which they were found, and it is not stated

* Read before the Royal Society of New South Wales, 1st October 1873.

whether they occurred in the present river-bed or in an ancient river-drift.

But we have a full account of the geology of the diamond-bearing district detailed in the above-mentioned paper by Messrs. Norman Taylor and Thomson, and from it we shall see that the Mudgee and Bingera districts have many points of resemblance.

The Mudgee diamond-workings are distant some 170 miles south of Bingera, on the Cudgegong River, which runs into the Macquarie River, and that again into the Darling River.

Diamonds were first discovered here in 1867 by the gold-diggers, who neglected them for some time, but in 1869 they were worked pretty extensively. The localities lie along the river in the form of outliers of an old river-drift, at varying distances from the river, and at heights of forty feet or so above it. These outliers are capped by deposits of basalt, hard and compact, and in some cases columnar. This basalt is regarded by Mr. Taylor as of Post-Pleiocene age, but this has not been determined directly by any fossil evidence.

The great denudation which the district has sustained is at once apparent from the drift, together with its protective covering of basalt, having been cut up into these isolated patches or outliers.

The remains of the drift can still be traced for some seventeen miles up the river, and in parts it still retains a thickness of seventy feet.

The patches which were worked, as enumerated in the above-mentioned paper, are as follows:—Jordan's Hill, 40 acres; Two-mile Flat, 70 acres; Rocky Ridge, 40 acres; Horseshoe Bend, 20 acres; Hassall's Hill, 340 acres. Total, 510 acres.

A peculiar deposit of crystalline cinnabar was found in one patch.

In the above localities the drift has invariably been met with in tunnelling under or sinking through the basalt; and in places where the basalt had been denuded away the drift has either disappeared or has been scattered over the neighbourhood.

No diamonds have been found in the river-bed, except in places where the diggers have discharged the drift into the river when washing for gold.

The basalt, when not resting on the drift, frequently lies upon metamorphic shales, slates, sandstones, or greenstone.

The general formation of the neighbourhood is regarded as Upper Silurian, with overlying outliers of undoubtedly Carboniferous age.

The rocks in the vicinity are nearly vertical, with a general strike of N.N.W., and consist of red and yellow coarse and fine-grained indurated sandstones; thin, white, platy, argillaceous shales; pink and brown fine-grained sandstone, banded with purple stripes; slates and hard metamorphic schists; hard brecciated conglomerate, containing limestone nodules, flint, and red felspar in a greenish siliceous base. And with these occur dykes and ejections of intrusive greenstone. The rocks are generally devoid of mica. For the most part the Older Pleiocene diamond-bearing drift is coarse and loose, but parts are cemented together into a compact conglomerate by a white cement of a siliceous nature, sometimes rendered green by admixture with silicate of iron; in other cases oxides of iron and manganese have been the agglutinating agents. Diamonds were proved to exist in this solid portion by a special experiment of Mr. Taylor's.

The drift is chiefly made up of boulders and pebbles of quartz, jasper, agate, quartzite, flinty slate, shale, sandstone, with abundance of coarse sand and more or less clay.

The quartz pebbles are white, like vein quartz, but often encrusted with films of iron or manganese oxides.

Many of the boulders and pebbles are remarkable for a most peculiar brilliant siliceous polish, which is evidently not due to friction, since the cavities are equally well polished. Silicified wood is common, and coal has been found in the river higher up; also Carboniferous fossils, such as *Favosites Gothlandica*, and others.

Amongst the minerals associated with the diamond are the following. This list, we shall see, is almost identical with that furnished by Bingera:—

1. Black vesicular pleonaste. This mineral has not yet been found at Bingera.
2. Topaz.
3. Quartz.
4. Corundum.
  a. Sapphire.
  b. Adamantine spar.
  c. Barklyite.
  d. A bluish-white variety, characteristic of Mudgee.
  e. Ruby.
  f. Rolled corundum, dirty-white and pink.
5. Zircon.
6. Tourmaline.
7. Black titaniferous sand.
8. Black magnetic iron sand.
9. Brookite.
10. Wood-tin, rare.
11. Garnets.
12. Iron, from tools.
13. Gold.
14. The diamond.

The largest diamond found was 5⅝ carats = roughly 16·2 grains. The average specific gravity was 3·44, and the average weight 0·23 carat, or nearly 1 carat grain each. The carat contains 4 carat grains, which are equal to 3·16 grains troy.

The Newer Pleiocene drift afforded a few diamonds, and being derived partly from the older drift, its materials are somewhat similar; but, in addition to the gems as enumerated, a few grains of osmiridium have been collected from it.

Diamonds have also been found in Victoria, but in no large quantities, and of but small size; but no report of their geological position appears to have been published.

We will now return to the more immediate subject of this note.

THE BINGERA DIAMOND WORKINGS.

The diamond-bearing deposits at present undergoing development are some seven or eight miles, more or less, to the south of Bingera, and are situated in a kind of basin or closed valley amidst the hills; this basin is about four miles long by three wide.

This, together with the surrounding district, is evidently of Devonian or Carboniferous age, but all attempts to procure fossils, in order to verify this, have hitherto failed. As before mentioned, the weather was too wet to allow me to make a proper search myself; in fact, it was only with very great difficulty that one could get about at all in the then state of the country. The weather was so thick, from the pouring rain and constant mists, that but meagre and unsatisfactory glimpses were obtained even of the country's general aspect. Nearly the whole of the basin seems to have been originally more or less covered with drift, parts of it having since been removed by denudation.

Running into the valley are various spurs of basalt, which apparently cover portions of the drift; but at present this is only a conjecture, since the workings have not yet been carried on sufficiently far to show whether this be the case or not, neither by tunnelling under it nor by sinking shafts; but I hope soon to receive information upon this head, for, when on the spot, I suggested that a shaft should be sunk, which will decide the question. Should the drift be proved to pass under the basalt, the known diamond-bearing area will be greatly increased. The probabilities are in favour that it does.

Both the basalt and the drift have undergone much denudation.

The drift is said to be traceable along the course of the river for some (30) thirty miles.

The drift is the forsaken bed of some river, and in all probability that of the Horton.

The rock upon which the diamond drift rests, or the "bed rock" of the minerals, is an argillaceous shale. Outcrops of this are seen in one or two places, but no good section is shown.

In other parts of the ground we see a compact, rather small-grained siliceous brecciated conglomerate, strongly agglutinated together by a ferruginous cement; occasionally the pebbles incorporated in this conglomerate are of rather large size.

In one part of the field the junction of the conglomerate with the argillaceous shale is clearly shown in the cutting formed by a small gully.

Both the shale and conglomerate beds appear to have undergone much disturbance; and at this particular spot diamonds are said to be plentiful on the conglomerate but not on the shale. The surface of the shale is here free from drift, but the conglomerate does not appear to be quite free from it. The miners regard the conglomerate as being of itself diamond-bearing, but this has not been put to any absolute proof.

Up to the present all the diamonds have been found within a foot or so of the surface; in fact, just at the grass roots. In no case have the workings been carried to greater depths than two or three feet; in some parts examined the drift itself is not thicker than that.

In the former sinkings made by the gold-diggers diamonds have occasionally been met with at depths of sixty feet, or even more; but as the men were working for gold, no great attention was paid to the diamonds, and it is quite likely that they fell in from the surface.

The method employed in the search for the precious stones is very simple. The drift is stripped off and carted to the puddling-machines, where it undergoes a great diminution in bulk by the removal of the clay and fine sand; the large pebbles are then screened off, and the clean gravel remaining is passed through one of Hunt's diamond-saving machines. But since this apparatus depends upon the principle of separation by difference in specific gravity, it does not, perhaps, afford the best method which could be devised; it may answer well enough for gold and other bodies of very high specific gravity, but must certainly answer very imperfectly for diamonds, on account of their comparatively low specific gravity, viz., 3·4 to 3·5, which is nearly equalled by most of the accompanying minerals, and exceeded by some.

I should be inclined to recommend the methods employed in Africa and Brazil, since they would probably prove more efficacious.

We may now pass on to consider in more detail the mineralogical nature of the drift, or "wash-dirt" as it is termed by the miners.

From Messrs. Westcott & M'Caw's claims I obtained three different specimens.

Wash-dirt No. 1.—This is a pale brown clay, binding together well-rolled pebbles, subangular and angular fragments of variously coloured jasper—red, green, brown, &c. Also black flinty slate, tourmaline, argillaceous sandstone, shale, and other rocks.

Wash-dirt No. 2.—This is rather darker in colour than No. 1, and the clay is more tenacious; the contained pebbles are of much the same character. The clay has a brecciated structure, and differs in colour in parts, fragments of it being nearly white. On the spot, when freshly dug out, portions of the clay are of a bright green colour, due to the presence of a ferrous silicate, which, by exposure to the air, absorbs oxygen, and passes into the reddish ferric silicate which imparts the red colour to the clay.

Wash-dirt No. 3.—This kind contains a larger proportion of pebbles than either No. 1 or No. 2; it is of a light colour, and much less indurated, being of a sandy nature. This also contains pebbles of argillaceous shale.

Unrolled blocks of the bed rock are met with in all the drift. In all three we find occasional minute crystals of selenite, probably of very recent origin.

During the process of extracting the diamond from the wash-dirt the material is sized as it passes through the machines; but as it is hardly necessary to consider these sands and gravels separately, it will be as well to consider their constituents merely, irrespective of the size, since they all contain nearly the same minerals, although not in the same proportions; but as the large pebbles and boulders which are removed immediately after the stuff is puddled do differ from the finer parts considerably, we shall take them by themselves.

Pebbles and Boulders.—These consist of masses of red, green, brown, and other coloured jaspers; white quartz, common agate, black flinty slate; fine sandstone, into which manganese and iron oxides have infiltered, leaving dendritic markings between the joints. Many of the pebbles are also coated externally in the same way. Nodules of magnesite and concretions of limonite or brown iron ore, of concentric structure—some of the magnesite still showing the limonite *in situ.* Rolled masses of hard, compact, brecciated conglomerate, often containing much manganese in the cement; masses of silicified wood (but this is not very common), cacholong, and greenstone. The rolled masses of sandstone, and especially the argillaceous sandstone, often assume long finger-shaped forms, and are accordingly termed " finger stones " by the diggers. The pebbles are not polished, as at Mudgee.

The list of gems, stones, and other minerals accompanying the diamond includes the following :—

1. *Tourmaline,* or "jet stone" of the miners, occurs as rolled prisms, usually from a $\frac{1}{4}$ to $\frac{3}{4}$ inch long. They usually retain the

trigonal section, but sometimes no trace of crystalline form is left, and they appear merely as more or less rounded black pebbles, often with a pitted surface, totally unlike the usual appearance of tourmaline. The blow-pipe decides their character at once, for they intumesce before it, and in other respects answer to the well-known tests. These "black jet stones" are invariably found with the diamond, and are regarded by the miners as one of the best indications of its presence.

2. *Zircon* occurs in small crystals of red and brown colours, also nearly colourless, but more commonly as rolled pieces of a brown shade. A cleavage plane is usually to be seen.

3. *Sapphire.*—Generally in small angular pieces, and usually of a pale colour; in many the blue tint does not overspread the whole of the fragment. The *Ruby* is present, but very rare. One fragment showed the faces of an acute hexagonal pyramid and basal pinacoid. The lower half of the crystal had been fractured; it was of a red colour, but possessed a purple-coloured central mass. The fragments of sapphire are far less in size than those found at Mudgee and in other places, and far less rolled; the majority appear to have undergone no rounding at all—thus presenting a broad distinction between the gems from the two places. A little corundum is found.

4. *Topaz.*—As rounded fragments, and sometimes with rough crystalline outline. They are generally of a dull yellowish colour, or colourless and transparent, small in size, and often apparently freely fractured.

5. *Garnet.*—In small, rough-looking, ill-formed crystals, of a dull red colour.

6. *Spinelle.*—Not very common, generally in small red or pinkish fragments.

7. *Quartz.*—Small prisms, capped with the pyramid, more or less rolled, transparent, and of a pale dirty red, also smoky; also small jasper pebbles, &c. &c. Amongst the jasper pebbles are some of pale mottled tints of yellow, pink, drab, brown, bluish-grey, &c.; these are termed "morlops" by the miners, and are regarded by them with much favour, as they say they never find one of them in the dish without diamonds accompanying it. Their average specific gravity, taken from a large number, is 3·25. As this is nearly the same as that of the diamond, we can readily understand their being found together. Many must be lost in the washing processes. They are oval in form, smooth, and rarely exceed $\frac{1}{4}$ inch in length. The miners can give no origin for the name, and it does not appear to be mentioned in any works on mineralogy.

8. *Brookite.*—Small flat fragments, very rare.

9. *Titaniferous Iron.*—Rather common.

10. *Magnetic Iron Ore.*—In small grains, showing an octahedral

form under the microscope, coated with hydrated sesquioxide of iron, easily removed by the magnet. Gold in small particles was often found attached to the grains of magnetite.

11. *Wood-tin.*—Rare, in small rolled particles.

12. *Gold.*—Fine grains and scales, present only in small quantity, and the greater portion attached to the magnetite; hence the magnet was found the most ready means of removing it.

13. *Osmiridium.*—In small brittle plates; rare.

14. *The Diamond.*—As already stated, they are for the most part small in size. Some are clear, colourless, and transparent, while others have a pale straw-yellow tint. One or two dark ones, very small, have been seen; also a greenish one. The specific gravity, as deduced from nineteen specimens, is 3·42 (the Mudgee being 3·44).

In some the crystalline form is well and distinctly shown, but others possess very much rounded faces. Some of the best crystals were those of the triakis-octahedron, the triakis-tetrahedron, the octahedron, the tetrakis-hexahedron (or four-faced cube), and the hexakis-octahedron.

No fractured specimens have been detected, but it is rather common to find them with very much pitted surfaces, and with internal black specks.

One of the companies, when prospecting the ground, found the drift to yield as follows:—

6 loads yielded	. . . .	41 diamonds
4½ „	. . . . .	143 „
6 „	. . . . .	88 „
6 „	. . . . .	125 „
6 „	. . . . .	163 „
6 „	. . . . .	89 „
Refuse from machines, &c.	. . . .	41 „
———		
34½		690 diamonds,

or an average of 20 diamonds per ton of stuff, regarding the load as equal to one ton. The above were obtained by the Gwydir Diamond-Mining Company.

The following is an account of the number obtained by Messrs. Westcott & M'Caw from the Eaglehawk claim, up to August 26th 1873 :—

400 diamonds, weighing	. . . . .	192 grains
420 „ „	. . . .	199 „
310 „ „	. . . .	153 „
200 „ „	. . . .	109 „
350 „ „	. . . .	150 „
———		
1,680		803 grains troy.

And, as examples of the number obtained per load of stuff, the following may be cited:—

5 loads yielded 86 diamonds, weighing 32 grains.
8 loads yielded 68 diamonds, weighing 30 grains.

Up to the present no large diamonds have been found, the largest hitherto met with being 1 only of 8 grains, 1 of 4 grains, 6 of 3 grains, 85 of 2 grains, and 1587 of less than 2 grains.

No mention is made of the kind of drift from which the above quantities were obtained; they, however, afford an opportunity of roughly estimating the yield.

"It is reported from Bingera, in the *Tamworth News* of September 26, that Mr. Gardiner has obtained 115 diamonds, and that the Gwydir Company are progressing vigorously. The Giant's Knob is rich in gems, the yield averaging about 140 to the machine full, when the dirt is taken from the diamond drift.

"A correspondent of the *Tamworth Examiner*, on the 12th instant, states that there have lately been large finds of diamonds in the district of Bingera. The Gwydir Diamond Company have prospected now twenty-one pieces of land, nineteen of which have proved to be more or less diamond-producing soil, containing *grupiara* or alluvial deposit, whose surface shows it to be the unused bed of a stream or river; *Burgalhas*, small angular fragments of rocks, bestrewing the surface of the ground; *cascalho*, fragments of rocks and sand mixed up with clay, and forming the bed of a river; and *takoa carza*, which are the above materials cemented together into a conglomerate mass. All the above, however, are known by the generic name of *Cascalho*. The masses of stones themselves, which rarely exceed a cubic foot in size, contain itacolumite, jasper, and peridots and granite. These are the known indications of the whereabouts of diamonds as trusted to and found to be correct both in the East Indies and the Brazils. The nineteen successful prospects of the Gwydir Company have produced each, on an average, thirty-five diamonds to every six loads (of one ton) of washdirt, and they have now by them some 11,000 * glistening pebbles, ready to transmit to Amsterdam, Paris, or some other European continental market, and are at present making extensive arrangements for the formation of three more dams and puddling apparatus on other parts of their land where good supplies of water are to be found. He also gives the following as the find of Messrs. Westcott & M'Caw:—Up to the week ending July 12, 100 diamonds; up to the week ending July 19, 113 diamonds; up to the week ending July 26, 119 diamonds—total 322."—*S. M. Herald*, August 21, 1873.

The only minerals found at Mudgee which have not yet been dis-

* Query 1100.

covered at Bingera are cinnabar and vesicular pleonaste. Bingera, in turn, seems to possess one or two characteristics, such as the magnesite containing the nodules of limonite—these are perfectly spherical at times—and the "morlops" form of jasper. As these are nothing more than small jasper pebbles, careful search would probably prove their presence in most river-drifts containing rolled jasper.

From the foregoing we see that the diamond in Australia is associated with sandstones, shales, conglomerates, and trap-rocks; and perhaps it would not be amiss just to see if this be the case in other countries. Thus, in the Brazils, in Bahia, the matrix of the diamond is said undoubtedly to be a Tertiary sandstone. Burton,* in his book on the Highlands of Brazil, states that it occurs in itacolumite, a metamorphosed palæozoic rock; but these statements require confirmation. This, however, is known indisputably—that they occur in the alluvial drifts of various kinds similar to those of New South Wales.

Diamonds found on the Cuddapah Hills in India are stated to occur in a conglomerate, and between Sangor and Mirzapore in a solid sandstone, and also in a ferruginous conglomerate; and in a gravel at Cuddapore containing pebbles of trap, granite, schist, quartz, jasper, sandstone, and also of the neighbouring limestone; basalt also is found near by.

And at Bangnapilly the diamond is said to have been found in a sandstone, together with corundum and magnetite, as well as in breccia; and the slates there are flinty. The district of Kumarea and Bridgepore is conglomeratic, and associated with sandstone beds. Other diamond-bearing localities of India, also, are conglomeratic.

In Russia, too, conglomerates seem to be the present receptacle of the diamond; iridium is there associated with it.

Borneo—again here we find it in a conglomerate containing quartz, &c., and associated with gold, platinum, and osmiridium.

Then, too, in Africa they are found in a drift, and usually within a few feet of the surface—from three to nine feet, and rarely down so far as thirty feet.†

Here, again, one of the main features of the district is the presence of sandstone, either of Upper Silurian or Devonian age; trap is also present, and a conglomerate or breccia containing boulders of granite, gneiss, mica schist, porphyry, sandstone, jasper, slate, agate, &c.

In conclusion, we are still as much in the dark respecting the origin of the diamond, or even its true matrix, for no good proof has yet been offered on this question. As we have seen, in nearly all cases

---

* See vol. ii. p. 144.

† This was before the workings had been so fully developed. Now the mineral matter found with the diamond at the Cape is seen to be of a totally different character from the Australian alluvial deposits.

it occurs in an ancient river-drift, and is usually associated with sand-stones, conglomerates, and trap rocks. The sapphires from Bingera seem to have undergone but little alteration, and consequently have not travelled far, so that perhaps we may soon light upon their source and that of the diamonds simultaneously. Bingera certainly seems the most hopeful locality to elucidate this point of any at present known.

Before closing this paper I must express my obligations to Messrs. Westcott & M'Caw, and to Mr. Dougherty of the Gwydir Diamond Company, for their great assistance in procuring and sending me suites of specimens illustrating the various rocks and minerals of the diamond workings, and wish them success in their endeavours to open up this industry, which I hope will prove to be a new source of wealth to New South Wales; and this result appears to be highly probable, since the whole of the above-mentioned valuable finds have been made by the exertions of but a few, perhaps not more than five or six workers.

REPORT ON THE DISCOVERY OF DIAMONDS AT BALD HILL, NEAR HILL END.

UNIVERSITY OF SYDNEY, *December* 5, 1873.

To the Hon. the Minister for Lands.

SIR,—In reply to your request of the 2d instant, I have the honour to furnish you with the following particulars relating to the mineral specimens from Bald Hill, near Hill End, which accompanied your communication.

Diamonds.—Three in number; the largest of them is in the form of a six-faced octahedron, rather flattened, owing to four of the groups of faces being more highly developed than the remaining four. The faces and edges are rounded somewhat, but this has not been caused by attrition. Diamonds often appear as if waterworn, but in reality this is seldom the case; the rolled and waterworn appearance is due to the fact that the diamond usually crystallises with curved faces and rounded edges. It is clear and colourless, and perfectly free from all visible internal flaws. The surface is likewise free from flaws; but scattered over some of the faces are a few minute and insignificant triangular markings, but these are quite superficial, and will disappear during the ordinary process of cutting. It possesses a specific gravity of 3·58, and weighs 9·6 grains (troy), *i.e.*, a little over 3 carats. It is generally calculated that diamonds lose one-half their weight during the process of cutting and polishing, and their true value cannot be ascertained until this has been done. The diamond next in size possesses the same crystallographic form as the one above men-

tioned, but is not so much compressed. It has a weight of 4·5 grains
(troy), or nearly one 1½ carat. It has a chip on one edge, and contains a speck of foreign matter. It is a straw-colour. The smallest
diamond weighs about half a grain; it has the form of a six-faced
tetrahedron, and possesses a high lustre, but is rather off colour.

Accompanying the diamonds were two small specimens of gem
sand.

GEM SAND No. 1.

In this the following substances were found to be present:—

1. *Corundum.*—When blue this is known as the *sapphire*, and
when red as the *ruby*.

  a. Common Corundum.—Present in small fragments of bluish,
       greenish, and grey tints.

  b. Sapphire.—In small particles of a blue colour, some so
       dark as to appear almost black, and others very light.
       Some of the fragments still show their crystalline form,
       viz., a hexagonal pyramid; but most of them do not,
       and are either much rolled, subangular, or angular in
       their outline.

The ruby is absent, but probably would have been present had the
sample of gem sand been larger.

2. *Zircon.*—Plentiful, usually in the form of much-rolled pieces.
Generally of a brown colour, sometimes red, and at others nearly
colourless. The small and nearly colourless crystals possess a very
high lustre, almost equal to that of the diamond, so that they might
readily, without careful examination, be mistaken for that gem.

3. *Quartz.*—Usually as small well-rolled grains, either colourless,
milky, or yellowish. Sometimes as hexagonal prisms, capped with
the hexagonal pyramid. Jasper of various colours, such as red, yellow,
grey, also occurs, together with black grains of flinty slate.

4. *Rutile.*—In angular fragments, still showing traces of crystallisation. Distinguished by its brown colour and metallic lustre, and
by the presence of numerous fine striæ on the faces of the prism. It
very much resembles tinstone in appearance. In composition it consists of titanic acid.

5. *Brookite* also occurs. This is another form of titanic acid.
Rutile crystallises in striated tetragonal prisms, whilst this crystallises
in tabular forms belonging to the rhombic system. It is present in
small quantity in the form of flat irregular plates, brown or grey in
colour.

6. *Topaz.*—Present in small rolled and angular fragments, colourless, and in pale tints of yellow and greenish blue. The latter coloured
topaz is often erroneously termed the aquamarine.

7. *Beryl* or *Emerald.*—Doubtful, but one or two very small fragments resembling it.

8. *Garnet.*—Small, rough, common garnets, of no value.

9. *Tourmaline.*—A few rounded pieces, but none showing the crystalline form, which is that of a three-sided prism.

10. *Gold.*—Present in the form of scales.

GEM SAND No. 2.

This consists of larger grains than No. 1; in fact, they are small pebbles.

1. *Quartz.*—Present principally in the form of jasper, of various colours—red, brown, green, yellowish, &c. &c.; also variegated. Colourless and yellow quartz pebbles are also found, together with black pebbles of flinty slate.

2. *Corundum.*—Present as common corundum, and as the sapphire.

3. *Brookite.*—Same as gem sand No. 1, only in larger pieces.

4. *Topaz.*—Clear and colourless; also tinted.

From the foregoing it will be seen that the Bald Hill gem sands very closely resemble those from Bingera and Mudgee.

None of the gems contained in the parcels submitted to me, with the exception of the diamonds, are of any commercial value, except for grinding and polishing purposes. Still, they are of great value as indications, for where such occur there is every prospect of finding others of larger size and better quality.

An examination of the original "wash-dirt" or "drift" might yield valuable information, and larger samples of the gem sand will probably be found to contain such minerals as iridium, titaniferous iron, tinstone, magnetite, &c., like the Bingera gem sand.

I may, perhaps, mention that in 1867 a brilliant of the first water, and without flaw, weighing one carat, was worth about £20; if weighing one and a half carat, about £45; and if two carats, about £80, and so on; but since that time the prices have probably undergone much change. According to the September number of the "British Trade Circular," the prices ruling for Cape diamonds, uncut, are in proportion much lower than the above.

I return the diamonds and gem sand per bearer.—I have, &c.,

A. LIVERSIDGE.

PRELIMINARY NOTES ON SOME ROCKS AND MINERALS FROM NEW GUINEA, &c.[*]

The specimens mentioned as having been obtained from the Fly River were collected by Signor D'Albertis, and those from Yule Island, Gulf of Papua, by the late Captain Onslow, R.N.

An account of these specimens was not published earlier because I have hitherto hoped to make chemical and microscopical analyses of the most important, but this work must be postponed for a more favourable time. I also expected to receive an account of the fossils which accompanied the rocks from Professor P. Martin Duncan, to whom they were sent for description.[†]

The fossils were mainly of a Mesozoic character, including *belemnites, pectens, ammonites, carcharodon teeth, coral,* &c., and seem to have been derived from Cretaceous rocks.

Signor D'Albertis states that all the specimens which are not otherwise noted are from the Fly River, in about lat. 5° 30′ S., some 250 miles in the interior.

Quartz.—In the form of white rolled pebbles with a few imperfect crystals in parts, apparently derived from a vein.

These were tested for gold by the dry way, but none was found to be present.

Some of the quartz pebbles were very brittle, and had evidently been subjected to the action of fire; probably they had been used for lining cooking-pits.

Quartzite.—Grey, almost white, tender, glistening, and under the microscope most of the grains present one or more crystal faces, and many are fairly complete hexagonal pyramids. So friable that it can be rubbed down between the fingers. These pebbles also have probably been used in fireplaces. They were found to be free from gold.

Also heavy, hard, and compact quartzite pebbles, containing minute disseminated particles of iron pyrites; the specific gravity is 2·599. The quantity of pyrites is very small; these were found to contain minute traces of gold, but only by operating upon large quantities.

---

[*] Read before the Royal Society of New South Wales, December 1, 1886.

[†] Professor Duncan has since informed me that they are too imperfect for description.

Some of the pebbles of quartzite were black, others only black in the interior; the former had probably not been burnt in the native fireplaces.

Flint.—In the form of light brown pebbles, somewhat like those from the Woolwich beds; others containing more impurity, and some resembling Egyptian jasper.

One flint pebble bears the imperfect cast of a *pecten*, and in others are markings somewhat spongiform in appearance. One is perforated by a tube about 2 inches long, containing a movable core of light porous siliceous matter about ¾ inch in diameter.

In one case the flint has a specific gravity of 2·586, and in another 2·570.

Conglomerate.—A pebble containing rolled fragments of white quartz, pale brown felspar, some showing twinned crystals, and black jasper or basalt, cemented together by a black ferruginous paste.

Basalt.—In the form of a pebble of porphyritic structure, and containing large crystals of augite. Another specimen, in the form of an almost spherical pebble, broke with a subconchoidal fracture, and showed weathering for about ⅛ inch from outer surface.

The specific gravity was low, being only 2·678.

Porphyry.—Small pebble of a red colour, an intimate mixture of red orthoclase felspar and quartz, a little hornblende and mica present. The exterior of the stone is quite black, and probably, like most of the others, was taken by Signor D'Albertis from a fireplace.

Iron Pyrites.—In dark liver-coloured masses, showing radiate structure when broken open (marcasite); in appearance very like that from the Lias clay. Rapidly oxidises to sulphate.

Was found on assaying to yield minute traces of gold.

Limonite.—In the form of nodules with a concentric structure. Some hollow inside; others have a loose nucleus or kernel inside, and in flat cake-like masses, of fair quality as an ore of iron.

Also as bright red and yellow ochres; one specimen was in a calabash, and the other wrapped in ti-tree bark, probably ready for use as pigments and personal adornment.

There are also specimens of ferruginous unctuous clays of red, brown, pink, and grey colours, from the banks of the Katau and Fly Rivers.

Limestone.—Compact, somewhat crystalline, no fossils, weathered outside; evidently from the sea-shore, since one piece has an oyster-shell attached.

A fragment of stalactitic aragonite, but blackened and saturated with tarry matter.

One stalagmitic mass was found to be hollow like a geode, and the cavity lined with small quartz crystals.

Amongst the calcareous specimens are pebbles made up of rolled masses of corals cemented with calcareous matter; in other cases made up of silt and volcanic ashes, also cemented together by calcium carbonate.

Also white and grey slightly crystalline limestones containing fossil corals. One piece of limestone has weathered into a peculiar spathulate form, and might almost have been mistaken for a native spoon.

There are several specimens of a yellow granular friable limestone containing fragments of various fossils, corals, &c., with waterworn surfaces and bearing attached oyster-shells in places.

Lignite.—Yarrn Island, three miles from the mainland of New Guinea. Of a brown colour, with lamellar structure where weathered, somewhat like dysodile; exhibits a woody structure, gives a brown streak or powder, breaks with a well-marked conchoidal fracture yielding a black lustrous surface. Burns with difficulty, without flame, emits but little smell, and leaves a dark grey ash, very bulky and resembling a wood ash.

It is not sectile, but flies off in powder before the knife.

Fossil Wood.—Many specimens were found in the natives' fireplaces, some of them presenting very curious and fantastic shapes; the woody structure is well preserved; on the outside they are brown, but black and porous within. The black portion when crushed and ignited on platinum foil glows feebly and leaves a white ash, but can hardly be said to burn. This mineralised wood—for it consists mainly of silica—is probably used by the natives merely to line their fireplaces, just as they use the rocks and pebbles, and not for any value it possesses as a fuel.

Pumice (from the Katau River).—In the form of a rolled mass of a light grey colour, very similar in appearance to the common white pumice of New Zealand and other Pacific sources.

Most of the stone implements, adzes, and club-heads, many of the latter being in the form of rings, stars, crescents, &c., are made of hard igneous crystalline rocks; some are apparently from modern lavas, but others of older basalts, diorites, and porphyries; but it was difficult to determine the rock with certainty in many cases, on account of the glaze-like black polish which they possessed, and which Signor D'Albertis did not wish removed; to obtain fresh fractures was, of course, out of the question without also destroying their value as ethnological specimens.

The egg-shaped stones are fashioned out of a variegated limestone, and the sharpening-stones or hone is a grey felsitic rock.

The following specimens were collected by the late Captain Onslow, R.N., in August 1875 :—

BAXTER RIVER, NEW GUINEA.

Rolled nodules of white vein-quartz, with a pebble of grey-coloured chalcedonic quartz; of a basalt and of a dark grey felsitic rock. Another specimen is a pebble of a dark-coloured very tough diabase rock.

Concretions of impure red hæmatite, like those occurring in the Waianamatta shale, and, like them, probably derived from a shale.

Various specimens of alluvial deposits obtained from the river-banks, passing from loose grey and black soils to stiff and tenacious red clays. One of these is laminated like a shale and contains mica scales. It is probably the source of the ironstone concretions.

The scales of mica apparently indicate the presence of older crystalline rocks in the interior.

Obtained from thirty-five miles up the river. Some of the soils are very full of vegetable matter, and should be very fertile.

Bole.—Of a red colour, also variegated red and white; adheres to the tongue, and scales off in fragments with a conchoidal fracture; gives shining streak, and is subtranslucent. Falls to pieces when placed in water with a slight crackling sound, and emits a rapid stream of minute air-bubbles.

YULE ISLAND.

The specimens from Yule Island consist simply of quartz pebbles. Some pieces of white more or less crystalline limestone, apparently from a coral reef; and others of a grey colour, one of which, evidently coral reef *débris,* enclosed rolled white quartz and other pebbles.

DARNLEY ISLAND.

With the above are a few specimens from Darnley Island. These consist of a black vesicular lava containing some carbonate of lime in the cavities; a rolled nodule of basalt, and a fragment of a buff-coloured tufaceous rock, containing a considerable amount of carbonate of lime in the form of recognisable fragments, but most of it disseminated in fine grains mixed with iron oxide.

On treating a fragment of the rock with hydrochloric acid, it dissolves in great part, and leaves a grey-coloured residue which, under the microscope, has the appearance of a volcanic ash; fragments of green augite, brown and colourless crystals, being abundant.

Chrysolite.—Small pale green rolled pebbles, fairly transparent; all external crystalline form worn off. Collected by the Rev. G. Brown in Samoa.

Aragonite.—In the form of stalactites from Tanna.

Sulphur, sulphates, and similar volcanic minerals are common in Tanna and some of the other islands.

As a whole, the variety of minerals at present obtained from the islands and New Guinea is very limited.

ROCKS FROM NEW BRITAIN AND NEW IRELAND.*

The specimens forming the subject of this notice were collected in the year 1875 by the Rev. George Brown, Wesleyan missionary, to whom my thanks are due for the opportunity to examine these and other specimens which he has brought from the islands from time to time.

SPECIMENS FROM NEW IRELAND.

Porphyry.—In the collection are several well-rounded pebbles of porphyry. In all of them the felspar crystals are small, but fairly well defined, embedded in a felspathic base or paste. In most cases the base is some shade of green, the colour varying from a light to a dark green; in one case the base is reddish-brown or chocolate-colour, and in this instance some of the felspar crystals are somewhat larger—one being about $\frac{1}{4}$ inch across.

Diorite.—Composed of a white felspar and quartz, with dark green hornblende, without mica.

Calcite.—In the form of veins, some nearly an inch thick, running through a dark brown rock, which is evidently of igneous origin, and possesses a slightly porphyritic structure in part, with obscure white felspar crystals. Almost colourless, and readily yielding cleavage rhombs.

Limestone.—In various forms—one somewhat crystalline, with reddish-brown streaks, might be described as a marble; does not appear to have been derived from recent coral rock.

A pebble of dark grey compact limestone, from the Mata-Kau River, but weathered almost to a white colour for about $\frac{3}{81}$ inch from surface.

A somewhat crystalline light grey limestone, from mountain 2500 feet high. In appearance it somewhat resembles a coral limestone; none or but very obscure indications of organic structure—not even on the external surface, which is much weathered.

Amongst the specimens are two rounded nodules of calcareous mudstone, containing some remains of branching corals, probably recent or living forms, but they are so much rolled that their structure is very obscure.

* Read before the Royal Society of New South Wales, December 8, 1880.

Pale brown calcareous mudstone, looks at first sight much like a sandstone—contains much volcanic ash.

A difficulty in properly describing some of the specimens is caused by the fact that they are merely small detached and rolled fragments.

Ancient Volcanic Ash.—Having the appearance of a dark-coloured conglomerate, hard and compact, made up of red, brown, black, and other pebbles embedded in a dark green felspathic base, which is porphyritic in parts, from the presence of disseminated white and grey crystals of felspar.

Jasper Pebbles.—One of a beautiful deep red in part, with patches of white quartz. A cavity on one side is filled with a porphyry made up of a dark green base with small disseminated white felspar crystals. The porphyry resembles that of the old volcanic ash conglomerate, and is probably part of it; the jasper pebble, however, does not look fused at all, but merely rolled.

Sandstone.—Pale brownish-grey, marked with thin dark parallel layers, evidently planes of stratification. The dark bands are rendered so by the presence of small hornblende or augite crystals, readily discernible under the microscope, being more or less transparent, and of a green colour, but not with the unassisted eye.

Epidote Rock.—A pebble apparently made up of a felspar with thin veins of epidote.

Decomposed Porphyry.—A red ferruginous pebble, breaking with an earthy fracture, darker mottlings in parts, and marked with a few white specks and very thin felspathic veins. Probably a decomposed ferruginous igneous rock.

Alluvial Deposit, from the river. Brown in colour, very fine grain, perforated with worm-burrows.

Another specimen is labelled " Stones and earth from the interior of New Ireland, 2500 feet." The earth is simply a light porous clay-coloured soil, but the stone looks like a much decomposed trachyte; under the microscope it is seen to retain traces of a crystalline structure, and green-coloured translucent crystals, of what appear to be hornblende, are abundant.

Lava, from river-bank. A dusky purple slate-coloured igneous rock, full of small amygdaloidal cavities. The cavities are, for the most part, about $\frac{1}{8}$ inch across, but some are nearly 1 inch long, but the width and depth not more than about $\frac{1}{8}$ inch to $\frac{1}{4}$ inch.

These cavities are arranged in fairly regular layers, and are drawn out in the direction of the flow of the once fluid lava.

Many of the cavities are filled with quartz; the central parts consist of small more or less perfectly developed transparent crystals, seated upon a lining of chalcedony. Some contain a thin velvety coating of crystallised chlorite, and others are completely filled with chalcedony.

Chemical Composition.

Loss at 100° C.	·402
Silica	67·664
Alumina	15·402
Iron sesquioxide	1·963
„ monoxide	3·491
Manganese monoxide	·762
Lime	2·963
Magnesia	trace
Carbon dioxide	„
Potash	1·220
Soda	6·010
	99·877

Tough, breaks with a fairly even fracture; the cleavage planes of elongated twin crystals of felspar, embedded in a granular paste, are well shown in places.

Specific gravity, in powder, at 17° C. = 2·694.

Amongst the New Ireland specimens is a rock with bright green mottlings, looking almost like a serpentine, but it is not serpentine, probably a decomposing igneous rock.

SPECIMENS FROM NEW BRITAIN.

Volcanic Conglomerate.—Composed for the most part of rounded fragments of a dark-coloured igneous rock, probably basalt, with lighter-coloured and greenish pebbles cemented together by black and dark green pastes. This specimen is very much less compact than those from New Ireland; the pebbles are so loosely bound together that they can be separated by the fingers, the paste being comparatively soft, and mixed with delessite (?) in parts.

Pumice.—Most of the specimens are black. One specimen is of a pale brown colour, and is rather more vesicular than the black pumice; this on analysis yielded the following results:—

Chemical Composition.

Loss at 100° C.	2·025
Combined water, by direct weighing	5·975
Silica	56·566
Alumina	17·820
Iron sesquioxide	2·910
„ monoxide	2·645
Manganese monoxide	·841
Lime	5·106
Magnesia	traces
Potash	2·610
Soda	3·094
	99·592

Specific gravity, 2·359 at 21·2° C.

Lava, from the volcano, New Britain. With a dark grey base, almost black, containing crystals of glassy felspar.

Some of the specimens are of low specific gravity, and very scoriaceous. Certain of the cavities contain a white powdery mineral partly soluble with effervescence in hydrochloric acid.

Chemical Composition.

Loss at 100° C.	·119
Loss at red heat	·390
Silica	57·465
Alumina	19·200
Iron sesquioxide	3·833
„ monoxide	3·223
Manganese monoxide	·974
Lime	9·353
Magnesia	·487
Soda	2·470
Potash	1·358
Carbon dioxide	trace
Sulphur trioxide	·225
	99·097

Specific gravity, 2·738 at 21·2° C.

Some of the small specimens of lava are vermiform or worm-like in shape, others are in the form of lapilli, &c.

Obsidian, or volcanic glass, from the volcano, New Britain. Some of it is black in colour, but greyish in parts; more or less parallel greyish bands also occur in it. One specimen of a pitchy black colour contains a few scattered felspar crystals, and another is, in addition, characterised by the presence of vesicular cavities.

Sulphur, from the crater of the volcano in Blanche Bay, in the form of small pieces, evidently broken off an incrustation, of about $\frac{1}{2}$ inch to $\frac{3}{4}$ inch in thickness, very clean and of a bright sulphur-yellow colour, probably very pure. The cavities in it are lined with small crystals, but for the most part it is somewhat friable and resembles flowers of sulphur.

Gypsum, also found in the crater with the sulphur, in the form of acicular crystals.

Aragonite.—In the form of nodular masses, seen on fracture to be built up of beautiful transparent columnar crystals arranged in a radiate manner. They look as if they had been set free from amygdaloidal cavities in igneous rocks.

Limestone.—White, granular.

Quartz.—Of a chalcedonic character.

Unfortunately none of the specimens contain fossils, so that they throw but little light upon the geological age of these islands.

The rolled pebbles of porphyry, epidote rock, and others of the crystalline rocks from New Ireland indicate the presence of much more ancient rocks in that island than do any of those examined from New Britain; they cannot, however, without further evidence, be assigned to any particular geological period, for such metamorphosed rocks may belong to almost any of the older series. Most of the specimens from New Britain, on the other hand, are apparently of comparatively recent geological origin; some of the specimens of limestone may even be of existing coral growth, but there is no trace of organic structure remaining to indicate the age of the rock, the structure being subcrystalline.

The igneous rocks are all doubtless modern volcanic products.

ON THE COMPOSITION OF SOME PUMICE AND LAVA FROM THE PACIFIC.*

Pumice.—Masses of pumice are frequently cast upon the beach along the coast of New South Wales, and at times are also found in the harbours, and they are not infrequently picked up within the Sydney Harbour.

The source of this pumice is, of course, a foreign one, and doubtless it is derived from more than one of the volcanic centres in the Pacific, but which of them does not as yet appear to be very clear.

It is always waterworn, and at times more or less coated with serpulæ, and has evidently been long in its travels across the sea. It is stated to be more abundant after an easterly gale, and is found more often on the north side of the inlets along the coast than in other situations; in size the pieces vary from quite small fragments to pieces 9 or 12 inches through.

Some of the specimens are black and others are white, or rather of a dirty white or grey colour.

Thinking that the chemical composition might throw some light on their source, analyses were made of a specimen of each variety, with the following results:—

Black Pumice.—Bondi Beach.

Chemical Composition.

Moisture	·147
Silica	63·630
Alumina	17·994
Iron sesquioxide	5·838
„ monoxide	traces
Manganese monoxide	·691
Lime	4·205
Magnesia	none
Soda	4·252
Potash	3·809
	100·566

Specific gravity = 2·307 at 15° C. in powder.

White Pumice.—Bondi Beach; much waterworn.

* Read before the Royal Society of New South Wales, December 1, 1886.

Chemical Composition.

Moisture	1·818
Silica	68·149
Alumina	16·493
Iron sesquioxide	3·255
„ monoxide	none
Manganese	·256
Lime	4·005
Magnesia	none
Soda	3·881
Potash	1·590
	99·447

Specific gravity = 2·107 at 18° C. in powder.

I have not been able to analyse any of the specimens of pumice from the islands, so that no direct comparison can be made between their composition and that of the drift pumice, but on comparing the above analysis with those of the Krakatoa pumice of the 1883 eruption, a very great resemblance is at once seen, although there are points in which they differ in composition, the most striking being the absence of any magnesia in the Bondi specimens. The Bondi specimens were, however, collected some years before the Krakatoa eruption; they may possibly have been drifted across from previous eruptions in that district.

It would be very interesting to trace the limits of the distribution of pumice along the Australian coast, and I trust that some one will undertake this duty.

Lava.—Chocolate-coloured, from the island of Tanna, very vesicular and almost a pumice in structure (see "Rocks from New Britain and New Ireland," A. Liversidge, F.R.S., *Journal of the Royal Society of New South Wales*, 1882); and in specific gravity it is just over 1·0, since it slowly sinks in water, but the powder has about the usual density of rocks, viz., specific gravity, 2·720 at 21·2° C. in powder.

It also contains small white glassy crystals of felspar. This specimen was collected by the late Commodore Goodenough.

Chemical Composition.

Moisture	·201
Silica	57·041
Alumina	19·512
Iron sesquioxide	5·499
„ monoxide	2·714
Manganese monoxide	2·053
Lime	8·157
Magnesia	none
Soda	2·831
Potash	2·375
	100·383

Lava.—From Port Resolution, island of Tanna. This is a black vesicular rock, with a pitchy or resinous lustre on the freshly fractured surfaces. The weathered surfaces are brown, and in the interior of some of the vesicles scattered through the lava are white glassy crystals of felspar.

Chemical Composition.

Hygroscopic moisture	·139
Silica	53·312
Alumina	9·007
Iron sesquioxide	17·339
„ monoxide	2·002
Magnesia	·727
Lime	9·058
Manganese monoxide	1·443
Soda	3·417
Potash	3·347
	99·791

Specific gravity = 2·686 at 15° C.

The proportion of iron to this rock is unusually large.

Lava.—From the island of Tanna. This is a black scoriaceous and vesicular lava; the freshly fractured surfaces are highly lustrous, like the former lava, it contains small isolated glassy crystals of felspar.

Chemical Composition.

Moisture	·241
Silica	56·755
Alumina	21·096
Iron sesquioxide	4·521
„ monoxide	3·021
Manganese	traces
Magnesia	„
Lime	9·014
Soda	2·804
Potash	3·272
	100·724

Specific gravity = 2·666 at 15° C.

Thus there is a considerable difference in the composition of these two lavas.

Comparative Table.

	White Pumice, Bondi.	Black Pumice, Bondi.
Moisture at 100° C. . . .	1·818	·147
Silica	68·149	63·630
Alumina	16·493	17·994
Iron sesquioxide . . .	3·255	5·838
„ monoxide . . .	...	traces
Manganese monoxide . .	·256	·691
Lime	4·005	4·205
Magnesia	none	none
Soda	3·881	4·252
Potash	1·590	3·809
	99·447	100·566
Specific gravity . . .	2·107	2·307

Description.	Krakatoa, 1883.			Chocolate Lava, Tanna.	Lava, Port Resolution.	Scoriaceous Lava, Tanna.
	1.	2.	3.			
Loss on ignition	2·17	2·74	2·12	·201	·139	·241
*Silica	63·30	65·04	68·06	57·041	53·312	56·755
Alumina	14·52	14·63	15·03	19·512	9·007	21·096
Iron sesquioxide	5·82	4·47	·28	5·499	17·339	4·521
„ monoxide		2·82	3·66	2·714	2·002	3·021
Manganese	·23	trace	trace	2·053	1·443	traces
Lime.	4·00	3·34	2·71	8·157	9·058	9·014
Magnesia	1·66	1·20	·81	none	·727	traces
Soda	5·14	4·23	4·25	2·831	3·417	2·804
Potash	1·43	·97	3·41	2·375	3·347	3·272
*Titanic acid	1·08	...	·38	...	...	...
	99·35	99·44	100·71	100·383	99·791	100·724
Specific gravity	...	...	...	2 720	2·686	2·666

No. 1 by Sauer, No. 2 by Renard, No. 3 by K. Oebbeke, *Journ. Chem. Soc.*, 1884, p. 974-5.

REMARKS ON THE NEW ZEALAND HOT SPRINGS.

At a meeting of the Chemical Section of the Royal Society of New South Wales in 1877 Professor Liversidge exhibited some interesting specimens of the siliceous and other deposits from some of the hot springs in New Zealand. He stated that he was not prepared with any paper descriptive of them, or of the geology of the localities whence they were obtained, for such would take up much more time than he could at present devote to the matter; and, moreover, such a paper would, perhaps, be more or less superfluous, after the many able descriptions of these springs which had been already published by various observers. He would only trouble them with a few remarks upon certain of the specimens, and would invite their attention to the series of photographs placed on the table, in lieu of any description of the place. Amongst the specimens were some samples of siliceous sinter from a spring opposite the hotel at Ohaiawai, on the road between the Bay of Islands and Hokianga, on the west coast; also of cinnabar from the mercurial springs near to the same place. Professor Liversidge mentioned that he was much struck by the general similarity between the "volcanic" phenomena at Ohaiawai and those presented by the burning coal-seam at Mount Wingen, the so-called "burning mountain" near to Scone—the chief differences being that the phenomena at the latter place are confined to a more limited area, and that water is absent; hence there are no hot springs or pools of warm water at Mount Wingen, as there would be if the jets of steam and heated vapour had to make their upward passage through water. The escaping gases at both places possess very much the same general character, and deposit similar sublimates of sulphur and certain volatile salts around the vents. At the hot mercurial springs near Ohaiawai the mercury occurs both in the native state and in the form of cinnabar; some of the cinnabar is apparently of recent deposition, since it was observed in one place to uniformly and completely fill certain small cracks and crevices existing in the shaly rock, but the greater part of it is evidently mechanically brought to the spot by the small stream which runs down to the lake. The presence of an extensive deposit of coal, shale, or pyrites undergoing a smothered combustion would be quite sufficient to account for the phenomena observed at Ohaiawai and other hot springs in New Zealand—even for the celebrated ones at Ohinemutu and at the Pink and White Terraces on the Lake Rotomahana. Respecting the beautiful bright blue colour of the water in the basins on the Pink and White Terraces at Rotomahana, a blue so extraordinarily beautiful that many travellers are unable to find words to express their admiration for it, Professor Liversidge explained that the blue colour was due to the reflection of the light from the innumerable minute particles of silica suspended in the water, just as the pale sky-blue colour of a mixture of milk and water is caused by the reflection of the light from the minute fat globules suspended in it. He was also inclined to attribute the equally beautiful blue colour of the lower layers of steam floating over the surface of the boiling waters to a similar cause, for he had but little doubt that the escaping steam bears minute particles of silica with it in its upward course. The colour of the steam does not appear to be due to a reflection of the colour of the water below it, any more than the colour of the water in the basin is due to a reflection of the sky. The beautiful opalescent blue of the water in the basin is very different in appearance from the clear but equally magnificent purple blue of one of the *ngwahas* or boiling springs near Ohinemutu. This latter blue water is remarkably transparent, and one can see down through it to very great depths, but the Terrace water is rendered too turbid by the silica in suspension for the eye to penetrate far down into it. Part of the silica brought up by the water is thrown down as a soft pulverulent deposit at the bottom of the basins, while another portion is more slowly precipitated as a hard and smooth stony material, known as siliceous sinter or geysirite, and it is of this that the terraces and basins are built up. So rapidly is this deposited that leaves and twigs become quickly invested, and it is stated that even dead birds become coated with it before the animal matter has time to decay and fall to pieces. The pink colour of the Pink Terrace is apparently due to the entanglement by the asperities on the stalactitic faces of the terraces of small quantities of red clay brought down by the water. A specimen of recently formed iron pyrites, which had been taken from a mass now forming upon some dead twigs in one of the hot springs (Jack Loffley's, the Taupo guide) at Lake Taupo, was also shown. On examination, this mass of mixed newly formed iron pyrites and dead vegetable matter was found to contain traces of gold.

ON THE OCCURRENCE OF CHALK IN THE NEW BRITAIN GROUP.*

In the following brief notice it is my wish to communicate to the Society a description of the physical properties and chemical composition of one of the geological specimens recently brought from the above group of islands.

The specimen which I now have the pleasure to lay before you is not only interesting in itself as an example of what is known as an organically formed rock, since it is built up almost entirely of the calcareous skeletal remains of organic forms, but it is interesting in a still higher degree, as it apparently indicates that a most important geological discovery has been made of the presence of chalk in a hitherto unknown and even unsuspected locality.

In October last the Rev. G. Brown, Wesleyan missionary, brought (amongst other specimens) from New Britain and New Ireland (New Britain group, lat. 4° S. and 150° E. long.) certain grotesque figures of men and animals which had been carved by the natives of the above islands out of a soft white somewhat pulverulent material, having much the appearance of plaster-of-Paris or chalk.

Some of these figures were deposited in the Museum, and a fragment broken off from one of them was placed in my hands for identification.

On examination, the remains of numerous foraminifera are at once detected, the forms of the larger ones being plainly visible even to the unaided eye; under the microscope the whole mass of the rock is seen to be almost entirely composed of the shells and fragments of shells of foraminifera, the remains of globigerina being most abundant.

To obtain the shells of the foraminifera free from the cementing calcareous matter, it is only necessary to gently rub the surface of the specimen with a soft tooth or nail brush under a stream of water, when the whole surface of the fragment submitted to the operation speedily becomes studded with the minute shells and fragments of shells of foraminifera, now left standing out in relief.

To obtain the foraminifera perfectly free from the accompanying powder, it is sufficient to dry the collected *débris*, and to place it upon

* Read before the Royal Society of New South Wales, July 4, 1877.

the surface of some clean water contained in a glass beaker or other vessel; the larger and more cavernous foraminifera float on the surface of the water, while the broken fragments, much of the amorphous powder, and many of the denser foraminifera are deposited at the bottom of the vessel as a sediment. The very light and finely divided parts are got rid of by decanting the milky supernatant liquid.

In the sediment the microscope reveals the presence of the smaller foraminifera, of a few sponge spicules, and minute grains of what are evidently siliceous and igneous rocks.

The further examination showed that the material is limestone, having a very close resemblance to chalk, both in chemical composition and in physical properties; in colour it is not the dazzling white of some chalk, but bears a closer resemblance to the light grey varieties.

Although it is essentially composed of carbonate of lime, still it is by no means pure; there are certain impurities present, in the form of alumina, iron, silica, manganese, and other substances; but reference will again be made to this question later on.

To ascertain whether my supposition that the rock might be regarded as chalk and not merely as a soft white friable recent limestone, or as a deposit such as is now forming over parts of the beds of the Atlantic and Pacific Oceans, I took an early opportunity, when writing, to enclose a portion of the material to Mr. H. B. Brady, F.R.S., of New-castle-on-Tyne, who has devoted himself to the study of foraminiferous deposits, and who is recognised as one of the first authorities upon these matters. I have since received a reply from him, in which he says :—

" First let me speak of your chalk from the New Britain group. I suppose you have ascertained that it is a Cretaceous chalk, and not a friable Tertiary limestone. All the foraminifera, or nearly so, are South Atlantic recent deep-sea species, *Globigerina bulloides, G. inflata, Pulvinulina Menardii* (a thick variety which I do not think is yet named), *P. Micheliniana,* and probably *P. Karsteni, Pullenia spheroides, Nonionia depressula, Bulimina Buchiana,* fragments of *Dentalina, Uvigerina,* &c. ; also a characteristic *pulvinulina* with thick shell and honeycombed surface, not yet described, of which I have quantities in the *Challenger* material. . . . The whole of the *Challenger* foraminifera have been handed over to me to work out."

In answer to a question as to the locality and mode of occurrence of the material used for the carvings, the Rev. G. Brown wrote to me as follows :—

" The chalk of which the figures are formed is, I am informed, only found on the beach after an earthquake, being cast up there in large pieces by the tidal wave; it is only found, so far as we know at present, in one district on the east side of New Ireland."

We have now to consider its chemical composition in somewhat closer detail, and to compare the results furnished by it on analysis with those yielded by specimens of typical or true chalk.

Chemical Composition of Specimen from New Ireland.

Hygroscopic moisture, *i.e.*, water driven off at 100° C.	1·202
Carbonic anhydride	35·337
Iron sesquioxide	1·597
Alumina	3·131
Silica	7·933
Phosphoric acid	minute trace
Manganese protoxide	·623
Lime	45·278
Magnesia	·476
Potash	·308
Soda	·260
Chlorine	·105
Combined water and loss	3·750
	100·000

Specific gravity = 2·199 at 59° Fahr.

The specific gravity was taken from a mass weighing about 78 grammes, which was allowed to soak in water for about one hour and a half; in fact, until all air-bubbles ceased to be evolved; a small quantity of the block scaled off when immersed in the water, a correction for which had to be made.

The above figures show that, in round numbers, about 81 per cent. of the specimen consists of calcium carbonate; thus it is undoubtedly a far less pure limestone than the ordinary white chalk, as the following figures indicate:—

Chemical Composition of Chalk from other Places.

A specimen of chalk from near Gravesend, which was analysed by Mr. W. J. Ward, yielded the following results:—

Calcium carbonate	98·52
Magnesium carbonate	·29
Calcium sulphate	·14
Manganese binoxide	·04
Phosphoric acid	traces
Organic matter	...
Insoluble matter, chiefly silica	·65
	99·64

Mr. David Forbes, F.R.S., also examined some specimens of chalk, the analyses of which are here cited. The first analysis shows the composition of a piece of white chalk from Shoreham, in Sussex, and the second of a piece of grey chalk from Folkestone.

	White Chalk.	Grey Chalk.
Calcium carbonate	98·40	94·09
Magnesium carbonate . . .	·08	·31
Phosphoric acid } Alumina and loss }	·42	trace
Sodium chloride	...	1·29
Water	...	·70
Insoluble rock *débris*	1·10	3·61
	100·00	100·00

(*Vide* "Geology of England and Wales," Woodward, p. 239.)

Another sample of chalk obtained from a well at Driffield was found by Mr. T. Hodgson to have the following composition:—

Moisture	5·20
Calcium carbonate	93·30
Magnesium carbonate	·15
Iron sesquioxide and alumina	·20
Silica	1·15
	100·00

The specimen from New Ireland closely resembles in chemical composition the chalk-like rock occurring in New Zealand.

Dr. Hector, C.M.G., F.R.S., Director of the Geological Survey of New Zealand, publishes in his *Annual Report* for 1875–6, the description and analysis of a limestone made by Mr. Skey, chemist to the Survey, as follows:—

"No. 1767. Chalk, contributed by Mr. H. Higginson, from South Canterbury; very closely resembles some taken from the same district by the Survey some time since. These samples, as to their physical and chemical character, also their general appearance, exactly represent the chalk of the Cretaceous formation as occurring in England."

Analysis.

Carbonate of lime	84·12
Carbonate of magnesia	2·10
Clay	12·57
Iron oxides and alumina, soluble in acid	1·21
	100·00

It is, however, far less impure than the "chalk mud" of the Atlantic for the analyses quoted by Professor Sir Charles Wyville Thomson, F.R.S., in his "Depths of the Sea," p. 469, show that the "chalk mud" contains merely some 60 per cent. of calcium carbonate, and with as much as from 20 to 30 per cent. of silica, and varying proportions of alumina, magnesia, iron, and other substances.

The same author mentions that the typical chalk is free from silica, and so it would appear to be from the above-quoted analyses; but the "insoluble rock *débris*" of the late talented David Forbes, F.R.S., probably consisted largely of silica.

The only locality for chalk in the Pacific Islands to which I can find any reference occurs in Professor Dana's work on "Corals and Coral Islands" (see p. 308). But this even is not true chalk; it is merely a recent limestone derived from disintegrated corals, and which resembles chalk.

Mr. Dana there says:—

"The formation of chalk from coral is known to be exemplified at only one spot among the reefs of the Pacific.

"The coral mud often looks as if it might be a fit material for its production. Moreover, when simply dried it has much the appearance of chalk, a fact pointed out by Lieutenant Nelson in his memoir on the Bermudas (1834), and also by Mr. Darwin, and suggested to the author by the mud in the lagoon of Honden Island. Still, this does not explain the origin of chalk, for, under all ordinary circumstances, this mud solidifies into compact limestone instead of chalk, a result which would be naturally expected. What condition, then, is necessary to vary the result and set aside the ordinary process?

"The only locality of chalk among the reefs of the Pacific, referred to above, was not found on any of the coral islands, but in the elevated reef of Oahu, near Honolulu, of which reef it forms a constituent part. It is 20 or 30 feet in extent, and 8 or 10 feet deep.

"The rock could not be distinguished from much of the chalk of England; it is equally fine and even in its texture, as earthy in its fracture, and so soft as to be used on the blackboard in the native schools.

"Some embedded shells look precisely like chalk fossils. It contained, according to Professor Silliman, 92·80 per cent. of carbonate of lime, 2·38 of carbonate of magnesia, besides some alumina, oxide of iron, silica, &c.

"The locality is situated on the shores, quite above high-tide level, near the foot of Diamond Hill. This hill is an extinct tufa cone, nearly 700 feet in height, rising from the water's edge, and in its origin it must have been partly submarine. It is one of the lateral cones of Eastern Oahu, and was thrown up at the time of an eruption through a fissure, the lava of which appears at the base. There was some coral on the shores when the eruption took place, as is evident from embedded fragments in the tufa; but the reef containing the chalk appeared to have been subsequent in formation, and afforded no certain proof of any connection between the fires of the mountain and the formation of the chalk.

"The fine earthy texture of the material is evidence that the deposit was not a subaerial sea-shore accumulation, since only sandstones and conglomerates, with rare instances of more compact rocks, are thus formed. Sand-rockmaking is the peculiar prerogative, the world over, of shores exposed to waves, or strong currents, either of marine or of fresh water. We should infer, therefore, that the accumulation was produced either in a confined area, into which the fine material from a beach may have been washed, or on the shore of a shallow quiet sea—in other words, under the same conditions, nearly, as are required to produce the calcareous mud of the coral island. But although the agency of fire in the result cannot be proved, it is by no means improbable, from the position of the bed of chalk, that there may have been a hot spring at the spot occupied by it.

"That there was some peculiar circumstances distinguishing this from other parts of the reef is evident.

"This, if a true conclusion, is to be taken, however, only as one method by which chalk may be made, for there is no reason to suppose that the chalk of the chalk formation has been subjected to heat; on the contrary, it is now well ascertained that it is of cold-water origin, even to its flints, and that it is made up largely of minute foraminifera, the shells of rhizopods.

"Professor Bailey found under his microscope no traces of foraminifera, or of anything distinctly organic, in the chalk."

The entire absence of any remains of foraminifera must, I venture to think, completely destroy any claim for the Oahu limestone to be regarded as chalk proper.

Neither can the Atlantic ooze, rich though it be in coccoliths and the shells of foraminifera, be regarded as chalk. It is true that it may in future geological ages fulfil Professor Wyville Thomson's prediction and become such, but even of that we cannot be certain. At present it is a soft calcareous mud, and a very impure one. When consolidated and converted into dry land, instead of forming a brilliant white chalk limestone, a hard compact argillaceous or siliceous slaty limestone may be the result.

The true white chalk so familiar to Englishmen is found over an area extending from the southern part of Sweden to Bordeaux, a distance in round numbers of 850 miles, and again from the northern part of Ireland to the Crimea, i.e., about 1140 miles.

I am, of course, referring to the extent of the soft white limestone known emphatically as chalk, not to the areas occupied by that great variety of rocks which are classed with the chalk, and which are collectively known as the rocks of the chalk or Cretaceous period from the fact that they contain certain fossils in common. These rocks have a very wide distribution, being found in Europe, Asia, Africa, America, and in Australia from Western Australia to Queensland, and New Zealand.

It may, perhaps, be mentioned as an argument in favour of the probability of the New Ireland limestone being properly regarded of Cretaceous age, that we have Cretaceous rocks in Queensland as far north as 11° S., and in New Guinea, still nearer to New Ireland, we have rocks which undoubtedly belong to the Mesozoic or Secondary period, for amongst the geological specimens brought by Signor D'Albertis from the Fly River and submitted to me for examination, there were *belemnites*, an *ammonite* (this ammonite bears a very close resemblance to a liassic form), and other fossils, such as *carcharodon teeth* and *pectens*, all of which may or may not belong to the Cretaceous age.

It would be by no means a startling thing to find that these Secondary beds had an extension to the New Britain group of islands, a distance of only a few hundred miles, which would comprise an area by no means equal to the extent of country occupied in Europe by the typical white chalk.

It should, however, be mentioned that no true white chalk has yet been found either in Queensland or in New Guinea.

In conclusion, it may be stated that the principal reasons in favour of the rock being regarded as chalk are, that physically it is almost indistinguishable from most typical specimens of that rock, and that it has had the same organic origin; the foraminifera alone are not, unfortunately, sufficient to rigidly determine the geological age of the specimen, because they are types which have been persistent from the Cretaceous period to the present time.

Note.—Since the above paper was written the following papers have been published referring to recent marine deposits in the Pacific Ocean:—

GUPPY: "Observations on the Recent Calcareous Formations of the Solomon Group made during 1882-84." (*Trans. Roy. Soc. Edin.*, vol. xxxii. p. 545.)

MURRAY: "On the Distribution of Volcanic Debris over the Floor of the Ocean —its Character, Source, and some of the Products of its Disintegration and Decomposition." (*Proc. Roy. Soc. Edin.*, vol. ix. p. 247.)

MURRAY AND RENARD: "On the Nomenclature, Origin, and Distribution of Deep-Sea Deposits." (*Proc. Roy. Soc. Edin.*, vol. xii. p. 495.)

NOTES ON SOME MINERALS FROM NEW CALEDONIA.*

The following notes and analyses have been made upon some minerals which were kindly placed at my disposal by Mr. Pryor, F.G.S., manager of the Balade Copper-mine, Ouegoa, New Caledonia, Mr. Rossiter of Noumea, Mr. Douglas Dixon, late of Sydney, and other friends. My thanks are especially due to Mr. Pryor, who has evidently been at very great pains to collect good and typical specimens for me, as well as to send trustworthy information as to localities, mode of occurrence, and other similar matters.

I should mention that this paper is intended to be merely a description of the particular specimens received, and as far as I have yet been able to work upon them; it is not meant to be a general account of the minerals of New Caledonia.

GOLD.

Disseminated in fine grains and particles through a mica-schist much stained with red oxide of iron; in parts of the rock pseudomorphous cubical cavities are abundant, apparently left by the removal of crystals of iron pyrites; the red colour of much of the schist is probably due to the decomposition of the pyrites, sesquioxide of iron has been formed, and the gold, which was doubtless held by pyrites, set free.

The bright-red-coloured schist is sometimes mistaken by miners for red oxide of copper and for gossan.

Gold is also met with in a talcose schist with quartz.

Locality.—Fern Hill Mine, Manghine, Diàhot River; also in auriferous pyrites at Niengneue.

COPPER.

Native Copper.—As irregular strings and thin plates, filling the joints in rotten and much-fissured quartz-veins; most of the fissures are about half an inch apart, and more or less at right angles to each other; the metallic copper is accompanied by a certain amount of the red oxide of copper or cuprite. Balade Mine.

Copper Gossan.—Of the usual character, consisting of friable earthy

* Read before the Royal Society of New South Wales, September 1, 1880, by A. Liversidge.

red oxide of iron, containing a trace of copper mixed with more or less quartz. Balade and Sentinelle Mines.

Cuprite.—The red oxide of copper, crystallised in the form of minute octahedra lining small cavities in a light-brown-coloured siliceous vein-stuff. Bouenoumala; Balade and Sentinelle Mines.

Tile Ore.—This, the earthy variety of red oxide of copper, occurs at the Balade and Sentinelle Mines. The specimens received from Mr. Pryor are mixed with streaks of the green carbonate of copper.

Tenorite.—The black oxide of copper occurs at the Balade Mine in the form of a loose black powder intimately associated with copper pyrites.

Sulphate of Copper.—From the Balade Mine, where it is met with on the outcrop of the lode in the form of beautiful pale blue crystals—some distinct; but in other specimens the crystals are very small and arranged in mammillated aggregations.

Malachite.—The green hydrated carbonate of copper. Most of the specimens from the Balade Mine merely show it as a coating, or sparingly diffused through the mica-schist; some are more massive, but friable and more or less earthy, but none sufficiently compact to present the usual characteristics of the typical malachite. It is also found at Goundolai, Diàhot River, associated with cuprite and other ores of copper; also Sentinelle Mine.

Chessylite.—The blue hydrated carbonate of copper occurs with other copper ores at the Sentinelle Mine, Diahot River, situated about two miles from the Port of Pam, and seventeen from the Balade Mine. Some of the chessylite is in the form of nodular crystallised masses, with a radiated internal structure, associated with a white kaolin-like clay, presenting very much the same appearance as the chessylite from the Cobar Mine, and apparently occurring under somewhat similar circumstances.

Redruthite.—Copper glance or the grey subsulphide of copper; massive, of very good quality, associated with cuprite. Balade Mine.

Bornite.—The variegated or purple sulphide of copper and iron also occurs at the Balade Mine; of a bronze colour, massive, and of good quality.

Chalcopyrites.—This, the common form of copper pyrites, appears to be very abundant at the Balade Mine, and of good quality. It occurs both massive and in the form of small strings and layers running through a mica-schist, in much the same way as we often find layers of quartz under similar circumstances; this mica-schist is often very much contorted, and in such a way as to present a very pretty wavy silky lustre.

The chalcopyrites is also occasionally met with lining fissures.. One or two of the specimens kindly sent to me by Mr. Pryor are fairly well crystallised, the form being in one case the sphenoid with curved faces; in the other specimens, also from the Balade Mine, the crystals are smaller, but better developed, consisting of groups of tetragonal pyramids combined with faces of the secondary prism and the basal pinacoid.

Most of the specimens of chalcopyrites received from the Balade Mine would be described by miners as peacock copper, on account of the iridescent tarnish tints which they present.

Associated with the copper pyrites are the minerals quartz, both opaque white and translucent, calcite, dolomite (ferruginous), passing into siderite, the carbonate of iron, chlorite, magnetite, pyrrhotine, iron pyrites, and others. One of the most interesting is, perhaps, the white fine-grained crystalline marble, closely resembling statuary marble in appearance. It is very unusual for marble to appear under such circumstances; calcite is, however, very commonly met with in mineral veins. Some of the fragments of marble are quite small, and are almost completely surrounded by the copper pyrites with sharply defined boundaries, just as if lumps of marble had fallen into the vein and had been subsequently surrounded by ore.

Mr. Pryor states that the calcite, siderite, chlorite, and dolomite are only found when the beds of quartz which lie between laminæ of the strata are in actual contact with the copper deposits; they then not infrequently form the upper or under boundaries of the ore deposits. At the point of contact he has observed also that the opaque white quartz generally gives way to the transparent variety.

In speaking of the deposits of copper ore at the Balade Mine Mr. Pryor says:—" The deposits consist of compact yellow copper ore almost free from any kind of gangue whatever. Quartz even is only found in quantity upon the coming in or going out of ore, forming as it were contact beds between it and the schist. It is also noteworthy that no cavities or hollows are found in these 'pipes' of mineral, and this circumstance accounts for there being no crystals of any kind, except a few very small ones of calcite and chalcopyrites which line the faces of some of the joints. These joints extend uninterruptedly across the strata and ore at the same angle, which varies from 40° N. in the smaller to vertical in the larger ones, and the two faults, while their strike is approximately the same, and may all, therefore, be referred to the same system of fracture. Another series of joints—merely divisions in the rock—occur, however, and these dip south, but are not represented in the deposits.

" To give you an idea of the geological formation of the district, I have collected thirty-four specimens illustrative of the various metamorphic rocks which are met with in ascending the mountain in a north or

S

transverse direction for a distance of about 2½ miles, *i.e.*, from its base
near where the deposits of copper crop out at surface.

"This chain of hills attains its greatest altitude at this point, where
it is 2500 feet above sea-level, and extends about thirty miles east and
ten west, with a general strike east 27° S. There are being worked at
present five distinct pipes of ore, S. 41½° W. with remarkable regularity
at an angle varying from 20° to 45°. On some future occasion I hope
to furnish you with a description of these singular formations and observations thereon, as well as the necessary reduced sections and diagrams,
without which it would be difficult to explain them. Garnets with
glaucophane and crystals of amphibole occur abundantly in these rocks,
but I have found none larger than those sent you. The crystals of
titaniferous iron and magnetite are not so plentiful apparently, at least
on the surface, from whence I procured nearly all the specimens of rocks.
About three-quarters of a mile to the east of this mine there appear to
be either intrusive masses or dykes of serpentine (judging from what can
be seen at surface), associated with which are small pieces of chrysolite
and asbestos, while the enclosing country is talcose schist."

Lead.

Galena.—The sulphide of lead; from Coumac, in masses with a finely
granular structure, reputed to be highly argentiferous. The specimen
given to me by Mr. Rossiter contained but a small quantity of silver.

Zinc.

Zinc Blende.—The sulphide of zinc; specimens from Coumac and the
Baie Lebris, said to be argentiferous, are black in colour, massive, with
granular structure, and in parts much stained with oxide of iron.

Antimony.

Antimonite.—The sulphide of antimony; a fine specimen of the massive variety from Nakety, on the east coast, with coarsely bladed structure like much of the Borneo ore, was contained in Mr. Rossiter's
collection.

The specimen is coated in part with yellow oxide of antimony, to
the thickness of about half an inch.

Titanium.

Rutile.—The dioxide of titanium TiO_2.

Crystallised in incompletely developed prisms, much striated, of a
dark hair-brown colour; in most cases the prisms are much flattened

and partly embedded, and not sufficiently well formed to admit of measurements being made with the goniometer.

In another specimen long, slender, translucent reddish-brown crystals of rutile penetrate through and through a mass of rock crystal. Some of the crystals are bent and broken.

Ouegoa, Diàhot River.

NICKEL.

M. Jules Garnier seems to have been the first to discover the existence of a nickel-bearing mineral in New Caledonia. He first met with it as far back as 1864, and made his discovery public in 1867,[*] but he did not, apparently, make any investigation into the chemical composition of the mineral in question. Afterwards M. Garnier placed some of the mineral in the hands of M. Jannettaz, Mineralogist to the Natural History Museum of Paris.

In a letter to the *Moniteur de la Nouvelle Calédonie* of January 6, 1875, upon the minerals of New Caledonia, translated and quoted by the late Rev. W. B. Clarke,[†] M. Garnier claims the priority of discovery as follows :—

" I have recorded this in my journal, 24th September 1864 :—' Continuing to ascend the river of Dumbéa; the rocks which I meet with are little variable'; they are amphibolitic, and often hold chromate of iron ; the rock is also accompanied by a green matter which sticks to the surface—nickel.' Moreover, it was one of the first steps in the country to announce nickel. I sent specimens of it to the Rev. W. B. Clarke, as he has had the goodness to state in his letter. I did not then give the descriptions, waiting for the definite work which I could only make in a place where I could be aided by the light of clever experiments, and also with instruments for investigation that I lacked in the colony. It was Mons. Jannettaz, Mineralogist at the Museum, who was so good as to analyse this green substance, which was thought might also be chrome in a certain condition, the oxidising salt, if we might judge by the abundance of chromate of iron in all the rocks. The analysis of Mons. Jannettaz gave me satisfaction. It was really that of nickel, and I was then able to say in my ' Geology of New Caledonia,' p. 85 (1867) :—' It would be highly interesting to study more completely the deposits of nickel, &c.' "

" In 1869 I again wrote:—' The serpentines, and in a general way of the rest, all the rocks which accompany them, are often covered with a coating of beautiful green, which is nothing but silicate of nickel, alumina,

* "Geol. Nouv. Cal.," p. 85, J. Garnier, 1867.
† *Jour. Roy. Soc. N. S. W.*, vol. ix. p. 47.

and magnesia. . . . The nickel in this condition is so abundant that we ought to hope to find one day a workable deposit of it ("Bulletin de l'Industrie Minérale," p. 301 tome xv.).'"

From the above it does not appear that M. Jannettaz made a complete analysis of the mineral; he merely ascertained its composition qualitatively.

Since my previous analysis of the nickel-bearing minerals from New Caledonia * I have had opportunities to examine a very large number of specimens from different deposits in New Caledonia, and especially of the one named *Noumeaite.* The variety known as *Garnierite* does not appear to be at all abundant, nor does it appear to be of much importance to the mineralogist.

Both varieties lose a portion of their combined waters when heated to 100° C.; the amount is variable in different specimens.

Noumeaite.—No crystallised specimens appear yet to have been met with; the mineral appears to be completely amorphous, not even a crystalline structure being recognisable, unless the fibrous appearance of some be regarded as such; it occurs in massive pieces, in botryoidal and stalactitic forms, as incrustations with smooth mammillated surfaces, in brecciated masses, as the cementing material of serpentine breccias, also as concretions, and in the massive form with a *petaloidal* structure, *i.e.* the mineral splits up into pieces with smooth polished concave-convex surfaces which fit into one another somewhat like the petals of an unopened flower-bud; this kind of structure is very often seen in mineral veins of all kinds, and in their walls also where there is a *slickenside* or *miroir.* Occasionally it is found invested with a drusy coat of small sparkling quartz-crystals.

In hardness and toughness it varies very much, sometimes being quite soft and brittle, crumbling between the fingers, and in other cases both hard enough and tough enough to be cut into ornaments. These harder varieties take a very fair polish, and rival malachite in beauty and effect. At the Paris Exhibition of 1878 Messrs. Christofle had some beautiful polished samples, including some columns veneered with noumeaite.

In colour it is met with of various shades of green, from the very palest tinge through apple-green to a full rich malachite-green; the very pale varieties apparently seem to be nothing more than a hydrated silicate of magnesia more or less charged with silicate of nickel. Some of the pale varieties, although not hard, are, from their great toughness extremely difficult to powder.

---

* "A New Nickel-bearing Mineral from New Caledonia," by A. Liversidge, *Quart. Jour. Chem. Soc.*, London, July 1874 ; "Nickel Minerals from New Caledonia," by A. Liversidge, *Jour. Roy. Soc. N. S. W.*, 1874.

One specimen of noumeaite from Mont d'Or passes on one side into a layer of pale green jade-like mineral, breaking with a splintering fracture and possessing a hardness of rather more than 6, and otherwise resembling jade. This layer had apparently been in contact with the walls of the vein, and had somewhat the appearance of a *slickenside*. I have not yet had time to examine the specimen further.

Some specimens have been found to contain minute traces of copper.

The following analyses, numbered from 1 to 7, were made from sets of specimens which I had carefully freed from the matrix. These sets were prepared so as to ascertain how far specimens resembling one another in colour and appearance, and from the same mine, differed from one another in chemical composition. Dr. Leibius, of the Mint, Sydney, was kind enough to undertake the analysis of one set of these specimens.

No. 1. A light-green-coloured specimen showing petaloidal structure ; from the Bel Air Mine, Ouaïlou, east coast.

Analyses.

	a.	b.	c.	d.
Water lost of 100° C.	10·01	10·95	12·38	14·47
Combined water at red heat	9·62	8·82	7·31 (by diff.)	6·77
Silica	48·90	48·25	49·36	44·96
Soluble	...	...	·17	·56
Alumina	} trace	·55	traces	·56
Iron sesquioxide				
Nickel protoxide	14·85	14·60	13·75	14·6
Magnesia	16·22	16·40	17·03	17·43
	99·60 (Leibius)	99·57	100·00	99·37

In *b*, *c*, and *d* the first portion of the water was driven off at 105° C. instead of at 100° C.

No. 2. A pale variety, very tough, from the same large block.

Analyses.

Water lost at 100° C.	12·71	11·28
Combined water	9·26	10·37 (by diff.)
Silica	48·85	50·15
Alumina	trace	·57
Iron sesquioxide	,,	trace
Nickel protoxide	11·50	10·20
Magnesia	16·81	17·43
	99·13 (Leibius)	100·00

No. 3. A dark translucent green, brittle, botryoidal or mammillated form from Boa Kaine Mine, Kanala.

Analysis.

Water lost at 100° C.	7·24
Combined, loss at red heat	12·92
Silica	35·50
Alumina	·85
Iron sesquioxide	trace
Nickel protoxide	29·65
Magnesia	13·44

99·60 (Leibius)

Another specimen of this, but of a lighter colour, had the following composition:—

Analysis.

Water lost at 100° C.	8·65
Combined water	8·95
Silica	36·79
„ soluble	·70
Alumina } Iron sesquioxide }	·11
Nickel oxide (NiO)	29·72
Magnesia	14·97

99·89

No. 4. Of a rich green colour, intermixed with lighter portions, brecciated and showing a striated and fluted surface, next to the walls of the vein, somewhat like a slickenside in appearance; the specimen of white hydrated silicate of magnesia (see No. 7) formed the boundary-wall or casing of the vein.   Bel Air Mine.

Analysis.

Water lost at 100° C.	11·15
Combined water lost at red heat	8·50
Silica	46·20
Alumina } Iron sesquioxide }	absent
Nickel protoxide	20·88
Magnesia	12·93

99·66 (Leibius)

No. 5. A translucent dark-green-coloured brittle variety, with mammillated surfaces, from Nakety.

Analyses.

	I.	II.
Water lost at 100° C.	6·70	6·44
Combined water lost at red heat	12·39	11·53 (by diff.)
Silica	39·75	38·35
Alumina	traces	·40
Iron sesquioxide	„	·15
Nickel protoxide	29·10	32·52
Magnesia	12·84	10·61

100·78 (Leibius) 100·00

No. 6. A translucent pale green variety from Ouaïlou.

Analysis.

Water lost at 100° C.	11·05
„ lost at red heat	9·70
Silica	48·00
Alumina	absent
Iron sesquioxide	...
Nickel protoxide	15·39
Magnesia	16·92
	101·06 (Leibius)

No. 7. The casing from the walls of a vein of dark green noumeaite at the Bel Air Mine, Kanala, consisted of a dazzling white very tough hydrated silicate of magnesia which in parts was quite free from nickel, and in others merely tinged with the palest green. The surface towards the vein was much grooved, striated, and polished, and had apparently formed part of a slickenside.

This mineral very closely resembles meerschaum in composition, in appearance, and in many of its properties. It is, however, much tougher than ordinary meerschaum, being as difficult to break apart as rock cork; it, moreover, presents in parts a more or less well-developed petaloidal structure. The specific gravity is 2·55, meerschaum being only about 1·3 to 1·6. There are occasional black dendritic markings within it.

One specimen possessing a very pale green tinge gave the following results:—

Analysis.

Water lost at 100° C.	14·30
„ combined	8·87
Silica	51·81
„ soluble	·13
Alumina) Iron sesquioxide)·	1·36
Nickel oxide (NiO)	2·32
Magnesia	21·35
	100·14

Two other specimens devoid of any green tinge gave the following results:—

Analyses.

	I.	II.
Water lost at 100° C.	11·77	13·30 (at 105° C.)
Combined water, at red heat	9·70	8·58 (by diff.)
Silica	53·80	53·80
Alumina	absent	·75
Iron sesquioxide	„	trace
Nickel protoxide	·24	·58
Magnesia	24·82	22·99 (Leibius)
	100·33	100·00

The above composition furnishes the formula, $2MgO,3SiO_2,5H_2O$, or $2MgO,3SiO_2H_2O$, if the water driven off at 100° C. be disregarded. The brownish or plum-coloured serpentine with which the noumeaite is often associated usually contains alumina, iron, &c.; hence (in cases where analyses show the presence of any considerable quantity of these) it may be, I think, assumed that the mineral has not been completely separated from its gangue, but that both have been taken together.

The following analyses of some of the dark green brecciated Kanala ores, from which the gangue had not been wholly removed, will serve to show this:—

Analyses.

Water lost at 105° C. . . .	8·765	8·016	8·65
„ combined, by difference .	9·034	6·550	8·95
Silica	47·041	38·108	36·79
„ soluble	...	...	·70
Alumina	1·376	2·584 ⎫	
Iron sesquioxide	·157	1·137 ⎭	5·36
Nickel oxide	14·544	31·853	24·72
Lime	absent	trace	...
Magnesia	19·033	11·752	14·97
	100·000	100·000	100·14

Another ore gave:—

Analysis.

Water lost at 100°, C.	9·46
„ combined, by difference	7·77
Silica	43·12
„ soluble	·93
Alumina . ⎫	1·31
Iron sesquioxide ⎭	
Nickel oxide	19·89
Magnesia	17·52
	100·00

The foregoing analyses confirm the statement made in 1874 that the mineral is of uncertain composition; it ranges from practically pure hydrated silicate of magnesia to what is also practically only hydrated silicate of nickel. Some specimens which are now being examined quantitatively contain but a very small quantity of magnesia.

Garnierite.—Since the receipt of the first specimen in 1874 I have obtained only one or two additional examples of this variety of the hydrated silicate of nickel and magnesia. It is at once distinguished from the more important mineral noumeaite by its adherence to the tongue, and by its falling to pieces when immersed in water, and (like halloysite) even when allowed to remain adherent to the tongue for a moment or so. It might be roughly described, apart from its composition, as being a green-coloured halloysite. Apart from these characters, there appears to be but little difference between the two varieties.

Kupfernickel, the arsenide of nickel, is reported from Kanala, but the statement requires confirmation. A specimen so labelled proved to be poor chalcopyrites encrusted with a thin film of impure green carbonate of copper. I expect this mineral will be found in New Caledonia, but up to the present I have not seen any authentic specimens.

COBALT.

Up to the present the only cobalt-bearing mineral from New Caledonia which I have had an opportunity to examine has all been of one kind, *viz.*, the variety known as earthy cobalt ore, asbolite or " wad," *i.e.*, an impure oxide of manganese containing cobalt oxide. It apparently occurs in the form of irregular deposits, and as more or less spherical concretionary nodular masses, with mammillated surfaces, embedded in an unctuous red clay. This clay is probably derived from the decomposition of the serpentine and other rocks of the district.

These nodules are black or bluish-black in colour, but usually superficially coated either with the red-coloured clay or with red oxide of iron. I understand that they are quite soft when first dug up; they readily stain the fingers, and yield to the knife at once, cutting like graphite, but with a blue-black shining metallic streak instead of a grey-black one.

Some of the nodules present a very vesicular structure, like certain kinds of lava. Even the apparently quite compact nodules often enclose patches of the clay, especially towards the centre.

Many of them present a very striking resemblance to the manganese nodules dredged up from the depths of the sea by the *Challenger*. I do not feel quite justified in throwing out any suggestions as to whether they were formed under similar conditions, since I have no personal knowledge of the conditions under which they are found; but, as far as an opinion can be formed from the specimens which I have had an opportunity to examine, I am inclined to think that they were not, but that the concretionary process has been set up subsequently; that is, the cobalt seems to have been originally disseminated throughout the clay, but has since segregated together, and assumed the nodular form.

Asbolite also occurs as dendritic markings in the kaolin from a locality on the River Leia.

Some nodules of the ore from Unia were examined, with the following results :—

Analyses.

	Specimen No. 1.	No. 2.	No. 3.
Water lost at 100° C.	8·68	10·19	10·54
„ combined	8·87	9·74	9·83
Silica	15·34	15·15	17·20
Alumina	8·86	8·70	7·65
Iron sesquioxide	10·41	10·26	5·51
Chromium sesquioxide	·52	·51	·87
Nickel oxide	traces	traces	traces
Cobalt „	15·67	15·43	13·59
Manganese peroxide (MnO₂)	11·52	9·57	12·05
Lime	traces	traces	traces
Magnesia	20·80	20·46	22·63
	100·67	100·01	99·87

The following contains but little magnesia—Coumac:—

Analysis.

Water lost at 100° C.	2·86
„ combined, by difference	16·57
Silica	1·06
Alumina	11·37
Iron sesquioxide	23·52
Chromium sesquioxide	traces
Manganese peroxide (MnO₂)	32·41
Cobalt oxide (CoO)	10·42
Lime	absent
Magnesia	1·79
	100·00

The following analysis of a specimen from Baie des Pirogues shows the presence of nickel in rather larger proportion than usual:—

Analysis.

Water lost at 100° C.	6·072
Combined water, by difference	13·759
Silica with traces of chrome iron	4·476
Alumina	21·529
Iron sesquioxide	18·396
Chromium sesquioxide	traces
Manganese peroxide (MnO₂)	27·588
Cobalt oxide	4·927
Nickel „	2·256
Lime	trace
Magnesia	·418
Potash	·123
Soda	·216
Phosphoric acid (P₂O₅)	·240
	100·000

From the foregoing analyses it will be seen that the earthy cobalt ore from New Caledonia differs considerably from those met with in other places; baryta is entirely absent, although often present in this mineral from other localities (see analyses given by Dana, "Descriptive Mineralogy," p. 182), but magnesia seems to have taken its place in the asbolite from some of the New Caledonian mines.

Specimen from Unia of poor quality.

Analysis.

Water lost at 100° C.	2·720
Combined water, by difference	22·901
Insoluble silica	1·699
Soluble	·230
Tin	trace
Iron sesquioxide	10·308
Alumina	37·421
Chromium sesquioxide	traces
Manganese peroxide (MnO₂)	16·598
Cobalt oxide (CoO)	3·387
Nickel ,, (NiO)	2·645
Lime	absent
Magnesia	1·311
Potash	·176
Soda	·328
Sulphuric oxide (SO₃)	·276
	100·000

IRON.

Magnetite.—In the form of small very perfect octahedra diffused through massive granular chlorite. Balade Mine, Ouegoa.

Red Hæmatite.—In a red-coloured micaceous schist, highly charged with oxide of iron, and in consequence often mistaken for the gossan of a copper-vein; the iron oxide has apparently been formed at the expense of iron pyrites, since much of the schist is dotted all over with small rectangular cavities, pseudomorphous after iron pyrites. Balade Mine.

Brown Hæmatite.—A specimen from Baie du Sud evidently of very good quality, curiously permeated by reticulating cavities.

Also met with in a talcose schist in the form of pseudomorphs after cubes of iron pyrites. Mr. Pryor mentions breaking open one of these, and finding a nucleus inside, about one-third of the size of the entire pseudomorph, made up of minute crystals of pale-green-coloured sulphur, &c.

Pyrrhotine.—The magnetic variety of iron pyrites (Fe₇S₈), collected by Mr. Pryor at the Balade Mine.

Massive, of a brown-yellow colour with metallic lustre, associated with copper pyrites and transparent quartz, which is diffused through the mass, just as is seen in some of the pyrrhotine from Bodenmais, in Bavaria.

One of the specimens presented a somewhat cylindrical concretionary form, surrounded by a kind of crust of mica-schist, composed of white mica and quartz with a few flakes of black mica.

Through the joints of some of the specimens layers of brown hæmatite were present, underneath which the pyrrhotine had a crystalline surface, but no distinct crystals could be found. It was thought that perhaps this mineral might contain nickel, like the pyrrhotine from Kelfva, in Sweden, and in the Gap Mine in Pennsylvania, but none could be found in the specimen examined.

Marcasite.—In the form of nodules, smooth externally, and converted into hydrated brown oxide of iron to the depth of about half an inch, but the interior still in part consisting of radiating crystals of marcasite. These nodules closely resemble those from the English chalk, evidently set free from a soft matrix, which was probably limestone.

Locality, Mount Tiebaghi.

Iron Pyrites.—The bisulphide of iron, FeS_2.

In the form of fairly well developed cubes, both isolated and twinned, embedded in a slaty matrix from the Balade Mine.

CHROMIUM.

Chromite.—The deposit of chromite or chromate of iron, commonly known as chrome iron ore, appears to be very extensively developed in New Caledonia, as well as of extremely rich quality.

The ore is met with in the form of alluvial deposits as well as *in situ* in the serpentine and other rocks. I am informed that some of these alluvial deposits are now being worked on a large scale.

The majority of the specimens are massive, with a crystalline, granular, or lamellar structure; also in the form of more or less distinct lustrous black octahedra, closely packed together; often the ore is, however, stained with oxide of iron and mixed with more or less steatitic matter. Some of the specimens yield as much as 66 per cent. of chromium sesquioxide.

One specimen made up of rather large imperfectly developed iron-grey crystals—some nearly half an inch in diameter—was found to have the following composition :—

Analysis.

Silica and insoluble matter *	3·54
Alumina	4·51
Chromium sesquioxide	66·54
Iron protoxide	10·85
Magnesia	15·03
	100·47

* Free from chromium.

The amount of chromium sesquioxide is unusually large ; this is due to much of the iron protoxide being replaced by magnesia, the difference being due to the lower equivalent of the latter.

The above numbers approximate to the usual formula, RO,R_2O_3.

Localities: Petit Mont d'Or, Coumac, Tiebaghi, Ouaghi, Ouaïlou, Baie du Sud.

NON-METALLIFEROUS MINERALS.

Coal.—A specimen of the so-called anthracite from Paita, near Noumea, came into my possession some time ago; it is in the form of a nodular mass, hard, earthy, of poor quality, and quite unfit either for ordinary domestic use or metallurgical operations.

Quartz.—In the form of fragments of colourless and transparent rock crystal, also as vein-quartz, both white and tinged with various colours from the admixture of impurities; also in the form of white pebbles cemented together with brown oxide of iron, and mixed with some more or less decomposed mica, similar in appearance to the conglomerate from the New England diamond drift. Collected by Mr. Pryor.

Chalcedony—In flat pieces as if set free from fissures, often white outside like chalk flints ; in colour various shades of brown and grey, also quite white, as in carnelian. Collected by Mr. Rossiter, from Bouenoumala, Coumac.

Chert.—Of various shades of grey through brown to black, and much fissured, from Pointe Nea (?), near Noumea, apparently weathered out of a limestone rock ; breaks with the usual square splintering fracture of chert, and is thus distinguished from flint and other forms of quartz which break with a conchoidal fracture.

Opal.—One of the hydrated forms of silica. Of various shades of pale translucent brown through grey to opaque white; some of the white varieties have a flesh-coloured tinge, and dendritic markings are common in all the specimens from Olande.

Calcite.—Calcium carbonate crystallising in the rhombohedral system. There are several specimens of this mineral, some collected by Mr. Rossiter from near Port la Guerre, which were mostly massive cleavage fragments. Mr. Pryor's collection from the Balade Mine contained a few specimens crystallised in rhombohedra, and associated with small quartz crystals, taken from the joints of the mica-schist near to the deposits of copper ore. Also others apparently from the lode, intimately associated with copper pyrites.

Aragonite.—The variety of carbonate of lime crystallising in the rhombic system ; occurs of a reddish colour, presenting a coarsely crystalline structure on the fractured surface. The Balade Mine.

Another variety is of a pure white colour, breaking with a fine crystalline fracture, and presenting much the appearance of alabaster, apparently derived from veins only a few inches across; where stained with iron oxide resembles somewhat the celebrated Algerian onyx marble. It apparently forms the vein-stuff of certain portions of the copper-veins.

Limestone.—Of a grey or dove colour, suitable for building or ornamental purposes; from the Baie de l'Oyselinat, Noumea, and from near Coumac. Collected by Mr. Rossiter. Mr. Pryor sent some specimens from an outcrop on the Diàhot River, near to the Balade Mine. M. Ratte speaks of the great extent of this limestone in his Catalogue of minerals from New Caledonia sent to the Paris Exhibition of 1878.

Dolomite.—Occurs in the veins with the copper ore intermingled with quartz. Balade Mine.

Ankerite.—A variety of this mineral of a pale brown colour was found by Mr. Pryor at the Balade Mine associated with quartz and copper pyrites; breaks readily into more or less lamellar pieces; contains manganese, as well as iron, lime, and magnesia.

Magnesite.—In the massive form, white, very dense, hard, and breaking with a conchoidal fracture; somewhat platy structure. A qualitative analysis shows it to be very pure.

A concretionary variety was contained in Mr. Rossiter's collection labelled "Barytes from Bouenoumala," but on testing for barium none could be detected. The specimen had the same peculiarly reticulated surface and mammillated form as the magnesite found on the New South Wales diamond-fields.

Garnet.—In some cases these are very well crystallised in the form of the rhombic dodecahedron, varying in size from $\frac{1}{20}$ to $\frac{5}{8}$ inch in diameter, most of them being $\frac{1}{4}$ inch.

Some are brick-red and more or less opaque, whilst others are of a rich more or less transparent red, similar to the varieties used for jewellery.

The matrix is of two kinds; the one is a hard and very heavy schistose rock, composed of quartz, glaucophane, and some epidote; the other matrix is the rather uncommon variety of hornblende known as glaucophane. The faces of the larger rhombic dodecahedral crystals occurring in the glaucophane matrix are, as it were, built up of plates, so that the edges of the garnets would present, if cut through, a step-like section.

Usually each face of the garnet crystal is covered or in contact with a plate of mica; these mica crystals often extend beyond the face of the garnet in one or more directions. When the garnet is detached a mould of it is left, beautifully lined with mica.

In some cases the garnets have crystallised in thin red films between the plates of mica; in other places the solid garnet crystals penetrate right through the layers of mica.

Specific gravity, 4·011.

An analysis of the garnets was made with the following results :—

Analyses.

	I.	II.	Mean.
Silica	38·10	38·21	38·15
Alumina	22·09	22·27	22·18
Iron protoxide	21·17	21·35	21·26
Manganese protoxide . . .	5·50	5·58	5·54
Lime	7·88	7·68	7·78
Magnesia	4·64	4·84	4·74
Loss on ignition	0·33	0·29	0·31
	99·71	100·22	99·96

which gives the following formula :—

$$3(FeO,MnO,CaO,MgO)_3, 2SiO_3 + Al_2O_3,SiO_2.$$

Mica.—From the Balade Mine, in the form of white silvery plates, some of which are about half an inch in diameter; but no well-developed crystals were present. By transmitted light the thicker plates present a dull greenish shade. Disseminated through some of the masses of mica are small red translucent crystals of garnet, and between the plates of mica films of garnet have occasionally crystallised out.

Plates of this silvery white mica are also found sparsely scattered through the glaucophane, especially in the glaucophane bearing the garnets; in other cases, again, the mica is in excess, the glaucophane playing a subordinate part.

Specific gravity, 2·938.

The following analysis was prepared on a very small quantity of the material, as it was only possible to collect a very limited amount of this silvery white mica; hence much importance cannot be attached to it :—

Analysis.

Water, combined	4·31
Silica	50·60
Alumina	25·28
Iron protoxide	3·47
Manganese protoxide	0·50
Lime	1·04
Magnesia	4·86
Potash	6·69
Soda	2·49
Loss	0·76
	100·00

Neither lithium nor fluorine were present.

The above results do not quite agree with any published analysis, nor do they afford a satisfactory formula, but it is apparently a variety of muscovite mica.

Another specimen of mica, apparently of the same kind, but of a rather darker colour and with less lustre, was examined, with the following results:—

Analyses.

	I.	II.	Mean.
Water, combined	4·42	4·50	4·46
Silica	51·22	51·23	51·23
Alumina	27·29	27·41	27·35
Iron protoxide	2·45	2·75	2·60
Manganese	·34	...	·34
Lime	1·25	...	1·25
Magnesia	3·82	...	3·82
Potash	...	6·93	6·93
Soda	...	1·27	1·27
			99·25

The above corresponds to $2(\frac{1}{2}RO,\frac{1}{2}R_2O_3)3SiO_2+H_2O$.

Hornblende.—In the form of black and fibrous schistose masses, associated with white silvery mica and minute garnets, the three in alternating layers. From the Balade Mine.

Pyroxene or *Augite.*—A rolled nodule, made up of confused masses of crystals. From Tonsjete Bay.

Glaucophane.—This rare variety of hornblende seems to be abundant in the neighbourhood of the Balade Mine, as Mr. Pryor's collection contained several specimens, some of which differ in colour, structure, and general appearance.

Only one specimen is crystallised; the crystals are in the form of dark-blue-grey silky-looking prisms, seated upon a base of a micaceous schist, composed of mica, glaucophane, and garnets, with some quartz.

The prisms are about ⅛ inch in diameter and from ½ to ¾ inch long; they present no distinct faces, both the lateral and terminal faces being more or less rounded; the prisms are, in fact, merely bundles of lamellar or capillary crystals. Some of the prisms are completely isolated from the rest, whilst others are more or less interlaced and superimposed.

All the other specimens are massive, with a fibrous crystalline structure, of a peculiar violet colour, passing into a dark slaty blue on the one hand, and into a pale greyish colour on the other—the lighter violet varieties have a very beautiful silky lustre. The streak is of a pale bluish-grey. Before the blowpipe it fuses, intumesces slightly, colours flame-yellow, yields a dark glass; with sodium carbonate yields indications of manganese. Partly soluble in acids. H. = 6–7.

This occurrence of glaucophane is of considerable interest, since it has hitherto only been met with in a few places, such as the island of Syra, one of the Grecian Islands, and at Zermatt.*

At Syra it is found associated with garnet, hornblende, and mica, in

---

* Since the above was written in 1880, additional localities have been given by Professor Bonney in the *Mineralogical Magazine*, vol. vii., 1886.

a mica-slate. The New Caledonia mineral is also associated with garnet and mica; in fact, it forms in some cases the matrix of these minerals. On analysis it was found to have the following composition :—

Analyses.

	I.	II.	Mean.
Water	1·42	1·34	1·38
Silica	52·71	52·88	52·79
Alumina	14·20	14·69	14·44
Iron protoxide	9·89	9·76	9·82
Manganese	traces	traces	traces
Lime	4·31	4·27	4·29
Magnesia	11·12	10·92	11·02
Potash	·95	·80	·88
Soda	5·15	5·38	5·26
	99·75	100·04	99·88

For the sake of comparison, I append the analysis of the mineral from the Isle of Syra, Dana's "Descriptive Mineralogy," p. 244 (Schnedermann, J. pr. ch. xxxiv. p. 238), also an analysis by Bodewig of a specimen from Zermatt.

Analyses.

	Syra.			Zermatt.
SiO_2 . . .	56·49	SiO_2 . . .	57·81
Al_2O_3 . . .	12·23	Al_2O_3 . . .	12·03
FeO . . .	10·91	Fe_2O_3 . . .	2·17
MnO . . .	·50	FeO . . .	5·78
MgO . . .	7·97	MgO . . .	13·07
CaO . . .	2·25	CaO . . .	2·20
Na_2O . . .	9·28	Na_2O . . .	7·33
K_2O . . .	traces	...	...
	99·63		100·39

Diallage.—A rolled nodule—no locality, probably from Tchio.

Serpentine.—The rock known as ophiolite or serpentine is very largely developed in New Caledonia, forming, in fact, mountain ranges; but the mineral known as noble or precious serpentine is not common.

Some of the serpentine has a very peculiar plum colour and plum-like bloom on its surface.

A specimen of the common massive serpentine was found to contain 0·78 per cent. of nickel oxide.*

Marmolite.—A foliated variety of serpentine of a green colour, translucent, in flat platy fibres, passing into an asbestiform variety, said to be associated with the chrome iron ore deposits at Tiebaghi, on the west coast.

Talc.—Of a white silvery or pale green colour, highly lustrous,

* "Nickel Minerals from New Caledonia," *Jour. Roy. Soc. N. S. W.,* 1874, p. 80.

T

possessing a schistose structure, and containing long semi-transparent interlacing crystals of actinolite. Balade Mine.

Steatite.—Of a white colour, translucent, mixed with some serpentine. Collected by Mr. Rossiter at Yate.

A green variety from Moira, also at the Balade Mine.

Chlorite.—In masses of the usual dark olive-green colour, breaking with a crystalline fracture, presenting rosette-like groups of crystals. Balade Mine.

Kaolin.—From Ombatche. Of a dazzling white colour, very friable, with a harsh feeling. A qualitative examination only was made of the specimen on its being found to be practically pure hydrated silicate of alumina, with but a trace of sesquioxide of iron.

With little preparation would probably be extremely well adapted for the manufacture of porcelain of the best quality.

Allophane.—From a small island to the south of New Caledonia.

As an incrustation, of a pale blue colour; hardness about 3, brittle, is readily cut with a knife, yields a shining streak, adheres somewhat to the tongue, translucent, resinous lustre, fracture flat conchoidal.

Before the blowpipe it loses colour somewhat and becomes more or less white and opaque, splits up, but does not intumesce or fall to a powder; at first it imparted a pale green tinge to flame, infusible, in closed tube gives off water, and with microcosmic salt a skeleton of silica. When strongly ignited with cobalt nitrate a blue mass is left. Gelatinises with hydrochloric acid.

Halloysite.—Of pale tints of grey, yellow, green, and brown; found in the crevices of the rocks at Yate.

In conclusion, my thanks are due to my friend Dr. Leibius, Senior Assayer of the Sydney Branch of the Royal Mint, for his kindness in making for me the seven analyses marked with his name, and to Dr. Helms for his assistance in analysing the chrome iron, the glaucophane, and its included garnets and mica.

ALPHABETICAL LIST OF MINERAL LOCALITIES.

It is probable that this list contains many inaccuracies—in spite of the care and labour devoted to it—for the reasons already given; but it is thought that the only way to attain accuracy in this matter is to publish as complete and as correct a provisional list of mineral localities as is possible with our present knowledge. The author hopes that those who are interested in the subject will assist in this matter by sending to him corrections and additions. It must be understood that in many cases the minerals mentioned do not actually occur at the town or other locality stated, but in the neighbourhood; the actual spot often has no name, and the finder has to give the name of the nearest town, settlement, or natural feature. Localities for coal and gold are not always given, since the areas over which they occur are best seen from the map.

A

Abercrombie Caves, co. Georgiana : marble.
Abercrombie Ranges, co. Georgiana : asbestos, blue and green carbonates of copper, cerussite.
Abercrombie River, co. Georgiana: apatite, barytes, chabasite, diamond, garnet, gold, graphite, sapphire, topaz, zircon.
Aberfoil, co. Clarke: antimonite, corundum, native tin, osmium-iridium, platinum, tinstone.
Abingdon, co. Hardinge : tinstone.
Adaminaby, co. Wallace : gold, tinstone.
Adelong, co. Wynyard : calcite, copper ores, galena, gold, iron pyrites, scheelite, silver ores, stilbite, zinc blende.
Adelong Creek, co. Wynyard : galena, gold, magnetic pyrites, silver ore.
Albion Mine, co. Gough: rock-crystal, tinstone, wolfram.
Albury, co. Goulburn : barytes, chrysolite, gold, iron pyrites, muscovite, orthoclase, tourmaline.
Alum Creek : alunogen.
Angular Creek, co. Murchison: chromite.
Ann River, co. Clarke : tinstone.
Anvil Creek, co. Northumberland : coal.
Appin, co. Cumberland : alunogen, epsomite.
Apsley, co. Bathurst: copper ores.
Apsley River, co. Vernon : copper ores, marmolite.

Araluen, co. St. Vincent: gold, silver ores.
Armidale, co. Sandon : antimony (antimonite, cervantite), bismuth (native, bismuthite), chromite, gold, iron pyrites, lead (cerussite, galena), manganese, mispickel, scheelite.
Arnprior, co. St. Vincent: chiastolite, marble.
Ashford, co. Arrawatta : coal, tinstone.
Ash Island, co. Northumberland : gypsum.
Attunga, co. Inglis : hornblende, tinstone.
Auburn Vale Creek, co. Hardinge : diamond, tinstone.
Avisford, co. Wellington : zoisite.

B

Babinda, co. Flinders: copper carbonates.
Back Creek, co. Bathurst: goethite.
Back Creek, co. Georgiana : cerussite, manganese (diallogite, manganite, oxide, rhodonite), silver ore.
Back Creek, co. Gloucester : gold, mispickel.
Bala, co. King : copper ores.
Bald Hills, co. Wellington : agate, corundum, diamond, emerald, manganese oxide, ruby, rutile, sapphire.
Bald Nob Creek, co. Gough : steatite.
Bald Rocks, co. Murchison : gold, tinstone.
Ballimore, co. Lincoln : goethite.
Ballina, co. Rous : diamond.

Balola, co. Hardinge: orthoclase, topaz, tourmaline.

Bangalore, co. Argyle: marble.

Bara, co. Wellington: cerussite, pharmacosiderite.

Barbigal, co. Lincoln: coal, iron ores.

Barmedman, co. Bland: schorl.

Barraba, co. Darling: chromite, copper ores, gold, kaolin, magnetite, pyroxene, serpentine, tripoli.

Barrier Range, cos. Yancowinna and Farnell: barytes, bismuth ore, copper ores, chrysolite, fibrolite, gypsum, iron (ironstone, oxide, pyrites, pyrrhotine), lead (cerussite, galena), cobaltiferous manganese oxide, mica, nickel, silver, (chloride, embolite, iodide), talc, tin ores.

Bathgate, co. Cook: torbanite.

Bathurst, co. Bathurst: amethyst, antimonite, barytes, copper (native and ores), diamond, epidote, garnet, gypsum, gold, iron pyrites, jasper, kupfer-nickel, marble, manganese oxide, mispickel, osmium-iridium, pyromorphite, silver ores, spinelle, staurolite, talc, tinstone, titaniferous iron, topaz, wad.

Belaira, Mount Gipps, co. Yancowinna: galena.

Belara, co. Bligh: cerussite, copper ores, silver ore, zinc blende.

Bellinger River, co. Raleigh: antimonite, cervantite.

Bell River, co. Wellington: eisenkiesel, sapphire, topaz, and other gemstones.

Belubula River, co. Bathurst: gemstones, gold, magnetite, marble.

Belubula and Lachlan Rivers (between), co. Bathurst: barytes.

Belwood, co. Georgiana: rhaetizite.

Ben Bullen, co. Cook: limestone.

Bendemeer, co. Inglis: antimonite, chromite, hornblende, manganese (oxide, rhodonite), mispickel, platinum, sapphire, tinstone, tourmaline.

Bengonover Mine, co. Hardinge: diamond, gemstones, tinstone.

Ben Lomond, co. Gough: iron pyrites, sapphire.

Bermagui River, co. Dampier: copper ores, gold.

Berrima, co. Camden: coal, gems, halloysite, iron (goethite, limonite), sapphire, torbanite, zircon.

Bethungra, co. Clarendon; galena.

Bibbenluke, co. Wellesley: barytes, specular iron.

Big River, Auburn Vale, co. Hardinge: diamond and gemstones.

Billabong, co. Clarendon: tinstone.

Billygoa, co. Robinson: gold, silver ore.

Bimbijong and Eumbi (between), co. Phillip: ruby.

Binalong, co. Harden: copper ores, galena, iron (brown hæmatite, magnetite, pyrites, red hæmatite).

Binbanang, co. Bathurst: silver ores.

Binda, co. Georgiana: galena, iron pyrites, silver ore, zinc blende.

Bingera, co. Murchison: adamantine spar, albite, antimony (antimonite, cervantite), bismuthite, bronzite, chromite, copper ores, diaclasite, diallage, diamond, eisenkiesel, galena, garnet, gold, iron (limonite, pyrites, titaniferous), jasper, kyanite, magnesite, mispickel, molybdenite, osmium-iridium, picrolite, rock-crystal, sapphire, serpentine, spinelle, tellurium, tinstone, topaz, tourmaline, zircon.

Bingera Creek, co. Murchison: chromite, gold.

Bischoff Mine, co. Gough: lode-tin.

Blackheath, co. Cook: coal, goethite, torbanite.

Black Ranges, co. Argyle: tourmaline.

Black Swamp, co. Clive: lithia-mica.

Blair Hill, co. Gough: tinstone.

Bland, co. Bland: chromite, opal serpentine.

Blayney, co. Bathurst: allophane, copper (native and ores), hæmatite, manganese oxide.

Bloomfield, near Orange: opal.

Blue Mountains, co. Cook: chert, graphite, iron (hæmatite, limonite), wad.

Bobbera, Jingery, and Pambula (between), co. Auckland: epidote.

Bocoble, co. Ashburnham: copper ores, gold.

Bogan District and Bogan River: gypsum, iron pyrites, magnetite.

Bogan and Lachlan Rivers (between): goethite, tin.

Bogenbung (two miles east of Wollongelong Run): arragonite, manganese oxide.

Bogolong, co. Harden: magnetite.

Bolitho Mine, co. Harden: tinstone.

Bolivia, co. Clive: chlorite, copper ores, iron pyrites, lead (cerussite, galena), mispickel, molybdenite, zinc blende.

Bombala, co. Wellesley: asbolite (wad), calcite, copper ores, galena, gold, mispickel.

Bongongolong: gold, iron pyrites.

Bony Gully, co. Gough: nickel sulphide.

Bookham, co. Harden: galena, hæmatite, marble.

Bookookoorara, co. Buller : tinstone.

Boona West, co. Blaxland : tinstone.

Boonoo Boonoo, co. Buller : gold, tinstone.

Boorolong, co. Sandon : antimonite, stream-tin.

Boorook, co. Buller : gold, silver (antimonial, chloride, native, sulphide), vivianite.

Borah Mine, co. Hardinge : diamond, tin.

Boro, co. Murray : cobaltiferous manganese oxide, copper ores, galena, gold, hæmatite, pyrites, zinc blende.

Boro Creek, co. Argyle : pisolitic iron ore, wad.

Bowenfels, co. Cook : coal, gold, hæmatite, limonite.

Bowling Alley Point, co. Parry : antimonite, bronzite, chromite, copper (native), diallage, dolomite, garnet, gold, iron carbonate, mica, serpentine, zircon.

Bowman River, co. Gloucester : calcite, gold, iron pyrites.

Bowning, co. Harden : manganese oxide.

Bowning Creek, co. King : copper and silver ores.

Bowning Hill, co. Harden : dioptase.

Bowra, co. Raleigh : antimonite.

Braidwood, co. St. Vincent : copper ores, gold, iron pyrites, lead (carbonate, galena), mispickel, opal, silver, zinc.

Branxton, co. Northumberland : cannel coal.

Bredbo, co. Beresford : calamine, galena, green carbonate of copper.

Breelong Creek, co. Cowen : agate, ironstone, opal.

Brewarrina, co. Clyde : lead (carbonate, galena).

Briar Park, co. Georgiana : asbestos, copper ores.

Bringelet parish, co. Bathurst : manganese oxide.

Brisbane Water, co. Northumberland : hæmatite, pisolitic iron ore.

Britannia Mine, co. Gough : diamond, sapphire, and other gems, tinstone.

Broad Gully, co. St. Vincent : gold.

Broadwater : muscovite.

Brogo and Twofold Bay (between), co. Auckland : bog-butter.

Broken Hill, co. Yancowinna : copper ores, lead carbonate, silver (chloride, chlorobromide).

Bromby, co. Wellington : calcite.

Brook's Creek, co. Murray : diamond, galena, silver ores.

Broughton Creek, co. Camden : coal, torbanite.

Broughton Vale, co. Camden : goethite.

Broulee, co. St. Vincent : iron ores, silver, zinc blende.

Brownlea : silver ores.

Brown's Creek, co. Bathurst : copper ores, gold, magnetite.

Bruce Mine, co. Gough : native bismuth.

Brungle Hill : copper carbonate.

Brush Creek, co. Arrawatta : websterite.

Buckinbar, co. Gordon : copper, silver ores.

Bukkulla, co. Arrawatta : asbestos, coal.

Bulladelah, co. Gloucester : alunite.

Bullabalakit, co. Pottinger : amethyst.

Bullanamang : tourmaline.

Bull-dog Range, Mitchell's Creek, co. Roxburgh : gold, lead ore.

Bulli,'co. Camden : alunogen, coal, hæmatite, limonite.

Bullio Flat, co. Argyle : molybdenite.

Bundanoon, co. Camden : coal, gold, iron ore, mispickel.

Bundarra, co. Hardinge : mispickel, tinstone.

Bundian : epidote, orthoclase.

Bungawalbyn, co. Rous : coal.

Bungendore, co. Murray : braunite.

Bungonia, co. Argyle : alunogen, andalusite, antimonite, calcite, cobaltiferous manganese oxide, copper ores, epidote, galena, gold, gypsum, iron (pharmacosiderite, pisolitic iron, pyrites), limestone, plumbago, tinstone.

Burnaby Creek, co. Argyle : silver ores.

Burnett, co. Burnett : coal.

Burra Burra, co. Ashburnham : iron (goethite, magnetite), stream-tin.

Burra Creek, co. Selwyn : tinstone.

Burraga, co. Wellington : copper (pyrites, &c.), galena, iron (green silicate, pyrites).

Burragorang, co. Camden : epsomite, galena, ironstone, limestone, torbanite, zinc blende.

Burramagoo, co. Westmoreland : copper ore.

Burramungie and Morowat Rivers (between), tourmaline.

Burrandong, co. Wellington : anatase, brookite, diamond, gold.

Burrangong, co. Monteagle : gold.

Burroba Creek, co. Wellington : asbestos.

Burrowa, co. King : copper ores (sulphide, &c.), galena, ironstone.

Burrowa Creek, co. King : silver ores.

Butchart Mine, co. Gough : tinstone.

Butler's Creek, co. Roxburgh : gold.

Buttar Ranges, co. Northumberland : limonite.

Byrne's Lode, co. Clive : native bismuth.

Byron's Plains, co. Gough : augite, cairngorm.

C

Cadell's Reef, co. Phillip : scorodite.

Cadia, near Orange : gold, marcasite.

Cadiangulong, co. Bathurst: copper (chalcotrichite, melaconite, blue silicate).

Cagillico, co. Gipps : ironstone.

Calabash Creek, co. Bland : diamond.

Callalia Creek : agalmatolite, pyroxene.

Caloola, co. Bathurst: asbestos, barytes, gold, hæmatite, manganese (braunite, pyrolusite, wad).

Calton Hill, co. Durham : mercury, (native, cinnabar), platinum, native silver.

Cambalong, co. Wellesley : barytes, galena.

Camberra, co. Murray : galena, goethite.

Cambewarra Ranges, co. Camden : torbanite.

Cameron's Creek, co. Hardinge : pyroxene.

Campbell's Creek, co. Wellington : jamesonite.

Canadian Lead, co. Phillip : gold.

Candelo, co. Auckland : barytes, galena.

Cangai, co. Drake : gold.

Canobolas, co. Ashburnham : barytes, copper (native and ores), gold, ironstone.

Canomidine Creek, co. Ashburnham : copper ores, marble.

Canowindra, co. Ashburnham : gold.

Capertee, co. Hunter : calcite, coal, copper (green carbonate, sulphide, cuprite), galena, gold, iron pyrites, mica, molybdenite, torbanite, tourmaline.

Capertee District : alunogen, antimonite, copper ore, cobaltiferous manganese oxide, torbanite.

Captain's Flat, Molonglo, co. Murray : barytes, gold, iron (pyrites, specular), lead (carbonate, galena, protoxide).

Carangera, near Orange : asbestos, copper ores, gold.

Carangula, co. Dudley : antimonite, cervantite.

Carcoar, co. Bathurst : barytes, chalcedony, copper (native and ores), eisenkiesel, galena, gold, halloysite, iron (hæmatite, magnetite, marcasite, pyrites), mispickel, opal.

Cargo, co. Ashburnham : copper ores, gold, gypsum.

Carlo's Gap (seven miles from), co. Wellington : torbanite.

Carlyle's Creek, co. Inglis : tinstone.

Carona, co. Farnell : copper ore.

Carrawabbity, co. Ashburnham : marble.

Carroll (six miles south-east of), co. Buckland : calcite, stilbite.

Carroll's Creek, co. Buckland : tinstone.

Carwary, co. St. Vincent : hæmatite.

Carwell, co. St. Vincent : calcite, hydrotalcite, iron (goethite, hæmatite), opal.

Castlereagh River, co. Leichhardt : carnelian, opal.

Cataract River, co. Cumberland : arragonite, calcite.

Cavan, co. Cowley : limestone.

Charlton (near Rockley) : copper sulphide.

Chichester River, co. Gloucester : galena.

Chonta, co. Auckland : lignite.

Clarence Heads (near), co. Clarence : gold.

Clarence River, co. Clarence : antimonite, apatite, coal, copper ores, iron (goethite, magnetite), resinite, serpentine, silver ores.

Clarence Siding, co. Cook : coal, hæmatite.

Clear Creek, co. Parry : magnetite.

Clyde River, Shoalhaven, co. St. Vincent : coal, torbanite.

Coal Cliff, co. Cumberland : coal, limonite.

Coal Gully, co. Arrawatta : coal.

Coast (along), co. Rous : gold.

Cobar, co. Robinson : copper (native and ores, atacamite, blue and green carbonates), native bismuth, gold, opal.

Cobargo, co. Dampier : barytes.

Cockle Creek, co. Northumberland : coal.

Collington, co. Beresford : lead (galena, carbonate).

Collingwood : gold, oligoclase.

Colo, co. Hunter : iron sulphate.

Colo Gates, co. Camden : coal.

Combing Park, co. Bathurst : copper ores.

Combullanarang : magnetite.

Condobolin, cos. Gipps and Cunningham : copper ores, lead (carbonate, galena).

Conical Hills : zeolites.

Coodrabidgee River, co. Cowley : arragonite.

Cookaboo River : agate.

Coolac, co. Clarendon : serpentine.

Coolah, co. Napier : coal, ironstone, ozokerite.

Coolamine Plain, co. Wellington : albite.

Coolongolook, co. Gloucester : antimonite, gold, iron pyrites.

Cooma, co. Beresford : copper pyrites, emerald, gold, gypsum, hypersthene, lead (carbonate, galena), muscovite, tourmaline, tremoline.

Coombing Creek, co. Bathurst : copper ores, garnet, kupfermanganerz.

Coonabarabran, co. Cowen : bitumen.

Cootamundra, co. Harden : gold, wad.

Cooyal, co. Phillip : gold, iron (goethite, magnetite), rock crystal, topaz.

Copabella, co. Goulburn : copper ores.

Copeland, co. Gloucester : gold.

Cope's Creek, co. Hardinge : agate, cairngorm, diamond, emerald, fluorspar, galena, hornblende, rock crystal, rutile, sapphire, tinstone, topaz, tourmaline.

Copperhannia, co. Bathurst : chlorite, copper ores, gold.

Copper Hill, co. Wellington : copper ores, silver ores.

Cordeaux River, co. Camden : graphite.

Coroo : chabasite, opal.

Costigan's Mount, co. Georgiana : cerussite, copper carbonate, ironstone.

Cotta River (near), co. Cowley : copper ores.

Cotter's River, co. Murray : arragonite, copper ores, eisenkiesel.

Courntoundra Range, co. Yunghulgra : copper ores.

Cowabee, co. Bourke : gold.

Cowarbee Mine, co. Wynyard : gold.

Cow Flat, co. Bathurst : actinolite, asbestos, copper ores, marble, steatite, zinc blende.

Cowra, co. Bathurst : copper (native and ores), gold, opal.

Cowridge Creek, co. Bathurst : agate, chalcedony, sahlite.

Crookwell, co. Georgiana : galena.

Crookwell River, co. Georgiana : copper and silver ores.

Crudine, co. Wellington : cervantite, copper pyrites, lead (carbonate, galena).

Crudine Creek, co. Wellington : antimonite, gold.

Cudal, co. Ashburnham : gold.

Cudgegong District : antimony (antimonite, cervantite), calcite.

Cudgegong River, co. Phillip : anatase, arragonite, brookite, diamond, gold, gypsum, jasper, iron (limonite, titaniferous), mercury (native, cinnabar), orthoclase, ruby, sapphire, spinelle, topaz, zircon sand, and other gemstones.

Cudgegong River, co. Bligh : cinnabar, diamond, gems, gold.

Cudgegong, co. Wellington : copper ore.

Cullen Bullen, co. Roxburgh : alunogen, copper ores, epsomite.

Curangora, co. Murchison : asbestos, native lead.

Currajong (twelve miles from), co. Ashburnham : copper ores, gold.

Currawang, co. Argyle : copper ores.

D

Dabee, co. Phillip : alunogen, epsomite, Thomsonite.

Dalton, co. King : gold, pyrites.

Dangar's Falls, co. Sandon : antimony ores.

Darbarra Parish, co. Buccleuch : tinstone.

Darby's Run, co. Hardinge : galena.

Dark Corner, co. Westmoreland : mispickel.

Darling River : gypsum.

Deepwater, co. Gough : copper ores, fluorspar, iron pyrites, lead (cerussite, galena, oxide), stream-tin ore, zinc blende.

Deepwater Creek, co. Gough : galena, gemstones, &c., iron pyrites, mispickel, tinstone, zinc blende.

Delegate, co. Wellesley : gold.

Dena River, co. Dampier : gold.

Denisontown (Talbragar), co. Bligh : copper pyrites, galena, iron pyrites, zinc blende.

Derringellon Creek, co. King : copper ores.

Dewelamble : albite.

Dewingbong Mountains : agate.

Diamond Hill, co. Bathurst : epidote.

Ding Dong, co. Gough : mispickel.

Doctor's Creek, co. Murchison : diamond.

Dogtrap (near) : goethite.

Dowagarang, co. Wellington : nepheline, smaragdite.

Drake, co. Drake : antimonite, cervantite.

Dubbo, co. Lincoln : agate, amethyst, chalcedony, jet, galena, iron pyrites.

Duck Creek, co. Gregory : native bismuth.

Duckmaloi : epidote, garnet, wollastonite.

Dundee, co. Gough : copper ores, emerald, mispickel, plumbago, sapphire, tinstone, topaz.

Dunedoo, co. Lincoln : gold.

Dungog, co. Durham : antimonite, calcite, gold, mispickel, serpentine.

Dungowan's Creek, co. Parry : copper ores, jasper, red hæmatite.

E

East Maitland, co. Northumberland : coal.

Eden, co. Auckland : antimonite, copper ore, gold, tinstone.

Ellenborough River, co. Vernon : wad.

Elsmore, co. Gough : beryl, bismuth (native and ores), copper ores, emerald, fluorspar, mispickel, molybdenite, muscovite, topaz, tinstone, wolfram.

Emmaville, co. Gough : cerussite, chlorite, copper ores, fahlerz, fluorspar, galena, iron (arseniate, oxide, pyrites, pyrrhotine), mispickel, silver ores, zinc (blende, red oxide), zircon.

Emu Creek, co. Gough : gold, prehnite, scolezite.

Eremeran, co. Blaxland : tinstone.

Errol : limonite, magnetite.
Eskbank, co. Cook : coal, hæmatite.
Essendon, co. Westmoreland : chrysocolla.
Essington, co. Westmoreland : copper ores.
Eugowra, co. Ashburnham : copper ore.
Euingar Creek, co. Drake : rock crystal.
Eumbi and Bimbijong (between), co. Phillip : ruby.
Euroka Creek, co. Bathurst : barytes.
Eurongilly, co. Clarendon : gold.
Europambela Station, co. Vernon : carbonates of copper.
Eurow Mountain, co. Ashburnham : copper ore.
Everton, Burrowa : galena, silver ores.

F

Fairfield, co. Drake : copper ores, gold, lead ores (carbonate, galena), mispickel, tinstone, zinc blende.
Fairy Meadows : wad.
Fish River, co. Westmoreland : calcite, garnet, gold, marble, mispickel, oligoclase, saltpetre.
Fish River Creek, co. Westmoreland : mispickel.
Five-mile Flat Creek : titaniferous iron.
Foley's Folly, co. Parry : gold, jasper, opal.
Forbes, co. Ashburnham : copper ores, garnet, iron ore, marble.
Ford's Creek (six miles south of Gulgong) : cervantite.
Forest Reefs, co. Bathurst : marcasite.
Fountain Head : chabasite.
Frog's Hole, co. Auckland : calcite, copper ores.
Furrucabad, co. Gough : mispickel.

G

Gara, co. Sandon : antimony (native, antimonite, cervantite), scheelite.
Germanton and Albury (between), co. Goulburn : barytes.
Gerringong, co. Camden : laumonite.
Gibraltar Rock, co. Camden : alunogen.
Gilghi, co. Gough : stream-tin ore.
Ginderbyne : copper pyrites.
Gineroi, co. Murchison : antimonite.
Girilambone, co. Canbelego : copper ores.
Glanmire, co. Roxburgh : barytes, iron ore, manganese (rhodonite, wad).
Glen Creek, co. Gough : chlorite, galena, iron ore, mispickel, molybdenite, tinstone, topaz, wood-tin, wolfram.

Glen Elgin River, co. Gough : antimony ore, gold, sapphire.
Glen Innes, co. Gough : bismuth (native and ores), copper carbonate, galena, gold, iron ores, pyrites, silver ores, tin ore.
Glenlyon, co. Phillip : rock crystal.
Gloucester, co. Gloucester : coal.
Gobandry : antimonite.
Gobolion, co. Ashburnhum : jasper.
Golden Age Mine, Boorook, co. Buller : silver ores (chloride), argentiferous pyrites.
Gooderich, co. Gordon : copper, gold, molybdenum.
Good Hope Mine, co. King : fluorspar, lead ores.
Goolagong, co. Forbes : copper ore.
Gordon Brook, co. Drake : anthracite, chromite, copper ores.
Gordon Creek, co. Drake : tinstone.
Goree, co. Phillip : epidote.
Gorianiwa (near Mundooran) : torbanite.
Gosford, co. Northumberland : goethite.
Goulburn District, co. Argyle : barytes, coal, galena, goethite, gold, green carbonate of copper, iron ores, manganese oxide with cobalt and nickel, melaconite, mispickel, silver ores.
Goulburn Plains, co. Argyle : limestone.
Govett's Leap, co. Cook : gold (traces), wad.
Gow's Creek, co. Cook : flourspar.
Grafton, co. Clarence : antimonite, chromite, coal, galena, gold, magnetite.
Grampian Hills, co. Gough : copper ores, fluorspar, iron pyrites, mispickel, silver ore, tinstone, toad's-eye tin, wolfram, zinc blende.
Granite Diggings, co. Tongowoko : gold, tin ore.
Grassy Creek, co. Clive : tin ore.
Great Mullen Creek, co. Phillip : chrysolite, ruby.
Green Swamp, co. Roxburgh : gold, iron pyrites.
Grenfell, co. Monteagle : amethyst, gold, iron pyrites, magnesite, pyromorphite, silver ores, tinstone, toad's-eye tin.
Gresford, co. Durham : antimonite, cervantite.
Greta, co. Northumberland : coal, torbanite.
Grey Ranges, co. Evelyn : gypsum.
Grose Valley, co. Cook : coal.
Grove Creek, co. Georgiana : agate, mercury.
Gudgeby River (near), co. Cowley : talc.
Gulgong, co. Phillip : agate, albite, analcime, asbestos, calcite, chalcedony,

chlorite, chondrodite, chromite, copper ores, epidote, garnet, gold, iron (chaly-bite, pyrites), kaolin, lead (galena, mimetite), magnesite, manganese oxide, marble, mispickel, opal, stilbite, topaz.

Gumble, co. Ashburnham : blue and green carbonates of copper, gold, lode tin, silver ore, bismuth ore.

Gumble Flat, co. Ashburnham : silver ores.

Gundagai, co. Clarendon : antimonite, asbestos, copper ores, dolomite, goethite, gold, hornblende, lead (native, minium), löllingite, manganese (braunite, oxide), marble, mispickel, serpentine, topaz, zinc blende.

Gundalmulda Creek, co. Murchison : chromite.

Gundaroo, co. Murray : galena.

Gunnedah, co. Pottinger : antimony (antimonite, cervantite), calcite, chalcedony, coal, gold.

Gunning, co. King : copper ores, gold.

Gunningbland : agate.

Guntawang, co. Phillip : gold, pyroxene.

Gwydir District : agate, chromite, coal, diamond, epidote, jasper, limestone, sapphire.

H

Hall's Creek, co. Harden : rose-quartz, wad.

Hanging Rock, co. Parry : agate, chromite, gold, native lead, serpentine.

Hardwick, co. King : garnet.

Hargraves, co. Wellington : gold, mispickel.

Hargraves (forty miles west of), co. Wellington : barytes.

Hargraves Falls, co. Wellington : antimonite.

Hartley, co. Cook : chert, galena, garnet, gold, hæmatite, heulandite, jet, torbanite.

Hastings River, co. Macquarie : antimonite, marmolite.

Havilah, co. Phillip : chalcedony, gold, marble, rock crystal, smoky quartz.

Hawkin's Hill, co. Wellington : chabasite, corundum, gold, muscovite, olivine, pyrrhotine.

Heathcote, co. Cumberland : coal.

Hell's Hole, Mudgee Line : antimonite.

Henry River, co. Gresham : tinstone.

Herding Creek, co. Buller : tinstone, gem-stones, &c.

Hermine : antimonite.

Hill End, co. Wellington : gold, muscovite, wad.

Hillgrove, co. Sandon : antimonite, gold, scheelite.

Hogue's Creek, co. Gough : wolfram.

Holander's River, co. Westmoreland : copper pyrites, galena, iron pyrites.

Home Rule, co. Phillip : felspar, gold, ochre, opal, orthoclase, rock crystal.

Honey's Creek, co. Hardinge : tinstone.

Honeysuckle Creek, co. Hardinge : tinstone.

Honeysuckle Range, co. Cook : iron ores, silver ores.

Hookanvil Creek, co. Parry : opal.

Horton River, co. Murchison : chromite.

Howlong, co. Hume : gold, iron pyrites, mispickel.

Humbug Creek, co. Bland : gold.

Hunter River District : agate, chalcedony, chrysolite, galena, jasper, limestone, molybdenite, opal.

Hunter's Creek, co. Georgiana : antimonite, cervantite.

I

Icely, co. Bathurst : asbestos, copper, epsomite, mispickel, soapstone, steatite.

Illawarra District, co. Camden : chabasite, chert, coal, copper carbonate, iron ores, oil shales.

Inverary, co. Argyle : magnetite.

Inverell, co. Gough : adamantine spar, agate, analcime, aragonite, augite, cobalt ore, copper ores, diamond, galena, gmelenite, olivine, opal, sapphire, silicified wood, stream-tin ore, tinstone, topaz, tourmaline, wolfram.

Inverleigh, co. Brisbane : coal.

Irawang : gypsum.

Ironbarks, co. Wellington : chromite, copper pyrites, gold, iron pyrites, mispickel.

Isis River, co. Brisbane : galena, silver ores.

J

Jamberoo, co. Camden : chert, coal, iron (chalybite, goethite).

Jegedzeric Hill, co. Wallace : actinolite, epidote, tourmaline.

Jervis Bay, co. St. Vincent : coal, goethite.

Jineroo Mount, co. St. Vincent : copper ores, galena.

Jingellic Creek, co. Goulburn : mispickel, scheelite, schorl, silver ore, tinstone.

Joadja Creek, co. Camden : coal, jet, torbanite.

Jones' Creek, co. Clarendon : asbestos, baltimorite, calcite, chromite, dolomite, epidote, pyrophyllite, serpentine, steatite.
Jones' Mount, co. Georgiana : copper ores.
Joppa, co. Argyle : goethite.
Jordan's Crossing, co. Camden : coal.
Jordan's Hill, co. Phillip : opal.
Jordan's Hill, co. Wellington : aragonite, chalybite.
Judd's Creek (near Burraga) : silver ore.
Jugiong Creek, co. Harden : galena, magnetite.
Jungemonia and Uranbeen, co. Phillip : hornblende, steatite, talc.

K

Kaizer Mine, co. Bathurst : copper ores, chessylite, gold.
Kangaloola Creek, co. Georgiana : kyanite.
Kangaroo Flat, co. Gough : asbestos, beryl.
Katoomba, co. Cook : coal, torbanite, wad.
Kayon, co. Richmond : pleonaste.
Kelly's Creek : diallage, picrolite.
Kempsey, co. Dudley : antimony (antimonite, cervantite), barytes, galena, magnesite, marble, mispickel.
Kempsey (near), co. Macquarie : marble.
Kennedy's Creek, co. Monteagle : chromite.
Kentucky Ponds, co. Hardinge : tinstone.
Kiama, co. Camden : agate, amethyst, native copper, laumonite, opal.
Kiandra, co. Wallace : emerald, copper ores, galena, gold, iron pyrites, lignite, molybdenite, muscovite, psilomelane, zircon.
Kimo, co. Clarendon : copper ores, gold.
Kingdon's Pond, co. Brisbane : wulfenite.
King's Creek, co. Westmoreland : psilomelane.
Kingsdale, co. Argyle : goethite.
Kingsgate, co. Gough : bismuth ore (native, carbonate, sulphide), gold, mispickel, molybdenum, silver ore, tinstone.
King's Plains, co. Bathurst : asbestos, gold, kaolin.
Kroombit : copper ores.

L

Lachlan River : apatite, asbestos, chabasite, chert, chlorite, diamond, gold, halloysite, iron (goethite, magnetite, titaniferous), lignite, magnesite, oligoclase, opal, pleonaste, topaz.
Lake Cobham, co. Yantara : agate, gypsum.
Lake Cowal, co. Gipps : gold.

Lake George, co. Murray : cobaltiferous manganese oxide, jasper.
Lake Macquarie, co. Northumberland : coal torbanite.
Lake Tank, co. Gunderbooka : gypsum.
Lambing Flat, co. Monteagle : gold, kaolin, tinstone.
Lamb's Paddock, co. Gough : fluorspar.
Lambton, co. Northumberland : coal.
Lanyon (west of Mount Tennant) : emerald, orthoclase.
Lawson (near), co. Cook : coal.
Lawson's Creek, co. Phillip : galena, opal, orthoclase, ruby, sapphire, and other gemstones.
Leadville, co. Yancowinna : lead ores.
Leconfield, co. Northumberland : coal.
Lewis Ponds Creek, co. Wellington : asbestos, copper ores, epidote, goethite, gold, lead carbonate, magnesite.
Lismore, co. Rous : manganese oxide, iron ores.
Lismore and Ballina (between), co. Rous : opalescent sandstone.
Lithgow, co. Cook : coal, iron (goethite, limonite), mispickel.
Liverpool Plains : aragonite, limestone.
Lob's Hole, co. Buccleuch : aragonite, calcite, copper ores.
Locksley, co. Argyle : diallage, hornblende, iron pyrites.
Long Gully, co. Argyle : tinstone, wad.
Louisa Creek, co. Wellington : arsenic, (native, löllingite, mispickel, realgar), brucite, cadmium, chrysolite, copper ores, gold, magnesite, opal, pyrolusite, scorodite, sulphur, zinc blende.
Lowee : chalcedony, eisenkiesel, opal, soapstone.
Lucknow, co. Wellington : asbestos, calcite, copper ores, gold, iron pyrites, metallic antimony, mispickel, picrolite, pyroxene.
Lunatic, co. Drake : antimonite, arsenic (native), copper ores, gold, pyrrhotine.

M

MacDonald River (head of), co. Inglis : antimonite.
MacIntyre River, co. Gough : agate, cairngorm, jasper, tin.
Macleay River : antimonite, cervantite, galena, marble, silver ores.
Macquarie River : chrysoberyl, diamond, jasper, spinelle, topaz, zircon.
Macquarie Valley, co. Camden : gold.
Made Hill, co. Argyle : pisolitic iron ore.

Made Hill, co. Hardinge : pisolitic iron ore.

Maitland West, co. Northumberland : agate, carnelian, chalcedony, coal, goethite.

Major's Creek, co. St. Vincent : copper pyrites, galena, gold, hornblende, iron pyrites, zinc blende.

Mallone Creek : copper ores.

Manar, co. Murray : actinolite.

Manaro District, co. Woore : barytes, copper ores, galena, gold.

Mandama West, co. Bourke : pyrites.

Maneero, co. Wallace : alunogen, epidote, epsomite.

Mangoplah, co. Mitchell : gold.

Manilla, co. Darling : calcite, copper ores, epidote, serpentine.

Manilla River, co. Darling : copper ores, iron pyrites.

Manilla Waters, co. Darling : copper ores.

Manly Beach, co. Cumberland : red and brown hæmatite.

Manner's Creek, co. Wallace : tinstone.

Manning River, co. Macquarie : antimonite, marble, marmolite, silver ores.

Mann River, co. Gough : emerald, iron pyrites, mispickel, quartz-crystal, sapphire.

Mann River (eight miles from), co. Gough : fahlerz.

Marulan, co. Argyle : antimonite, cervantite, coal, lead (carbonate, galena), marble, mispickel.

Maryland Creek, co. Buller : tinstone, gemstones, &c.

Meadow Flat : cassiterite.

Melrose Ranges and River : copper ores, lead (carbonate, galena), silver ores.

Merimbula, co. Auckland : galena, gold, lignite, silver ore.

Merinoo, co. Wellesley : barytes.

Meroo Creek, co. Wellington : gold.

Merool Creek : stream-tin.

Merrendee, co. Wellington : actinolite, hornblende.

Merriwa, co. Brisbane : calcite.

Meryla, co. Camden : torbanite.

Middle Creek, co. Gough : cairngorm, diamond, fluorspar, galena, pyroxene, sapphire and other gems, tinstone.

Middle Creek, co. Westmoreland : serpentine, williamsite.

Mihi Creek, co. Sandon : cobaltiferous manganese oxide.

Milalong, co. Cook : coal.

Milburn Creek, co. Bathurst : copper ores.

Milparinka, co. Evelyn : lead carbonate.

Minumurra Creek, co. Camden : coal, limestone.

Mitchell River, co. Gresham : gold, tinstone.

Mitchell's Creek, co. Lincoln : copper, gold.

Mitchell's Creek (Sunny Corner), co. Roxburgh : barytes, copper ores, fluorspar, gold, iron (magnetite, pyrites), lead (carbonate, galena, pyromorphite), marble, silver ores, wad, zinc blende.

Mittagong, co. Camden : anthracite, coal, diamond, gems, gold, iron (bog iron ore, goethite), sapphires.

Mittagong, co. Mitchell : gold.

Modbury Creek, co. Murray : chiastolite.

Mole River, co. Clive : tinstone, gemstones, &c.

Mole Tableland, co. Clive : tinstone, gemstones, &c.

Molong, co. Ashburnham : copper ores, jasper, lead, prehnite.

Molong Creek, co. Ashburnham : copper ores, smaragdite.

Molonglo River, co. Murray : silver ores, wulfenite.

Monaltrie, co. Richmond : chalcedony, quartz, siliceous sinter.

Money Ranges (between Yass and Gundagai) : galena.

Monga, co. St. Vincent : gold, silver ores.

Monkey Hill, co. Wellington : diamond.

Montagu Island : adularia.

Montreal, co. Dampier : gold.

Mookerawa Creek, co. Wellington : gold, mercury.

Moonbi (fifteen miles from) : copper pyrites, iron pyrites, mispickel.

Moredun Creek, co. Hardinge : tinstone.

Moree (forty miles north-east of), co. Burnett : stream-tin.

Morullan : epidote.

Moruya, co. Dampier : arsenic, arsenical antimonial silver ore, gold, mercury, zinc blende.

Moruya River, co. Dampier : galena, gold, mispickel, silver ores.

Mount Agate, co. Brisbane : agate.

Mountain Creek, co. Murray : silver ores.

Mount Arrowsmith, co. Evelyn : copper ores.

Mount Browne, co. Evelyn : galena, gold, gypsum.

Mount Budawang, co. St. Vincent : torbanite.

Mount Clarence, co. Cook : goethite.

Mount Dixon, co. King : albite.

Mount Dromedary, co. Dampier : gold, ironstone.

Mount Edgecombe : iron ore, torbanite.

Mount Ephraim (near Hanging Rock), co. Parry : gold.

Mount Gipps, co. Yancowinna: bismuth carbonate, copper (green carbonate), garnet, iron carbonate, lead carbonate, silver (chloride, &c.).

Mount Gipps (eighteen miles north of), co. Farnell : tin ore.

Mount Grosvenor, co. Bathurst: galena, mispickel, silver ores.

Mount Hope, co Blaxland : copper ores.

Mount James, Rocky River (Armidale district) : bismuth (native), zircon.

Mount Keira, co. Camden: coal, goethite, graphite.

Mount Kembla, co. Camden : coal, torbanite.

Mount King George, co. Cook : goethite.

Mount Lambie, co. Cook: fluorspar, garnet, iron (goethite, limonite, magnetite).

Mount Lawson, co. Bathurst : asbestos.

Mount Lindsay, co. Nandewar : eisenkiesel, orthoclase.

Mount Lowry Creek, co. Darling : tinstone.

Mount Lyell, co. Mootwingee : copper ores.

Mount MacDonald (near), co. Bathurst : mispickel.

Mount Macquarie, co. Bathurst : asbestos.

Mount Megalon, co. Cook : coal.

Mount Misery, co. Parry : chalcedony, serpentine, zircon.

Mount . Mitchell, co. Clarke : antimonite, tin ore.

Mount Murulla, co. Brisbane : wulfenite.

Mount Ovens, co. Roxburgh : mispickel.

Mount Pleasant, co. Camden : coal, iron ore.

Mount Pleasant, co. Parry: copper ores, jasper.

Mount Tennant, co. Cowley: emerald, epidote, tourmaline.

Mount Tomah, co. Cook : goethite.

Mount Trooper (near), co. Wallace : copper carbonate, lead (carbonate, galena, protoxide).

Mount Victoria, co. Cook : chert, hæmatite, wad.

Mount Walker, co. Cook : orthoclase.

Mount Walsh, co. Sandon : rutile.

Mount Werong, co. Georgiana : pleonaste, sapphire.

Mount Westmacott : coal.

Mount Wilson, co. Cook : orthoclase.

Mount Wingen, co. Brisbane : agate, alunogen, iron (limonite, magnetite), orthoclase, sulphur, wulfenite.

Mount York, co. Cook : coal, torbanite.

Mowembah, co. Wallace : actinolite, tinstone.

Mudgee, co. Wellington : alunogen, antimony, gold, manganese oxide, mispickel, torbanite.

Mud Wells : natron.

Mullion Range : lydian-stone.

Mullon : specular iron.

Mummurra, co. Bligh : wulfenite.

Munga Creek, co. Dudley : antimonite, cervantite.

Muntabilli River : pleonaste.

Murrumbateman, co. Murray : marble.

Murrumbidgee River, co. Murray : albite, chromite, galena, gold, jasper, marble, marmolite, serpentine, silver ores, tourmaline.

Murrumburrah, co. Harden : galena, gold.

Murrurundi, co. Brisbane : agate, carnelian, pyroxene, torbanite, zeolites.

Murrurundi Tunnel, co. Brisbane : apophyllite, chabasite, gismondine, gmelenite, halloysite, natrolite.

Muswellbrook, co. Durham : chabasite, De Lessite, limestone, zeolites.

Muttama, co. Harden : galena, gold, iron pyrites.

Myall Creek, co. Murchison : calcite, limestone, tinstone.

Mylora, co. Harden : galena gold, hæmatite.

Myralla : apophyllite.

N

Naas Valley, co. Cowley : muscovite, orthoclase.

Nambucca, co. Raleigh : galena, iron pyrites, zinc blende.

Nambucca River and Trial Bay, co. Raleigh : cobaltiferous manganese oxide.

Namoi River, co. Darling : agate, iron pyrites, sapphire.

Nana Creek, co. Fitzroy : gold, iron pyrites.

Nangahra Creek, co. Darling : tinstone.

Narellan Creek, co. Monteagle : goethite.

Narrabri, co. Nandewar : carnelian, coal, obsidian, zeolites.

Narrabri (near), co. Nandewar : torbanite.

Narragal, co. Gordon : copper sulphide.

Narrandera, co. Cooper : goethite.

Narrangarie, co. Napier : silver (chlorides, &c.)

Native Dog Creek, co. Westmoreland : copper ores, galena, gold, iron (pyrites, spathic ore), sapphire and other gemstones, zinc blende.

Nattai, co. Camden : coal, iron (goethite, hæmatite, limonite, pisolitic ore), pleonaste.

Nattai River (near Colo) : torbanite.

Nemingah Flat, co. Inglis : marble.

Nerriga, co. St. Vincent : copper pyrites, torbanite.

Nerriga District, co. St. Vincent : antimonite, lead carbonate, mispickel.

Nerrigundah, co. Dampier : gold.

Nerrimunga, co. Argyle : copper ores, gold.

Newbridge, co. Bathurst : goethite, manganese oxide.

Newbridge (twelve miles south of) : talc.

New Bridge, co. Cook : goethite.

Newcastle, co. Northumberland : coal, manganese oxide.

New Chum Hill, co. Wallace : gold.

Newstead, co. Gough : chalcedony, chalybite, diamond, jasper, limestone, molybdenite, muscovite, pyroxene, rock crystal, sapphire, tin, wolfram.

New Summer Hill, co. Bathurst : argentiferous galena.

New Valley, co. Gough : graphite, tinstone.

Norwood, co. Argyle : goethite.

Nowra, co. St. Vincent : chalcopyrites, iron pyrites, lead (carbonate, galena).

Nuggetty Gully Creek, co. Wellington : jamesonite.

Nundle, co. Parry : agate, antimonite, axinite, black marble, chrysolite, copper ores, diallage, gold, hornblende, hyalite, iron (pyrites, pyrrhotine), jasper, mispickel, opal, serpentine, tourmaline, wad, zircon.

Nundle Creek, co. Parry : chromite, gold, sapphire.

Nurembla : agalmatolite.

Nymagee, co. Mouramba : chlorite, copper ores, lead carbonate, vivianite.

Nymboi River, co. Gresham : carnelian, gold.

O

Oaky Creek, co. Clarence : chromite.

Oban, co. Clarke : cairngorm. limonite, molybdenite, orthoclase, sapphire, spinelle, tinstone, topaz, tourmaline.

Oberon, co. Westmoreland : chlorite, copper ores, diamond, epidote, galena, gold, iron pyrites, mispickel, pyroxene.

Obley, co. Gordon : lead carbonate.

O'Connell, co. Westmoreland : copper carbonate, hæmatite, iron oxide, lead (carbonate, galena), opal.

Old Razor Back, co. Roxburgh : antimonite.

Ophir, co. Wellington : copper sulphide, emerald, galena, gold, iron (pyrites and titaniferous), platinum, silver chloride.

Orange, cos. Wellington and Bathurst : asbestos, bismuth ore, calcite, chlorite, copper ore, iron pyrites, marble, mispickel, muscovite, opal, serpentine, silver ores, wad, zinc blende.

Orara, co. Clarence : epidote, tourmaline.

Ororal : muscovite, opal.

Orunbimbie, co. Vernon : coal.

Ottery Lodes, co. Gough : mispickel, pyrrhotine, tinstone, zinc blende,

Oura, co. Clarendon : muscovite, schorl, spodumene.

Ournie, co. Selwyn : gold, mispickel.

P

Page River, co. Brisbane : galena, limestone, silver ores.

Palmer's, co. Roxburgh : antimonite.

Palmer's Oakley, co. Roxburgh : calcite, copper pyrites.

Pambula, co. Auckland : graphite.

Paradise Creek, co. Gough : chlorite, emerald, manganese sulphide, tinstone.

Parkes, co. Ashburnham : calcite, copper oxide, epidote, gold, iron oxide, pyrites.

Parkes Parish, co. Gough : copper pyrites, iron pyrites, zinc blende.

Parramatta River, co. Cumberland : clays, lignite, zeolites.

Paterson River, co. Durham : antimonite, cervantite, silver ores.

Peabody Mine, co. Ashburnham : copper (native and ores).

Peel, co. Roxburgh : galena.

Peel River, co. Inglis : copper ores.

Peel River, co. Parry : antimonite, aragonite, copper-nickel, gold, lead (native and galena), marble, marmolite, rock crystal, sapphire, serpentine.

Peelwood, co. Georgiana : antimonite, copper (native and ores), ironstone, lead (carbonate, galena, minium, stolzite), molybdenite, scheelite.

Pennant Hills, co. Cumberland : asbestos, calcite, zeolites.

Pentecost Island : manganese oxide.

Perico, co. Auckland : silver ore.

Pheasant Creek, co. Clive : mispickel, lodetin.

Picton, co. Camden : salt.

Piesse Knob, co. Yancowinna : copper ores.

Pigeon House, co. St. Vincent : pyroxene.

Pine Bone Creek : hæmatite.

Pink's Creek, co. Roxburgh : copper ore, ribbon-jasper, sapphire.

Pinnacles, co. Yancowinna: copper carbonate.

Pipeclay Creek (near Mudgee) : calcite.

Piper's Flat, co. Cook : coal, limonite.

Pittwater, co. Cumberland : alunogen.

Plattsburg, co. Northumberland : coal.

Plumbago Creek, co. Drake : graphite.

Pond's Creek, co. Gough : bismuthite, garnet, tin, topaz.

Ponsonby Parish, co. Bathurst : copper ores, hæmatite.

Port Hacking, co. Cumberland : aragonite, goethite.

Port Macquarie (near), co. Macquarie : braunite, copper ores, gold.

Port Stephens, co. Gloucester : coal, marble, torbanite.

Pretty Plains, co. Ashburnham : pyroxenee.

Pucka Parish, co. Clarence : chromite.

Pudman Creek, co. King : galena.

Pullitop Creek, co. Mitchell : gold, tinstone, wolfram.

Pye's Creek, co. Clive : antimonite, bismuth (metallic), copper ores, epsomite, garnet, iron (pyrites, pyrrhotine), lead (carbonate, galena, oxide), mispickel, lode-tin, zinc blende.

Pyramul, co. Wellington : antimonite, cervantite, gold.

Pyramul Creek, co. Wellington : diamond, gold.

Q

Queanbeyan, co. Murray : chlorite, marble, silver ores.

Queanbeyan River, co. Murray : copper ores, marbles.

Queanbeyan and Braidwood (dividing range between), co. Murray : barytes.

Quedong Mount, co. Wellesley : copper ores, galena.

Quialago Creek : limestone.

Quirindi Creek, co. Buckland : iron ore, manganese oxide, tinstone.

R

Ranger's Valley, co. Gough : cairngorm, steatite, tinstone.

Rat's Castle Creek, co. Phillip : chalcedony, hornblende, ruby, wavellite.

Razorback, co. Roxburgh : antimonite, cervantite.

Redgate, co. Gough : bismuth (native).

Redhead, co. Northumberland : coal.

Reedy Creek, co. Bathurst : diamond.

Reedy Creek, co. Murchison : chromite, copper, elaterite, galena, halloysite, prehnite, tinstone.

Reedy Creek, co. Murray : galena.

Richmond River, co. Richmond : cimolite, gold, jasper, meerschaum, opal.

Richmond River (on the coast), co. Richmond : gold, platinum.

Richmond River, co. Rous : coal.

Ringwood : coal.

Rix Creek, co. Durham : coal.

Rock Flat : aragonite.

Rockley (near), co. Georgiana : asbestos, copper ores, galena, hornblende, manganese oxide, marble, silver ore.

Rocksbury, co. Westmoreland : fibrolite.

Rocky Bridge Creek, co. Georgiana : barytes, chalybite, opal.

Rocky Ridge : kaolin.

Rocky River, co. Hardinge : antimonite, cervantite, gold, iron (magnetite and titaniferous), tinstone, zircon.

Rose Valley, co. Gough : sapphire, tinstone.

Round Hill (near), co. Yancowinna : galena.

Ruby Creek, co. Buller : tinstone, gemstones, &c.

Ruby Tin-mine, co. Buller : diamond, tinstone, gemstones, &c.

Rylstone, co. Roxburgh : albite, antimonite, cinnabar, manganese (braunite, manganblende).

S

Saddleback, co. Camden : torbanite.

Sally's Flat, co. Wellington : diamond, gold.

Salt Lake, co. Werunda : salt.

Sandy Creek, co. Hardinge : tinstone.

Sandy Mount, co. Clive : tinstone.

Sandy Swamp, co. Hardinge : galena.

Sara River, co. Clarke: osmium-iridium, sapphires and gemstones, tinstone.

Scone, co. Brisbane : mispickel.

Scrubby Gully, co. Gough : beryl, hyalite, pleonaste, sapphire, smoky quartz, topaz, tourmaline, zircon.

Scrubby Rush : silver ore.

Severn River, co. Clive : fahlerz, iron (pyrites, red oxide), lead (anglesite, carbonate, galena), mispickel, spinelle, stream-tin, wolfram.

Sewell's Creek, co. Georgiana : asbestos, copper carbonate, lead (carbonate, galena), sapphire, steatite.

Sewell's Creek, co. Westmoreland : asbestos, gold, specular iron.

Seymour, co. Wallace : copper ores.

Shannon River and Valley: co. Gough: tinstone.

Sharpening-stone Creek, co. King: antimonite.

Sheep Station Creek, co. Gough: tinstone.

Shell Harbour, co. Camden: gold.

Shellmalleer: silver ores.

Shepherd's Hill, co. Northumberland: hæmatite.

Shoalhaven River, co. St. Vincent: alunogen, antimonite, chrysolite, coal, copper ores, diamond, emerald, epidote, galena, gold, lignite, magnesite, marcasite, mispickel, platinum, tinstone, zircon.

Sidmouth Valley, co. Bathurst: copper ores, epidote, garnet, gold, manganese oxide, rock crystal.

Silent Grove Creek, co. Gough: bismuth (native), tinstone.

Silverdale, co. King: antimony, copper ores, fluorspar, lead (carbonate, galena, pyromorphite), marble, wad, zinc blende.

Silverton, co. Yancowinna: copper carbonate, garnet, gold, ironstone, lead (carbonate, galena, vanadiate), silver (kerargyrite, iodargyrite, embolite, &c.)

Singleton, co. Northumberland: coal, goethite, gypsum, thinolite.

Slate Creek, co. Argyle: copper sulphide, lead (carbonate, galena).

Slaughterhouse Creek, co. Wellesley: barytes.

Smashem's Creek, co. Hardinge: tinstone.

Smith's Paddock, co. Ashburnham: copper ores.

Snowy Mountain, co. Wallace: idocrase.

Snowy River, co. Wallace: gold, iron pyrites, lead (carbonate, galena), sapphire.

Sofala, co. Roxburgh: antimonite, gold.

Solferino, co. Drake: antimonite, arsenic (metallic), chlorite, copper sulphide, gold, iron (magnetite, pyrites), lead (carbonate, galena), rock crystal.

Sounding Rock: silver ore.

South Creek, co. Cumberland: coal.

South River Range: antimonite.

Spring Creek, co. Argyle: tinstone.

Spring Creek, co. Gough: tinstone.

Spring Creek, co. Wellington: gold, tinstone.

Stannifer, co. Gough: diamond, rose-quartz, tinstone.

Stanthorpe (east of), co. Buller: lode-tin.

Stockyard Creek, co. Gough: tinstone.

Stony Batta, co. Sandon: manganese oxide.

Stony Batta Creek, co. Hardinge: chromite, serpentine.

Stony Creek, co. Wynyard: copper ores, gold, halloysite, torbanite.

Strathbogie, co. Gough: apophyllite, copper pyrites, mispickel, lode-tin.

Stroud, co. Gloucester: gold.

Sturt's Meadows, Barrier Range: manganese oxide.

Sugarloaf Hill, co. Wellington: mimetite.

Summer Hill, co. Bathurst: hæmatite.

Summer Hill Creek, co. Bathurst: gold.

Sunny Corner (see Mitchell's Creek, co. Roxburgh).

Sutton Forest, co. Camden: chabasite, hæmatite, halloysite, manganese oxide.

Swallow's Nest: calcite, copper ores.

Swan Bay, co. Rous: coal.

Swanbrook, co. Gough: sapphire, zircon.

Swan Creek, co. Gough: tinstone.

Swinton Parish, co. Hardinge: tinstone.

T

Tacking Point, co. Macquarie: silver ores.

Talbragar River, co. Bligh: analcime, apophyllite, cervantite, chabasite, coal, titaniferous iron.

Tallawang, co. Bligh: calcite, gold.

Tallebung Mountains (twenty miles from Forbes): hornblende.

Talwal Creek, co. St. Vincent: galena.

Tambaroora, co. Wellington: gold, opal.

Tamworth District, co. Inglis: bronzite, calcite, chromite, cobaltiferous manganese oxide, diallage, garnet, gmelenite, goethite, gold, gypsum, hyalite, laumonite, manganese oxide, sapphire, serpentine, shale, stilbite, topaz, zeolites, zircon.

Tarana (near), co. Roxburgh: copper pyrites, galena, garnet, iron (pyrites, pyrrhotine), mispickel, molybdenite,.

Tarcutta, co. Wynyard: alunogen, gold, sulphur, tourmaline.

Tarraba: torbanite.

Tarrabandra, co. Wynyard: galena, iron pyrites, marble, mispickel, zinc blende.

Tarrago Creek, co. Argyle: iron pyrites, lead (carbonate, galena), marble.

Tea-tree Creek, co. Clarence: calcite, gold, tinstone.

Teesdale, co. Bathurst: silver.

Temora, co. Bland: antimonite, copper ores, galena, gold, ochreous ironstone.

Ten-mile, co. Clive: copper ores, iron pyrites.

Tenterfield, co. Clive: antimonite, bismuth (native), gold, hornblende, lead (carbonate, galena), magnetite, stilbite, lode-tin.

Teralba, co. Northumberland: coal.

Thackaringa, co. Yancowinna : barytes, copper ore, galena, garnet, iron pyrites, silver ore.

Thackaringa (near), co. Yancowinna : mispickel.

The Alps : tinstone.

The Gulf, co. Gough : beryl, bismuth (native and ores), chalcedony, mica, scheelite, lode-tin, stream-tin.

The Lagoons, co. Phillip : hypersthene.

The Pinnacle, co. Forbes : gold, lead (carbonate, galena), nepheline.

Thompson's Creek, co. Georgiana : copper ores.

Three-mile Flat, co. Wellington : copper carbonate.

Tiabundie Creek, co. Darling : tinstone.

Tighe's Hill, co. Northumberland : coal.

Timbarra, co. Drake : gold, vivianite.

Tingha, co. Hardinge : beryl, bismuthite, copper ores, galena, garnet, mispickel, sapphire, spinelle, talc, lode-, ruby-, and stream-tin, zinc blende.

Tintin Hull : asbestos, garnet, hornblende.

Tomingley, co. Narromine : gold, manganese oxide, copper carbonate.

Tooloom, co. Drake : gold.

Tooloom River, co. Buller : coal, gold.

Towamba, co. Auckland : epidote.

Trial Bay, near co. Raleigh : cobaltiferous manganese oxide.

Trunkey, co. Georgiana : actinolite, agate, asbestos, diamond, copper pyrites, galena, gold, opal, magnetite, marble, steatite, talc.

Tuena, co. Georgiana : copper ores, gold, lead carbonate, williamsite.

Tuggerah Beach Lake, co. Northumberland : yenite.

Tumberumba, co. Wynyard : gold, ruby, sapphire, stream-tin.

Tumberumba Creek, co. Selwyn : gold.

Tumut, co. Wynyard : gold, hæmatite, magnesite, stream-tin.

Tumut River, co. Selwyn : cachalong.

Turalla Creek, co. Argyle : lignite.

Turon River, co. Roxburgh : diamond, epsomite, gold, tinstone.

Tweed River : carnelian.

Twofold Bay, co. Auckland : amethyst, antimonial copper ore.

Two-mile Creek, co. Hardinge : iron ores.

Two-mile Creek, co. Roxburgh : chromite.

Two-mile Flat, co. Murchison : chrysolite.

Two-mile Flat (near Mudgee), co. Phillip : brookite, carbonaceous earth, chromite, coal, diamond, epsomite, gold, grossularite, halloysite, jasper, magnesite,

marble, muscovite, orthoclase, osmiumor osmi-iridium, pleonaste, rock crystal, sapphire, tinstone, titaniferous iron, topaz and other gemstones, torbanite.

Tyagong, co. Monteagle : gold.

U

Ulladulla, co. St. Vincent : coal.

Umberumba Creek, co. Yancowinna : galena.

Umberumberka, co. Yancowinna : galena, silver ore.

Undercliff, co. Buller : tinstone.

Upper Adelong, co. Wynyard : mispickel.

Upper Bellinger River, co. Raleigh : galena, mispickel, silver ore.

Uralla, co. Sandon : adamantine spar, antimonite, aragonite, chromite, copper carbonate, diamond, garnet, goethite, gold, hornblende, kaolin, mispickel, opal, rock crystal, rutile, sapphire, schorl, spinelle, tinstone, titaniferous iron, topaz, zircon.

Urana, co. Urana : gold.

Uranbeen, co. Phillip : hornblende, steatite, talc.

V

Vale of Clwydd, co. Cook : coal, gold, iron pyrites.

Vegetable Creek, co. Gough (see Emmaville) : aragonite, bismuth (native), diamond, mispickel, monazite, red oxide of zinc, sapphire, stream-tin, zeolites.

W

Wagga Wagga, co. Wynyard : gold, ironstone, tinstone, titaniferous iron.

Wagonga, co. Dampier : gold, iron pyrites, mercury.

Waianamatta, co. Camden : torbanite.

Waibong : limestone.

Walcha, co. Vernon : antimonite, chalcedony, gold, manganese oxide, torbanite.

Walker's Crossing, co. Cook, iron pyrites.

Wallabadah, co. Buckland : copper, galena, marble, zeolites.

Wallerawang, co. Cook : alunogen, andradite, antimonite, chert, coal, epsomite, garnet, iron, goethite, limonite, magnetite, limestone, marble.

Wallsend, co. Northumberland : coal.

Wambrook (near Cowra), co. Bathurst : tinstone.

Wangajong, co. Forbes : gold.

Wantiool, co. Clarendon : gold.

Waratah, co. Northumberland : coal, hydro-carbon (albertite).

Warialda, co. Burnett : diallage, serpentine.

Waroo, co. Bathurst : galena.

Warrambungall Mountains : aragonite, manganese ore, tinstone.

Warrell Creek, co. Raleigh : antimony, argentiferous mispickel, iron pyrites, lead ores, magnesite, ochres, zinc blende.

Warrill, co. Dudley : antimonite, silver ores.

Washpool Creek, co. Drake : antimonite, garnet.

Watson's Creek, co. Inglis : stream-tin.

Wattle Flat, co. Roxburgh : mispickel.

Wear's Creek, co. Parry : nickel ore.

Weddin Mountain : copper carbonate, lead carbonate.

Weeho, co. King : gold.

Wellbank, co. Wellington : copper ores.

Wellingrove, co. Gough : copper ores, silver ores.

Wellington, co. Wellington : agate, chalcedony, copper (native and ores), galena, gold, iron (pyrites and titaniferous), manganese (braunite, wad), marble, opal.

Wellington Caves (near), co. Wellington : copper ores, marble.

Wentworth, co. Bathurst : gold, magnetite, mispickel, picrolite.

Wentworth, co. Wentworth : asbestos.

Werris Creek, co. Buckland : stilbite.

Wheeo, co. Beresford : chrysocolla, copper ores, muscovite.

Whet Creek, co. Bathurst : gold.

Whipstick Flat, co. Wallace : gold.

Willeroi Station, co. Murray : goethite.

Williams River, co. Gloucester : gold, soapstone.

Wilson's Downfall, co. Buller : molybdenum oxide, tinstone.

Windellama (near), Shoalhaven River : cobaltiferous manganese oxide.

Windellama Creek, co. Argyle : pisolitic iron ore.

Windeyer, co. Wellington : gold.

Windindingerie Cataract : epidote, haüyne, orthoclase.

Windsor, co. Cumberland : pisolitic iron ore.

Wingecarribee River, co. Camden : coal, sapphire, zircon.

Wingello, co. Camden : alunogen.

Winter & Morgan's Mine (Sunny Corner), co. Roxburgh : barytes, copper ores, gold, galena, lead (pyromorphite), silver (native and embolite, &c.)

Winterton, Mitchell's Creek, co. Roxburgh, arsenic (native), zinc blende.

Wiseman's Creek, co. Westmoreland : antimonite, asbestos, copper (native and ores), fahlerz, galena, gold, iron pyrites, ironstone, manganese (kupfermanganerz, manganblende, wad), platinum, silver ores, zinc blende.

Wiseman's, South, co. Westmoreland : copper ores, fahlerz, fluorspar, gold, lead (anglesite, carbonate).

Wolgan, co. Cook : coal, gold, torbanite.

Wollombi River, co. Northumberland : corundum, sapphire, zircon.

Wollondilly, co. Camden : limestone, marble.

Wollondilly River, co. Argyle : copper ores.

Wollongong, co. Camden : calcite, coal, limestone.

Wombat, co. Harden : gold.

Wood's Flat, co. Bathurst : copper carbonate, gold, limonite.

Woodstown, co. Bathurst : gold.

Woolgarloo : chessylite, fluorspar, galena, silver ores.

Woolomi, co. Parry : chromite, jasper.

Wylie Creek, co. Buller : tinstone, gemstones, &c.

Wyndham : copper ores, talc.

Y

Yalwal Creek, co. St. Vincent : galena, gold, iron (pyrites, magnetite), platinum, zinc blende.

Yarrahappini, co. Raleigh : silver ores.

Yarralumla, co. Argyle : albite.

Yarrangobilly, co. Buccleuch : copper ores, gold.

Yarrangun : muscovite.

Yarrow (near), co. Gough : bismuth, bismuthite, &c., tinstone.

Yarrowford (twenty miles from), co. Gough : galena, mispickel.

Yass, co. King : antimony, chlorite, chromite, copper ores, galena, gypsum, magnetite, rock crystal, wad.

Yass Plains, co. King : marble.

Yass River, co. King : silver ores.

Yetholme (near), co. Bathurst : mispickel.

Young, co. Monteagle : gold, lead (carbonate, galena), tourmaline.

Young (twenty miles from) : chromite.

Y Waterholes, co. Gough : cassiterite, iron (limonite and titaniferous), spinel, zircon.

U

MINERAL LOCALITIES

ARRANGED ACCORDING TO COUNTIES.

Co. Argyle.

Bangalore.......... Marble.
Black Ranges..... Tourmaline.
Boro Creek........ Pisolitic iron ore, wad.
Brook's Creek..... Galena, diamonds.
Bullio Flat........ Molybdenite.
Bungonia........... Alunogen, andalusite, antimonite, calcite, copper ores, epidote, galena, gold, gypsum, iron (pharmacosiderite, pisolitic iron ore, pyrites), limestone, plumbago, tinstone.
Bungonia (near).. Cobaltiferous manganese oxide, copper ore.
Burnaby Creek... Silver ores.
Currawang......... Copper ores.
Goulburn (near).. Barytes, black oxide of copper, cobaltiferous manganese oxide with nickel, galena, goethite, gold, iron ores, malachite, manganese ore, mispickel, silver ore.
Goulburn Plains.. Limestone.
Inverary........... Magnetite.
Joppa.............. Goethite.
Kingsdale.......... Goethite.
Locksley........... Diallage, hornblende, iron pyrites.
Long Gully........ Tinstone, wad.
Made Hill......... Pisolitic iron ore.
Major's Creek..... Gold, pyrites, zinc blende.
Marulan........... Antimonite, cervantite, coal, lead (carbonate, galena), marble, mispickel.

Nerrimunga .... Copper ores, gold.
Norwood............ Brown hæmatite.
Slate Creek........ Copper [sulphide, lead (carbonate, galena).
Spring Creek...... Tinstone.
Tarrago Creek.... Iron pyrites, lead (carbonate, galena), marble.
Turalla Creek..... Lignite.
Windellama Creek Pisolitic iron ore.
Wollondilly River Copper ores.
Yarralumla........ Albite.

Co. Arrawatta.

Ashford .... Coal, tinstone.
Brush Creek...... Websterite.
Bukkulla........... Asbestos, coal.
Byron's Plains.... Augite, cairngorm.
Severn River...... Fahlerz, iron (pyrites, red oxide), lead (anglesite, carbonate, galena), mispickel, spinelle, streamtin, wolfram.

Co. Ashburnham.

Billabong.......... Tinstone.
Bocoble............ Copper ores, gold.
Burra Burra....... Iron (goethite, magnetite), tin.
Canobolas.......... Barytes, copper (native and ores), gold, ironstone.

Canomidine Creek Copper ores, marble.
Cargo.............. Copper ores, gypsum, gold.
Carrawabbity...... Marble.
Currajong (twelve Copper ores, gold.
miles from)
Eugowra............ Copper ore.
Eurow Mountain Copper ore.
Forbes.............. Copper ores, iron ore.
Forbes (near)...... Marble.
Gobolion............ Jasper.
Gumble............ Bismuth ore, copper ores, gold, lode-tin, silver ore.
Gumble Flat...... Silver ores.
Molong............ Copper ores, jasper, lead ores, prehnite.
Molong (near)..... Copper ores.
Molong Creek..... Copper ores, smaragdite.
Orange (ten miles Chlorite, iron pyrites.
from, on Cargo
Road)
Parkes.............. Calcite, copper ores, gold, iron oxide.
Parkes (near)...... Epidote, iron pyrites.
Peabody Mine.... Copper ores.
Pretty Plains...... Pyroxene.
Smith's Paddock Copper ores.

Co. Auckland.

Bobbera, Jingery, Epidote.
and Pambula
(between)
Brogo and Two- Bog Butter.
fold Bay (be-
tween)
Candelo (near).... Barytes, galena.
Eden................ Antimonite, gold.
Eden (thirty-six Copper ore.
miles north-
west of)
Frog's Hole........ Calcite, copper ores.
Merimbula......... Galena, gold, lignite, silver ore.
Pambula............ Graphite.
Perico.............. Silver ore.
Towamba .... Epidote.
Twofold Bay...... Amethyst, antimonial copper ore.

Co. Bathurst.

Apsley.............. Copper ores.
Back Creek........ Goethite.

Bathurst District Amethyst, antimonite, barytes, copper, copper-nickel, diamond, epidote, garnet, gold, gypsum, iron (pyrites, titaniferous), jasper, manganese ore, marble, mispickel, osmium-iridium, pyromorphite, silver ores, spinelle, staurolite, talc, tinstone, topaz, wad.
Belubula River... Gemstones, gold, magnetite, marble.
Belubula and Barytes.
Lachlan Rivers
(between)
Binbanang......... Silver ores.
Blayney............ Allophane, copper (native and ores), hæmatite, manganese ore.
Bringelet Parish Manganese oxide.
Brown's Creek.... Copper ores, gold, magnetite.
Bull Dog Range.. Gold, lead ores.
Cadiangulong...... Copper ores.
Caloola............. Asbestos, barytes, gold, hæmatite, manganese (braunite, pyrolusite, wad).
Carcoar............ Barytes, chalcedony, copper ores, eisenkiesel, galena, gold, halloysite, iron (hæmatite, magnetite, marcasite, pyrites), mispickel, opal.
Coombing Creek.. Copper ores, garnet, kupfermanganerz.
Copperhannia...... Chlorite, copper ores, gold.
Cow Flat........... Actinolite, asbestos, copper ores, galena, marble, steatite, zinc blende.
Cowra.............. Copper ores, gold, opal.
Cowridge Creek... Agate, chalcedony, sahlite.
Diamond Hill..... Epidote.
Euroka Creek..... Barytes.
Forest Reefs ...... Marcasite.
Icely.............. Asbestos, copper, epsomite, mispickel, steatite.
Kaizer Mine ...... Copper ores with gold.
King's Plains...... Asbestos, gold, kaolin.
Milburn Creek.... Copper ores.
Mount Grosvenor Galena, mispickel, silver ore.
Mount Lawson... Asbestos.
Mount MacDon- Mispickel.
ald

Mount Macquarie　Asbestos.
Newbridge......... Brown hæmatite, manganese.
New Summer Argentiferous galena.
Hill
Orange............. (See co. Wellington).
Ponsonby Parish. Copper ores, hæmatite.
Reedy Creek...... Diamond.
Rockley(five miles Hornblende.
west of)
Sidmouth Valley.. Copper ores, epidote, garnet, gold, manganese, rock crystal.
Summer Hill...... Hæmatite.
Teesdale............ Silver ore.
Waroo............... Galena.
Wentworth......... Gold, magnetite, mispickel, picrolite.
Wood's Flat....... Gold, limonite, malachite.

Co. Beresford.

Bredbo.............. Calamine, galena, malachite.
Collington .... Lead (carbonate, galena).
Cooma.............. Copper pyrites, gold, gypsum, hypersthene, lead (carbonate, galena), muscovite, tremolite, tourmaline.
Wheeo.............. Chrysocolla, copper ores, muscovite.

Co. Bland.

Barmedman (seventeen miles north-west of) Schorl.
Bland............... Chromite, opal.
Calabash Creek... Diamond.
Temora............. Antimonite, copper ores, galena, gold, ochreous ironstone.

Co. Blaxland.

Boona West....... Tinstone.
Eremeran.......... Tinstone.
Mount Hope...... Copper ores.

Co. Bligh.

Belara .... Copper ores, lead carbonate, silver ore, zinc blende.

Cudgegong River Diamond, gold, gems.
Denisontown...... Copper pyrites, galena, iron pyrites, zinc blende.
Munmurra......... Wulfenite.
Talbragar River.. Analcime, apophyllite, cervantite, chabasite, coal, titaniferous iron.
Tallawang......... Calcite, gold.

Co. Bourke.

Mandama West... Pyrites.

Co. Brisbane.

Isis River.......... Galena, silver ores.
Kingdon's Ponds Wulfenite.
Merriwa............ Calcite.
Mount Agate...... Agate.
Mount Murulla.... Wulfenite.
Mount Wingen... Agate, alunogen, limonite, magnetite, orthoclase, sulphur, wulfenite.
Murrurundi....... Agate, carnelian, pyroxene, torbanite, zeolites.
Murrurundi Tunnel Apophyllite, chabasite, gismondine, gmelenite, halloysite, natrolite.
Page River......... Galena, limestone, silver ores.
Scone............... Mispickel.

Co. Buccleuch.

Darbarra Parish.. Tinstone.
Lob's Hole......... Aragonite, calcite, copper ores.
Yarrangobilly..... Copper ores, gold.

Co. Buckland.

Carroll (six miles Calcite, stilbite.
south-east of)
Carroll's Creek. .. Tinstone.
Quirindi Creek.... Iron ore, manganese ore, tinstone.
Wallabadah........ Cimolite, copper, galena, marble, zeolites.
Werris Creek...... Stilbite.

Co. Buller.

Bookookoorara.... Tinstone.
Boonoo Boonoo ... Gold, tinstone.
Boorook............ Gold, silver ores, vivia-
                 nite.
Golden Age Mine Silver ores (chloride, &c.),
                 argentiferous pyrites.
Herding Creek.... Tinstone, gemstones, &c.
Maryland Creek.. Tinstone, gemstones, &c.
Ruby Creek....... Tinstone, gemstones, &c.
Ruby Tin-mine ... Diamond, gemstones, &c.,
                 tinstone.
Tooloom River.... Coal, gold.
Undercliff.......... Tinstone.
Wilson's Downfall Molybdenum oxide, tin
                 ore.
Wylie Creek...... Tinstone, gemstones, &c.

Co. Burnett.

Moree (forty miles Stream-tin ore.
   north-east of)
Warialda........... Diallage, serpentine.

Co. Camden.

Berrima............ Coal, gems, halloysite, iron
                 geothite, hæmatite, li-
                 monite), torbanite, sap-
                 phire, zircon.
Broughton Creek Torbanite, coal.
Broughton Vale... Goethite.
Bulli.............. Coal, limonite.
Bundanoon......... Coal, gold, iron ore, mis-
                 pickel.
Burragorang....... Epsomite, galena, iron-
                 stone, limestone, tor-
                 banite, zinc blende.
Cambewarra   Torbanite.
   Ranges
Cordeaux River... Graphite.
Gerringong......... Laumonite.
Gibraltar Rock·... Alunogen.
Illawarra District Chabasite, chert, copper
                 carbonate, oil shales.
Jamberoo.......... Chalybite, chert, coal,
                 goethite.
Jamberoo Moun- Goethite.
   tains
Joadja Creek...... Coal, jet, torbanite.
Kiama.............. Agate, amethyst, copper,
                 laumonite, opal.
Meryla............. Torbanite.
Minumurra Creek Coal, limestone.
Mittagong......... Anthracite, coal, dia-
                 mond, gemstones, gold,
                 iron (bog iron ore,
                 goethite).

Mount Keira...... Coal, goethite, graphite.
Mount Kembla... Coal, torbanite.
Mount Pleasant... Coal, iron.
Nattai.............. Coal, iron (goethite, hæ-
                 matite, limonite, pisoli-
                 tic iron ore), pleonaste.
Picton.............. Salt.
Saddleback........ Torbanite.
Sutton Forest...... Chabasite, halloysite, hæ-
                 matite, manganese ox-
                 ide.
Waianamatta...... Torbanite.
Wingecarribee   Sapphire, zircon.
   River
Wingello........... Alunogen.
Wollondilly....... Limestone, marble.
Wollongong....... Calcite, coal, limestone.

Co. Canbelego.

Girilambone .... Copper ores.

Co. Clarence.

Clarence River.... Antimonite, apatite, coal,
                 copper ores, gold, iron
                 (goethite, magnetite),
                 resinite, serpentine, sil-
                 ver ores.
Grafton............ Antimonite, chromite,
                 coal, galena, gold, mag-
                 netite.
Oaky Creek........ Chromite.
Orara.............. Epidote, tourmaline.
Pucka Parish...... Chromite.
Tea-tree Creek.... Calcite, gold, tinstone.

Co. Clarendon.

Bethungra.......... Galena.
Coolac............. Serpentine.
Gundagai District Antimonite, arsenic (löl-
                 lingite, mispickel), as-
                 bestos, copper ores, do-
                 lomite, goethite, gold,
                 hornblende, lead (native,
                 minium), manganese
                 (braunite,oxide),marble,
                 serpentine, topaz.
Gundagai (eight Zinc blende.
   miles west of)
Jones' Creek....... Asbestos, baltimorite, cal-
                 cite, chromite, dolomite,
                 epidote, pyrophyllite,
                 serpentine, steatite.
Junee District..... Gold, limestone.
Kimo............... Copper ores, gold.
Oura............... Muscovite, schorl, spodu-
                 mene
Sebastopol Reef... Galena, gold.

Co. Clarke.

Aberfoil River.... Antimonite, corundum, iridosmine, platinum, tin (native, cassiterite).
Ann River........ Tinstone.
Mount Mitchell... Antimonite, tin.
Oban .... Cairngorm, limonite, molybdenite, orthoclase, sapphire, spinelle, tinstone, topaz, tourmaline.
Sara River........ Osmium-iridium, sapphires and gemstones, tinstone.

Co. Clive.

Black Swamp...... Lithia-mica.
Bolivia.............. Chlorite, copper ores, iron pyrites, lead (carbonate, galena), mispickel, molybdenite, zinc blende.
Byrne's Lode...... Native bismuth.
Grassy Creek...... Tin ore.
Mole River........ Tinstone, gemstones, &c.
Mole Tableland... Tinstone, gemstones, &c.
Pheasant Creek... Mispickel, lode-tin.
Pye's Creek........ Antimonite, bismuth (metallic), copper ores, epsomite, garnet, iron (iron pyrites, magnetic pyrites), lead (carbonate, galena, oxide), lode-tin, zinc blende.
Sandy Mount...... Tinstone.
Ten-mile........... Copper ores, iron pyrites.
Tenterfield........ Antimonite, bismuth (native), gold, hornblende, lead (carbonate, galena), magnetite, stilbite, lode-tin.

Co. Clyde.

Brewarrina........ Lead (carbonate, galena).

Co. Cook.

Bathgate........... Torbanite.
Ben Bullen........ Limestone.
Blackheath........ Coal, goethite, torbanite.
Blue Mountains... Chert, graphite, iron (hæmatite, limonite), wad.
Bowenfels......... Coal, gold, limonite.
Clarence Tunnel.. Coal, hæmatite.
Eskbank........... Coal, hæmatite.
Govett's Leap..... Gold (traces), wad.
Gow's Creek....... Fluorspar.

Grose Valley...... Coal.
Hartley............ Chert, galena, garnet, gold, hæmatite, heulandite, jet, torbanite.
Honeysuckle Range   Iron ores, silver ores.
Katoomba.......... Coal, torbanite, wad (see Blue Mountains).
Lithgow............ Coal, iron ores (goethite, limonite, &c.), mispickel.
Mount Clarence... Goethite.
Mount King George   Goethite.
Mount Lambie......Fluorspar, garnet, iron (hæmatite, limonite, magnetite).
Mount Tomah..... Goethite.
Mount Victoria... Chert, torbanite.
Mount Walker.... Orthoclase.
Mount Wilson..... Oligoclase.
Mount York....... Coal, torbanite.
Mudgee Road..... Alunogen.
New Bridge........ Goethite.
Piper's Flat........ Coal, iron (limonite, magnetite), limestone.
Vale of Clwydd... Coal, gold, iron pyrites.
Walker's Crossing Iron pyrites.
Wallerawang...... Alunogen, andradite, antimonite, chert, coal, epsomite, garnet, iron (goethite, limonite, magnetite), limestone, marble.
Wolgan............ Coal, gold, torbanite.

Co. Cooper.

Narrandera......... Goethite.

Co. Cowen.

Breelong Creek... Agate, ironstone, opal.
Coonabarabran.... Bitumen.

Co. Cowley.

Cavan............... Limestone.
Coodrabidgee River   Aragonite.
Cotta River (near) Copper ores.
Gudgeby River (near)   Talc.
Mount Tennant... Emerald, epidote, tourmaline.
Naas Valley........ Muscovite, orthoclase.

Co. Cumberland.

Appin.............. Alunogen, epsomite.
Bulli................. Alunogen, coal, hæmatite.
Cataract River.... Aragonite, calcite.
Coal Cliff.......... Coal, limonite.
Manly Beach...... Hæmatite.
Parramatta River Hæmatite, lignite, zeo-
 lites.
Pennant Hills..... Asbestos, calcite.
Pittwater........... Alunogen.
Port Hacking...... Aragonite, goethite.
Sydney.............. Hæmatite.
Windsor............ Pisolitic iron ore.

Co. Dampier.

Cobargo............ Barytes.
Bermagui River... Copper ores, gold.
Moruya............. Arsenic, gold, mercury,
 zinc blende.
Moruya River..... Galena, gold, mispickel,
 silver ores.
Mount Drome- Gold, ironstone.
 dary
Wagonga........... Gold, iron pyrites, mer-
 cury.

Co. Darling.

Barraba............ Chromite, copper ores,
 gold, kaolin, magnetite,
 pyroxene, serpentine,
 tripoli.
Manilla............. Calcite, copper ores, epidote, iron pyrites, ser-
 pentine.
Mount Lowry Tinstone.
 Creek
Namoi River...... Agate, iron pyrites, sapphire.
Nangahra Creek.. Tinstone.
Tiabundie Creek Tinstone.

Co. Drake.

Drake............... Antimonite, cervantite.
Euingar Creek.... Rock crystal.
Fairfield............ Copper ores, gold, lead
 (carbonate, galena).
Gordon Brook..... Anthracite, chromite, cop-
 per ores, mispickel, tin-
 stone, zinc blende.
Gordon Creek..... Tinstone.
Lunatic .... Antimonite, arsenic (na-
 tive), gold, pyrrhotine.
Lunatic District.. Copper ores.

Plumbago Creek.. Graphite (plumbago).
Solferino........... Antimonite, arsenic (metallic), chlorite, copper,
 gold, iron (magnetite,
 pyrites), lead (cerussite
 galena), rock crystal.
Timbarra........... Gold, vivianite.
Washpool Creek.. Antimonite, garnet.

Co. Dudley.

Caragula........... Antimonite, cervantite.
Kempsey........... Antimony (antimonite,
 cervantite), barytes,
 galena, marble, mis-
 pickel.
Munga Creek...... Antimonite, cervantite.
Warrill .... Antimonite, silver ores.

Co. Durham.

Calton Hill......... Mercury (cinnabar, native), platinum, silver
 (native).
Dungog .... Calcite, gold, mispickel,
 serpentine.
Gresford .... Antimonite, cervantite.
Muswellbrook..... Chabasite, De Lessite,
 limestone, zeolites.
Paterson River... Antimonite, cervantite,
 silver ores.
Rix Creek.......... Coal, iron.

Co. Evelyn.

Grey Ranges...... Gypsum.
Milparinka........ Cerussite.
Mount Arrow- Agate, gypsum.
 smith
Mount Browne Galena, gypsum.
 District

Co. Farnell.

Carona.............. Copper ore.
Mount Gipps Tin ore.
 (eighteen miles
 north of)

Co. Fitzroy.

Nana Creek........ Gold, iron pyrites.

Co. Flinders.

Babinda............ Carbonates of copper.

Co. Forbes.

Goolagong .... Copper ore.
"The Pinnacle ".. Gold, lead (carbonate, galena), nepheline.

Co. Georgiana.

Abercrombie Caves — Marble.
Abercrombie Ranges — Asbestos, copper ores, lead carbonate.
Abercrombie River — Apatite, barytes, chabasite, diamond, garnet, gold, graphite, sapphire, topaz, zircon.
Back Creek........ Lead carbonate, manganese (diallogite, manganite, oxide, rhodonite), silver ore.
Belwood............ Rhætizite.
Binda............... Galena.
Briar Park......... Asbestos, copper ores.
Crookwell.......... Galena.
Crookwell River.. Copper and silver ores.
Costigan's Mount Copper carbonate, ironstone, lead carbonate.
Grove Creek....... Agate, mercury.
Hunt's Creek...... Antimonite, cervantite.
Jones' Mount..... Copper ores.
Kangaloola Creek Kyanite.
Mount Werong... Pleonaste, sapphire.
Peelwood.........:.. Antimonite, copper ores, lead (cerussite, galena, minium), molybdenite.
Rockley .... Asbestos, fahlerz, galena, marble, silver ores.
Rocky Bridge Creek — Barytes, chalybite, opal.
Sewell's Creek.... Asbestos, copper carbonate, lead (carbonate, galena).
Thompson's Creek Copper ores.
Trunkey............ Actinolite, agate, asbestos, copper pyrites, diamond, galena, gold, magnetic iron ore, marble, opal, steatite, talc.
Tuena............... Cerussite, copper ores, gold, williamsite.
Tuena (near)...... Cerussite, copper carbonate.

Co. Gipps.

Cagillico (thirty miles east of) — Ironstone.

Condobolin......... Copper ores, lead (cerussite, galena).
Forbes (fifty miles west of) Iron ores.

Co. Gloucester.

Back Creek........ Gold, mispickel.
Bowman River... Calcite, gold, iron pyrites.
Bulladelah......... Alunite.
Chichester River Galena.
Coolongolook...... Antimonite, gold, iron pyrites.
Dungog (twenty miles north of) Antimonite.
Port Stephens..... Coal, marble, torbanite.
Williams River.... Gold, soapstone.

Co. Gordon.

Buckinbar .:....... Copper, silver ores.
Gooderich.......... Copper, gold, molybdenum.
Narragal............ Copper sulphide.
Obley............... Cerussite.

Co. Gough.

Albion Mine...... Rock crystal, tinstone, wolfram.
Bald Nob Creek.. Steatite.
Ben Lomond...... Iron pyrites, sapphire.
Bischoff Mine..... Lode-tin.
Blair Hill.......... Tinstone.
Bony Gully........ Nickel sulphide.
Britannia Mine... Diamond, sapphire and other gemstones, tinstone.
Bruce Mine........ Native bismuth.
Butchart Mine.... Tin ores.
Deepwater......... Copper ores, fluorspar, iron pyrites, lead (carbonate, galena, oxide), streamtin ore, zinc blende.
Deepwater Creek Galena, gemstones, &c., iron pyrites, mispickel, tinstone, zinc blende.
Ding Dong......... Mispickel.
Dundee............. Copper ores, emerald, mispickel, plumbago, sapphire, tinstone, topaz.
Elsmore............ Beryl, bismuth (native) copper ores, emerald, fluorspar, mispickel, molybdenite, muscovite, tinstone, wolfram.

Emmaville (near) Chlorite, copper ores, galena, iron (oxide, pyrites, pyrrhotine), mispickel, silver ore, tinstone, zinc (blende, oxide), zircon.

Emu Creek......... Gold, prehnite, scolezite.

Furrucabad........ Mispickel.

Gilghi.............. Stream-tin ore.

Glen Creek......... Chlorite, galena, iron ore, mispickel, molybdenite, tinstone, wood-tin, topaz, wolfram.

Glen Elgin River Antimony ore, gold, sapphire.

Glen Innes.......... Bismuth (native and ores), copper carbonate, galena, gold, iron ore, silver ore, tin ore.

Glen Innes......... Iron pyrites, tin ore.

Grampians......... Copper ore, fluorspar, iron pyrites, mispickel, silver ore, tinstone, toad's eye tin, wolfram, zinc blende.

Hogue's Creek.... Wolfram.

Inverell .... Adamantine spar, agate, analcime, aragonite, augite, cobalt ore, copper ores, diamond, galena, gmelenite, olivine, opal, sapphire, silicified wood, stream-tin ore, tinstone, topaz, wolfram.

Kangaroo Flat.... Asbestos, beryl.

Kingsgate.......... Bismuth (native, carbonate, sulphide), gold, mispickel, molybdenum, silver ore, tinstone.

Lamb's Paddock.. Fluorspar.

MacIntyre River.. Agate, jasper, cairngorm, tin.

Mann River....... Emerald, fahlerz, iron pyrites, mispickel, quartz crystal, sapphire.

Middle Creek.... Cairngorm, diamond, fluorspar, galena, pyroxene, sapphires and other gems, tinstone.

Newstead.......... Chalcedony, chalybite, diamond, jasper, limestone, molybdenite, muscovite, pyroxene, rock crystal, sapphire, tin, wolfram.

New Valley....... Graphite, tinstone.

Ottery Lodes...... Mispickel, pyrrhotine, tinstone, zinc blende.

Paradise Creek... Chlorite, emerald, manganese sulphide, tinstone.

Parkes Parish..... Copper pyrites, iron pyrites, zinc blende.

Pond's Creek..... Bismuthite, garnet, tin, topaz.

Ranger's Valley.. Cairngorm, steatite, tinstone.

Redgate........... Native bismuth.

Rose Valley....... Sapphire, tinstone.

Scrubby Gully.... Beryl, hyalite, pleonaste, sapphire, smoky quartz, topaz, tourmaline, zircon.

Shannon River.... Tinstone.

Sheep Station Creek Tinstone.

Silent Grove Creek Native bismuth, tinstone.

Spring Creek...... Tinstone.

Stannifer.......... Diamond, rose-quartz, tin ore.

Stockyard Creek.. Tinstone.

Strathbogie........ Apophyllite, copper pyrites, mispickel, lode-tin.

Swanbrook......... Sapphire, zircon.

Swan Creek....... Tinstone.

The Gulf........... Beryl, bismuth (native and ores), chalcedony, iron oxide, mica, lode-tin, stream-tin, scheelite.

Vegetable Creek (Emmaville) Aragonite, bismuth (native), diamond, mispickel, monazite, sapphire, red oxide of zinc, tinstone, zeolites.

Wellingrove....... Copper ores, silver ores.

Yarrow River..... Bismuth ores, tinstone.

Y Waterholes..... Cassiterite, iron (limonite, titanic), spinelle, zircon.

Co. Goulburn.

Albury.............. Gold, iron pyrites.

Albury (seventy miles above) Chrysolite.

Albury and Germanton (between) Barytes.

Copabella.......... Copper ores.

Jingellic Creek.... Mispickel, scheelite, schorl, silver ores, tinstone.

Co. Gregory.

Duck Creek....... Native bismuth.

Co. Gresham.

Henry River.......	Tinstone.
Mitchell River....	Gold, tinstone.
Nymboi River.....	Carnelian, gold.

Co. Gunderbooka.

Lake Tank.........	Gypsum.

Co. Harden.

Binalong...........	Copper ores, galena, iron (brown hæmatite, magnetite, pyrites, red hæmatite).
Bogolong...........	Magnetite.
Bookham...........	Galena, hæmatite, marble.
Bowning...........	Dioptase, manganese oxide.
Cootamundra......	Gold, wad.
Hall's Creek.......	Rose-quartz, wad.
Murrumburrah....	Galena, gold.
Muttama...........	Galena, gold, iron pyrites.
Mylora.............	Galena, gold, hæmatite.

Co. Hardinge.

Auburn Vale Creek	Diamond, tinstone.
Balola...............	Orthoclase, topaz, tourmaline.
Bengonover Mine	Diamond and other gemstones, tinstone.
Big River...........	Diamond and gemstones.
Bolitho Mine......	Tinstone.
Borah Mine	Diamond, tin.
Bundarra...........	Mispickel, tinstone.
Cameron's Creek	Pyroxene.
Cope's Creek	Agate, cairngorm, diamond, emerald, fluorspar, galena, hornblende, rock crystal, rutile, sapphire, tinstone, topaz, tourmaline.
Darby's Run......	Galena.
Honey's Creek....	Tinstone.
Honeysuckle Creek	Tinstone.
Kentucky Ponds..	Tinstone.
Made Hill.........	Pisolitic iron ore.
Moredun Creek...	Tinstone.
Rocky River	Antimonite, cervantite, gold, iron (magnetite, titaniferous), tin, zircon.
Sandy Creek......	Tinstone.
Sandy Swamp.....	Galena.

Smashem's Creek	Tinstone.
Stony Batta Creek	Chromite, serpentine.
Swinton Parish...	Tinstone.
Tingha.............	Beryl, bismuthite, copper ores, galena, garnet, mispickel, sapphire, spinelle, talc, tin (rubytin ore, stream-tin), zinc blende.
Two-mile Creek...	Iron ores.

Co. Hume.

Howlong...........	Gold, iron pyrites, mispickel.

Co. Hunter.

Capertee...........	Calcite, coal, copper ores, galena, gold, iron pyrites, mica, molybdenite, torbanite, tourmaline.
Colo...............	Iron sulphate.

Co. Inglis.

Attunga............	Hornblende, tinstone.
Bendemeer........	Antimonite, chromite, hornblende, manganese (oxide, rhodonite), mispickel, platinum, sapphire, tinstone, tourmaline.
Carlyle's Creek...	Tinstone.
MacDonald River	Antimonite.
Nemingah Flat...	Marble.
Peel River.........	Copper ores.
Tamworth	Bronzite, calcite, chromite, cobalt ore, diallage, garnet, gmelenite, goethite, gold, gypsum, hyalite, laumonite, manganese, sapphire, serpentine, shale, stilbite, topaz, zeolites, zircon.
Tamworth District	Manganese ore.
Watson's Creek...	Stream-tin.

Co. King.

Bala...............	Copper ores.
Bowning Creek...	Copper and silver ores.
Burrowa...........	Copper ores, galena, silver ores.
Dalton.............	Gold, iron pyrites.
Derringellon Crk.	Copper ores.

Good Hope Mine Fluorspar, lead ores.
Gunning............ Copper ores, gold.
Hardwicke......... Garnet.
Mount Dixon...... Albite.
Pudman Creek.... Galena.
Sharpening-stone Antimonite.
Creek
Silverdale.......... Antimony, copper ores, fluorspar, lead (cerussite galena, pyromorphite), marble, wad, zinc blende.
Yass................ Antimony, chlorite, chromite, copper ores, galena, gypsum, magnetite, rock crystal, wad.
Yass Plains........ Marble.
Yass River......... Silver ores.

Co. Leichhardt.

Castlereagh River Carnelian, opal.

Co. Lincoln.

Ballimore........... Goethite.
Barbigal............ Coal, iron ores.
Dubbo.............. Agate, amethyst, chalcedony, jet.
Mitchell's Creek.. Copper, gold.

Co. Macquarie.

Hastings River... Antimonite, marmolite.
Kempsey........... Marble.
Manning River... Antimonite, marble, marmolite, silver ores.
Port Macquarie Braunite, copper ores, District gold.
Tacking Point.... Silver ores.

Co. Mitchell.

Mangoplah........ Gold.
Mittagong .... Gold.
Pullitop Creek.... Gold, tinstone, wolfram.

Co. Monteagle.

Grenfell............ Amethyst, gold, iron pyrites, magnesite, pyromorphite, silver ores, tinstone.
Kennedy's Creek Chromite.
Lambing Flat..... Gold, kaolin, tinstone.
Narellan Creek... Goethite.
Young.............. Cerussite, galena, gold, tourmaline.

Co. Mootwingee.

Mount Lyell....... Copper ores.

Co. Mouramba.

Nymagee........... Chlorite, copper ores, lead carbonate, vivianite.

Co. Murchison.

Angular Creek.... Chromite.
Bald Rocks........ Tinstone, gold.
Bingera............. Adamantine spar, albite, antimony (antimonite, cervantite), bismuth, bismuthite, bronzite, chromite, copper ores, diaclasite, diallage, diamond, eisenkiesel, galena, garnet, gold, iron (limonite, pyrites, titaniferous), magnesite, mispickel, molybdenite, osmo-iridium, picrolite, rock crystal, sapphire, serpentine, spinelle, tellurium, tinstone, topaz, tourmaline, zircon.
Bingera Creek..... Chromite, gold.
Curangora.......... Asbestos, native lead.
Doctor's Creek.... Diamond.
Gineroi............. Antimonite.
Gundalmulda Chromite.
Creek
Horton River...... Chromite, serpentine.
Myall Creek....... Calcite, limestone, tinstone.
Reedy Creek....... Chromite, copper, elaterite, galena, halloysite, prehnite, tinstone.
Two-mile Flat..... Chrysolite.

Co. Murray.

Boro.. .... Cobaltiferous manganese ore, copper ores, galena, gold, hæmatite, iron pyrites, zinc blende.
Brooks' Creek..... Silver ores.
Braidwood and Copper ore.
Queanbeyan (dividing range between)
Bungendore........ Braunite.
Camberra.......... Goethite.
Camberra Plains.. Galena.

Captain's Flat..... Barytes, gold, iron (pyrites, specular), lead (carbonate, galena, protoxide).

Cotter's River..... Arragonite, copper ores, eisenkiesel.

Gundaroo........... Galena.

Lake George...... Cobaltiferous manganese oxide, jasper.

Manar.............. Actinolite.

Modbury Creek... Chiastolite.

Molonglo River... Silver ores, wulfenite.

Mountain Creek.. Silver ores.

Murrumbateman Marble.

Murrumbidgee Galena, gold, silver ores.
River

Queanbeyan....... Chlorite, marble, copper and silver ores.

Reedy Creek...... Galena.

Willeroi Station.. Brown hæmatite.

Co. Nandewar.

Mount Lindsay... Eisenkiesel, orthoclase.

Narrabri........... Carnelian, coal, obsidian, zeolites.

Narrabri (near)... Kerosene shale.

Co. Napier.

Coolah.............. Coal, ironstone, ozokerite.

Narrangarie....... Silver ores (chloride, &c.)

Co. Narromine.

Tomingley......... Gold, manganese oxide.

Tomingley (two Copper carbonate.
miles from)

Co. Northumberland.

Ash Island........ Gypsum.

Branxton.......... Cannel coal.

Brisbane Water... Hæmatite, pisolitic iron ore.

Buttar Ranges.... Limonite.

Gosford............ Goethite.

Greta............... Coal, torbanite.

Lake Macquarie.. Coal, torbanite.

Newcastle......... Coal, limonite, manganese oxide.

Shepherd's Hill... Hæmatite.

Singleton.......... Coal, goethite, gypsum, thinolite.

Tuggerah Beach Titaniferous iron, yenite.
Lake

Waratah........... Chessylite, coal, hydrocarbon.

West Maitland... Agate, carnelian, chalcedony, coal, goethite.

Wollombi River... Corundum, sapphire, zircon.

Co. Parry.

Bowling Alley Pt. Antimonite, bronzite, chromite, diallage, dolomite, garnet, gold, iron carbonate, mica, native copper, serpentine, zircon.

Clear Creek....... Magnetite.

Dungowan......... Copper ores.

Dungowan's Ck. Red hæmatite.

Foley's Folly...... Gold, jasper, opal.

Hanging Rock... Chromite, gold, native lead, serpentine.

Hookanvil Creek Opal.

Mount Misery.... Chalcedony, gold, serpentine, zircon.

Mount Pleasant... Copper ores, gold, jasper.

Nundle .... Agate, antimonite, axinite, black marble, chrysolite, copper ores, diallage, gold, hornblende, hyatite, iron pyrites, jasper, mispickel, opal, pyrrhotine, serpentine, tourmaline, wad, zircon.

Nundle Creek..... Chromite, gold, sapphire.

Peel River......... Antimonite, aragonite, copper-nickel, gold, lead (native, galena), marble, marmolite, rock crystal, sapphire, serpentine.

Wear's Creek..... Nickel ore.

Woolomi........... Chromite, jasper.

Co. Phillip.

Cadell's Reef...... Scorodite.

Cooyal.............. Goethite, gold, magnetite, rock crystal, blue topaz.

Cudgegong River Anatase, aragonite, brookite, diamond, gold, gypsum, jasper, iron (limonite, titaniferous), mercury (cinnabar, native), orthoclase, ruby, sapphire, spinelle, topaz, zircon, and other gemstones.

Cudgegong Dist. Antimonite, cervantite.

Dabee................ Alunogen, epsomite, Thomsonite.

Glenlyon............ Rock crystal.
Goree................ Epidote.
GreatMullenCrk. Chrysolite, ruby.
Gulgong............ Agate, albite, analcime, asbestos, calcite, chalcedony, chalybite, chlorite, chondrodite, chromite, copper ores, epidote, garnet, gold, iron pyrites, kaolin, lead (galena, mimetite), magnesite, manganese oxide, marble, mispickel, opal, stilbite, topaz.
Guntawang........ Gold, pyroxene.
Havilah............... Chalcedony, gold, marble, smoky quartz.
Home Rule........ Felspar, gold, ochre, opal, orthoclase, rock crystal.
Jordan's Hill...... Opal.
Lawson's Creek... Galena, opal, orthoclase, ruby, sapphire, and other gemstones.
Rat's Castle Creek Chalcedony, hornblende, ruby, wavellite.
Tallawang......... Calcite, gold.
The Lagoons...... Hypersthene.
Two-mile Flat (near Mudgee) Brookite, carbonaceous earth, chromite, coal, epsomite, gold, grossularite, halloysite, jasper, magnesite, marble, muscovite, orthoclase, osmium-iridium, pleonaste, rock crystal, tinstone, titaniferous iron, topaz, torbanite.
EumbiandBimbijong (between) Ruby.
Jungemonia and Uranbeen (between) Hornblende, steatite, talc.

Co. Pottinger.

Gunnedah......... Antimony (antimonite, cervantite), calcite, chalcedony, coal, gold.

Co. Raleigh.

Bellinger River... Antimonite, cervantite.
Bowra............... Antimonite.
Macleay River District Galena, silver ores.
Nambucca......... Galena, iron pyrites, zinc blende.
Nambucca River and Trial Bay (between) Cobaltiferous manganese oxide.

Upper Bellinger River Galena, mispickel, silver ore.
Warrell Creek..... Antimony, argentiferous mispickel, iron pyrites, lead ores, magnesite, silver ores, zinc blende.
Yarrahappini...... Silver ores.

Co. Rankin.

Cobar (seventy miles west of) Fibrous malachite.

Co. Richmond.

Kayon............... Pleonaste.
Monaltrie......... Chalcedony, quartz, siliceous sinter.
Richmond River.. Cimolite, gold, jasper, meerschaum, opal, platinum.

Co. Robinson.

Billygoa............ Gold, silver ore.
Cobar......... ..... Bismuth (native), blue and green carbonates of copper and other copper minerals, gold, opal.

Co. Rous.

Ballina (near)..... Diamond.
Lismore............ Manganese oxide.
Lismore and Ballina (between) Opalescent sandstone.

Co. Roxburgh.

Cullen Bullen..... Alunogen, copper ores, epsomite.
Glanmire.......... Barytes, iron ore, manganese (oxide, rhodonite).
Green Swamp..... Gold, iron pyrites.
Mitchell's Creek.. Barytes, copper ores, fluorspar, gold, iron (magnetite, pyrites), lead (carbonate, galena, pyromorphite), marble, silver ores, wad, zinc blende.
Mount Ovens...... Mispickel.
Old Razorback.... Antimony ore.
Palmer's............ Antimonite.
Palmer's Oakley.. Calcite, copper pyrites.
Peel................. Galena.
Pink's Creek...... Copper, ribbon - jasper, sapphire.
Razorback........ Antimonite, cervantite.
Rylstone (near)... Albite, antimonite, cinnabar, manganese (braunite, mangan-blende).
Sofala............... Antimonite, gold.

Sunny Corner..... (See Mitchell's Creek).
Tarana (near)...... Copper pyrites, galena, garnet, iron pyrites, mispickel, magnetic pyrites, molybdenite.
Turon River....... Diamond, epsomite, gold, tinstone.
Two-mile Creek... Chromite.
Wattle Flat....... Mispickel.
Winter & Morgan's Mine (Sunny Corner, Mitchell's Crk.) Barytes, copper ores, galena, gold, pyromorphite, silver (native, embolite).
Winterton......... Native arsenic, zinc blende.
Yetholme (near).. Mispickel.

Co. Sandon.

Armidale........... Antimony (antimonite, cervantite), bismuth (native, bismuthite), gold, iron pyrites, lead (carbonate, galena), manganese, mispickel.
Armidale (near)... Chromite, scheelite.
Ben Lomond...... Iron pyrites, sapphire.
Boorolong......... Antimonite, stream-tin.
Gara............... Antimony (antimonite, cervantite, native), scheelite.
Hillgrove.......... Antimonite, gold, scheelite.
Mihi Creek........ Cobaltiferous manganese ore.
Mount James...... Native bismuth, zircon.
Mount Walsh..... Rutile.
Stony Batta....... Manganese ore.
Uralla.............. Adamantine spar, antimonite, aragonite, chromite, copper carbonate, diamond, garnet, gold, hornblende, iron (goethite, titaniferous), kaolin, mispickel, opal, rock crystal, rutile, sapphire, schorl, spinelle, tinstone, topaz, tourmaline, zircon.

Co. Selwyn.

Burra Creek....... Tinstone.
Ournie............. Gold, mispickel.
Tumut River...... Cachalong.

Co. St. Vincent.

Araluen........... Gold, silver ores.
Arnprior........... Chiastolite, marble.

Braidwood........ Copper ores, gold, iron pyrites, lead (carbonate, galena), mispickel, opal, silver, zinc.
Broulee........... Iron ores, silver, zinc blende.
Carwary........... Hæmatite.
Carwell........... Calcite, goethite, hæmatite, hydrotalcite, opal.
Clyde River, Shoalhaven Coal, torbanite.
Jervis Bay........ Goethite.
Jineroo Mount.... Copper ores, galena.
Major's Creek..... Copper pyrites, galena, gold, hornblende, iron pyrites, zinc blende.
Monga............. Gold, silver ores.
Mount Buddawang Torbanite.
Pigeon House..... Pyroxene.
Nowra .... Chalcopyrites, iron pyrites, lead (carbonate, galena).
Shoalhaven River Alunogen, antimonite, chrysolite, coal, copper ores, diamond, emerald, epidote, galena, lignite, magnesite, marcasite, mispickel, platinum, tinstone, zircon.
Talwal Creek..... Galena.
Yalwal Creek..... Galena, gold, iron pyrites, platinum, zinc blende.

Co. Tongowoko.

Granite Diggings Gold, tin.

Co. Urana.

Urana............. Gold.

Co. Vernon.

Apsley River...... Copper, marmolite.
Ellenborough River Wad.
Europambela Station Carbonates of copper.
Moonbi (five miles south-east of) Axinite.
Walcha............ Antimonite, chalcedony, gold, manganese oxide.
Walcha (near)..... Torbanite.

Co. Wallace.

Adaminaby........ Gold, tinstone.
Jegedzeric Hill ... Actinolite, epidote, tourmaline.

Maneero............ Alunogen, epidote, epsomite.

Manner's Creek... Tinstone.

Mount Trooper (near) Copper carbonate, lead (cerussite, galena, protoxide).

Mowembah....... Actinolite, tinstone.

Seymour............ Copper ores.

Snowy Mountain Idocrase.

Snowy River...... Gold, iron pyrites, lead (carbonate, galena), sapphire.

Co. Wellesley.

Bibbenluke........ Barytes, specular iron.

Bombala,.......... Copper, galena, gold.

Cambalong .... Barytes, galena.

Merinoo........... Barytes.

Quedong Mount.. Copper ores, galena.

Slaughterhouse Creek Barytes.

Co. Wellington.

Avisford........... Zoisite.

Bald Hills......... Agate, corundum, diamond, emerald, manganese, ruby, rutile, sapphire.

Bara................ Arseniate of iron, lead carbonate.

Bell River........ Sapphire, topaz, and other gemstones.

Bell River and Guano Hill (between) Eisenkiesel.

Bromby............ Calcite.

Burraga .... Copper, galena, iron pyrites.

Burraga (four miles from) Green silicate of iron.

Burrandong....... Anatase, brookite, diamond, gold.

Burroba Creek.... Asbestos.

Campbell's Creek Jamesonite.

Coolamine Plain.. Albite.

Copper Hill....... Copper ores, silver ores.

Crudine........... Cervantite, copper pyrites, lead (carbonate, galena).

Crudine Creek.... Antimonite, gold.

Cudgegong........ Copper ore.

Dowagarang....... Nepheline, smaragdite.

Hargraves .... Gold, mispickel.

Hargraves Falls.. Antimonite.

Hargraves (forty miles west of) Barytes.

Hawkins Hill..... Chabasite, corundum, gold, olivine, muscovite, pyrrhotine.

Hill End........... Gold, muscovite, wad.

Ironbarks .... Chromite, copper pyrites, gold, iron pyrites, mispickel.

Jordan's Hill...... Aragonite, chalybite.

Lewis Ponds Creek Asbestos, brown iron ore, copper ores, epidote, gold, lead carbonate, magnesite.

Louisa Creek...... Arsenic (native, löllingite, mispickel, realgar, scorodite), brucite, cadmium, chrysolite, copper ores, gold, magnesite, opal, pyrolusite, sulphur, zinc blende.

Lucknow........... Antimony (metallic), asbestos, calcite, copper ores, gold, iron pyrites, mispickel, picrolite, pyroxene, serpentine.

Merrendee........ Actinolite, hornblende.

Monkey Hill...... Diamond.

Mookerawa Creek Gold, mercury.

Mudgee............ Antimony, gold, manganese oxide, mispickel, torbanite.

Nuggetty Gully Creek Jamesonite.

Ophir .... Copper sulphide, emerald, galena, gold, iron pyrites, platinum, silver chloride, titaniferous iron.

Orange (partly in co. Bathurst) Asbestos, bismuth ore, calcite, copper ore, iron pyrites, marble, mispickel, muscovite, opal, serpentine, silver ores, wad, zinc blende.

Pyramul........... Antimonite, cervantite, gold.

Pyramul Creek... Diamond, gold.

Sally's Flat........ Diamond, gold.

Spring Creek...... Tinstone.

Sugarloaf Hill.... Mimetite.

Tambaroora....... Gold, opal.

Three-mile Flat... Copper carbonate.

Two-mile Flat.... Gold, diamond, | sapphire, and other gemstones.

Wellbank.......... Copper ores.

Wellington........ Agate, chalcedony, copper ores, galena, gold, iron (pyrites, titaniferous), manganese (braunite, wad), marble, opal.

Wellington Caves (near) Copper ores, marble.

Co. Wentworth.

Wentworth......... Asbestos.

Co. Werunda.

Salt Lake........... Salt.

Co. Westmoreland.

Burramagoo .... Copper ore.
Dark Corner....... Mispickel.
Essendon........... Chrysocolla.
Essington........... Copper ores.
Fish River......... Garnet, gold, mispickel, oligoclase.
Fish River Creek Calcite, marble, mispickel, saltpetre.
Holander's River Copper pyrites, galena, iron pyrites.
King's Creek...... Psilomelane.
Middle Creek...... Serpentine, williamsite.
Native Dog Creek Copper ores, galena, gold, iron (pyrites, spathic iron ore), zinc blende.
Oberon.............. Chlorite, copper ores, diamond, epidote, galena, gold, iron pyrites, mispickel, pyroxene.
O'Connell District Copper carbonate, lead (carbonate, galena), hæmatite, iron oxide, opal.
Rocksbury .... Fibrolite.
Sewell's Creek.... Asbestos, gold, specular iron.
Wiseman's Creek Antimonite, asbestos, copper ores, fahlerz, galena, gold, iron pyrites, ironstone, manganese (kupfermanganerz, manganblende, wad), silver ores, zinc blende.
Wiseman's Creek, Copper ores, fluorspar,
South   gold, lead (anglesite, carbonate).

Co. Woore.

Monaro District.. Copper ores, galena.
Monaro............. Barytes, galena, gold.

Co. Wynyard.

Adelong............ Calcite, copper ores, galena, gold, iron pyrites, scheelite, silver ores, stilbite, zinc blende.

Adelong Creek ... Galena, gold, magnetic pyrites, silver ores.
Cowarbee Mine... Gold (leaf).
Stony Creek ...... Copper ores, gold, halloysite, torbanite.
Tarcutta........... Alunogen, gold, sulphur, tourmaline.
Tarrabandra....... Galena, marble, mispickel, iron pyrites, zinc blende.
Tumberumba...... Gold, ruby, sapphire, stream-tin.
Tumut.............. Gold, hæmatite, magnesite, stream-tin.
Upper Adelong... Mispickel.
Wagga Wagga.... Gold, ironstone, titaniferous iron.

Co. Yancowinna.

Belaira,   Mount Galena.
Gipps
Broken Hill....... Copper ores, lead carbonate, silver ores (chloride, embolite, iodide, &c.)
Leadville........... Lead ores.
Mount Gipps...... Bismuth carbonate, garnet, iron carbonate, ironstone, lead carbonate, silver (kerargyrite, &c.)
Pinnacles (fifteen Copper carbonate.
miles east of)
Round Hill (near) Galena.
Silverton .... Copper carbonate, garnet, gold, ironstone, lead (carbonate, galena, vanadiate), silver ores (chloride, chlor-iodide, embolite, &c.).
Thackaringa....... Barytes, copper ore, galena, garnet, iron pyrites, mispickel, silver ore.
Umberumba   Galena.
Creek
Umberumberka... Galena, silver ore, native silver, &c.

Co. Yantara.

Lake Cobham..... Agate, gypsum.

Co. Yunghulgra.

Courntoundra   Copper ores.
Range

INDEX OF MINERALS.

ALPHABETICAL LIST OF MINERALS FOUND IN NEW SOUTH WALES.

THE END.

PRINTED BY BALLANTYNE, HANSON AND CO.
EDINBURGH AND LONDON.

Printed in the United States
By Bookmasters